This volume brings together scholars from several countries and a variety of disciplines to address the question of the influence that Epicureanism and Stoicism, two philosophies of nature and human nature articulated during classical antiquity, exerted on the development of European thought from ancient times to the Enlightenment. Although certain aspects of the influence of these philosophies on European culture have often been noted – for example, the influence of Stoicism on the development of Christian thought and the influence of Epicureanism on modern materialism – the essays in this volume contribute a new awareness of the degree to which these philosophies and their continued interaction informed European intellectual life well into early modern times.

The influence of the Epicurean and Stoic philosophies on European literature, philosophy, theology, and science are considered. Many thinkers continued to perceive these philosophies as significant alternatives for understanding the human and natural worlds. Having become incorporated into the canon of philosophical alternatives, Epicureanism and Stoicism exerted identifiable influences on scientific and philosophical thought until at least the middle of the eighteenth century.

Even though Aristotelian philosophy may have been left behind after the Renaissance, Reformation, and Scientific Revolution, these alternative ancient philosophies continued to play an active role in Western intellectual life. The history of early modern thought, rather than simply being regarded as the triumph of the moderns, can be understood at least in part as the substitution of one set of ancient models for another.

Atoms, *pneuma*, and tranquillity

Atoms, *pneuma*, and tranquillity

EPICUREAN AND STOIC THEMES IN EUROPEAN THOUGHT

Edited by

MARGARET J. OSLER
The University of Calgary

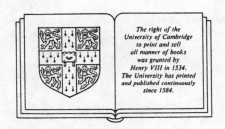

The right of the
University of Cambridge
to print and sell
all manner of books
was granted by
Henry VIII in 1534.
The University has printed
and published continuously
since 1584.

CAMBRIDGE UNIVERSITY PRESS

CAMBRIDGE
NEW YORK   PORT CHESTER   MELBOURNE   SYDNEY

CAMBRIDGE UNIVERSITY PRESS
Cambridge, New York, Melbourne, Madrid, Cape Town, Singapore, São Paulo

Cambridge University Press
The Edinburgh Building, Cambridge CB2 2RU, UK

Published in the United States of America by Cambridge University Press, New York

www.cambridge.org
Information on this title: www.cambridge.org/9780521400480

First published 1991
This digitally printed first paperback version 2005

A catalogue record for this publication is available from the British Library

Library of Congress Cataloguing in Publication data
Atoms, pneuma, and tranquillity: epicurean and stoic themes in
European thought/edited by Margaret J. Osler.
p.   cm.
Includes index.
ISBN 0-521-40048-1
1. Stoics.   2. Epicurus – Influence.   3. Philosophy, Ancient.
4. Philosophy, Medieval.   5. Philosophy, Modern.   I. Osler,
Margaret J., 1942–
B181.A76   1991
187–dc20   90-45068
                                                        CIP

ISBN-13 978-0-521-40048-0 hardback
ISBN-10 0-521-40048-1 hardback

ISBN-13 978-0-521-01846-3 paperback
ISBN-10 0-521-01846-3 paperback

For Marsha, exemplary friend and colleague

CONTENTS

CONTRIBUTORS

Peter Barker is Professor of Philosophy and Director of the Graduate Program in Science and Technology Studies at Virginia Polytechnic Institute and State University. He has written several articles on the influence of Stoicism on early modern science.

B. J. T. Dobbs is Professor of History at Northwestern University. Her research has primarily concerned the role of alchemy and chemistry in seventeenth-century natural philosophy. She has written two books on Newton's alchemical and theological researches, *The Foundations of Newton's Alchemy, or "The Hunting of the Greene Lyon"* (Cambridge: Cambridge University Press, 1975) and *The Janus Faces of Genius: The Role of Alchemy in Newton's Thought* (Cambridge: Cambridge University Press, forthcoming).

Louise Fothergill-Payne is Professor of Spanish at the University of Calgary. She has published widely on sixteenth- and seventeenth-century Spanish drama and has edited *Parallel Lives: Spanish and English Drama, 1580–1680* and *Prologue to Performance: Spanish Classical Theater Today* (Lewisburg, Pa.: Bucknell University Press, 1990). Her recent book *Seneca and Celestina* (Cambridge: Cambridge University Press, 1988) has expanded her field of inquiry to fifteenth-century intellectual history.

Thomas M. Lennon is Dean of the Faculty of Arts and Professor of Philosophy at the University of Western Ontario. He has published many articles and reviews, primarily on the history of seventeenth-century philosophy. His major work involves translation of and commentary upon the philosophy of Nicholas Malebranche. He has recently finished a book-length manuscript, *The Battle of the Gods and Giants: A History of Philosophy, 1655–1715.*

Maristella de P. Lorch is Professor and Chair of the Department of Italian at Barnard College, Columbia University. Her research and writing concerns the Italian Renaissance. Among her books are *A Defense of Life: Lorenzo Valla's Theory of Pleasure* (Munich: Fink, 1985) and (with Ernesto Grassi) *Folly and Insanity in Renaissance Literature* (Binghamton, N.Y.: Medieval and Renaissance Texts and Studies, 1986).

J. J. MacIntosh is Professor of Philosophy at the University of Calgary. He has wide-ranging interests in philosophy and in the history of philosophy, particularly in the seventeenth century. He is presently conducting a major study of Robert Boyle's philosophical views.

Calvin G. Normore is Professor of Philosophy at the University of Toronto and at Ohio State University. He is particularly interested in medieval philosophy and in modal and tense logic. Among his many publications is the article on "Future Contingents" in the *Cambridge History of Later Medieval Philosophy*.

Margaret J. Osler is Associate Professor of History at the University of Calgary. Her research and publications have focused on seventeenth-century philosophies of nature and their relationship to theology. She is presently completing a book, provisionally entitled *Theology and the Mechanical Philosophy: Gassendi and Descartes on Divine Will and the Philosophy of Nature*.

Letizia A. Panizza is Senior Lecturer and Head of the Italian Department at Royal Holloway and Bedford New College, University of London. Her research has focused on Italian humanists and ancient philosophical schools, especially Stoicism and Platonism, on Lorenzo Valla, and on debates between rhetoric and philosophy.

Nicholas G. Round is Stevenson Professor of Hispanic Studies at the University of Glasgow. His research interests include fifteenth-century Castilian cultural and political history and the theory and history of translation.

Lisa Tunick Sarasohn is Associate Professor of History at Oregon State University. She is currently writing a book on the moral and political philosophy of Pierre Gassendi. She is also conducting research on the French intellectual community of the 1630s and 1640s.

M. A. Stewart is Senior Lecturer and Head of the Department of Philosophy at the University of Lancaster, as well as Chairman of the British Society for the History of Philosophy. His research interests include English philosophy in the seventeenth century and the intellectual history of Scotland and Ireland in the eighteenth century.

Gerard Verbeke is Permanent Secretary of the Royal Belgian Academy of Sciences, Letters, and Fine Arts in Brussels. He was formerly Professor of Philosophy at the Catholic University of Louvain (Belgium). Author of many publications on ancient and medieval philosophy, he is particularly interested in the relationship between philosophy and Christianity in Western thought.

John P. Wright, Associate Professor of Philosophy at the University of Windsor, is author of *The Sceptical Realism of David Hume* (Minneapolis: University of Minnesota Press, 1983). He is engaged in an extended study of the conceptions of mind, soul, and the body in seventeenth- and eighteenth-century medicine.

ACKNOWLEDGMENTS

This volume and the conference from which it emerged were made possible by the efforts of many individuals and several institutions.

The Calgary Institute for the Humanities sponsored the conference on "Epicureanism and Stoicism," held in Calgary on 24–27 March 1988. I am grateful to Professor Harold Coward, Director of the institute, for his counsel and support, both material and moral. Gerry Dyer, the institute's administrator, did immense service in organizing and making arrangements for the conference as well as performing a multitude of administrative duties. I am grateful for her tact and good cheer as well as for all the labor she contributed. I would also like to acknowledge Cindy Atkinson, the institute's secretary, for a variety of services rendered.

The conference was made possible by a grant from the Social Sciences and Humanities Research Council of Canada and by a University of Calgary Conference Grant. Additional contributions came from the Faculty of Humanities, the Faculty of General Studies, the Department of Philosophy, and the Development Office, all of the University of Calgary. I gratefully acknowledge their support.

My colleagues at the University of Calgary, Louise Fothergill-Payne and J. J. MacIntosh, shared fully in the planning and organization of the conference. Decisions about the list of participants and the themes that we wanted the conference to address resulted from a number of fruitful planning sessions. Our wide range of scholarly interests produced a fertile mix of individuals at the conference. Many participants remarked on the exciting and stimulating character of the discussions that resulted. My two co-planners have also been exceedingly helpful during the process of editing this volume. I could not have carried out this project without their contributions.

Professor Gerard Verbeke assumed the role of unofficial general commentator at the conference. In addition to his paper, which gave the conference an auspicious beginning, he contributed extensive comments during the discussions following each of the other papers. His graceful erudition and lucid remarks were appreciatively noted by everyone.

I thank Peter Barker for his help in finding suitable illustrations for the dust jacket.

Suggestions solicited from two anonymous referees by Cambridge University Press greatly improved the quality and coherence of this volume. Helen Wheeler, History of Science Editor, Edith Feinstein, Production Editor, and the other extremely friendly and competent people at the New York office of Cambridge University Press have made the production of this volume a real pleasure. I also thank Christie Lerch for her heroic efforts in copyediting.

I am grateful to my parents, Abraham G. Osler and Sonia F. Osler, who have always supported my academic pursuits. In particular, I wish to thank my mother, who contributed her time and energy to check the page proofs.

My friend Betty Flagler has been a constant source of encouragement, for which I am grateful.

I owe a special debt of thanks to Marsha P. Hanen, to whom this volume is dedicated. Now President of the University of Winnipeg but then a close colleague at Calgary, she possesses the gift of retaining her humanity despite the relentless demands of administrative duties. She has generously shared her friendship and wisdom, which have constantly served to remind me that it is possible to approach the various demands of academic life with a sense of balance, understanding, and even humor. Her guidance at critical stages of this project was instrumental in bringing it to fruition.

Introduction

MARGARET J. OSLER AND LETIZIA A. PANIZZA

This volume is the product of a conference on "Epicureanism and Stoicism," sponsored by the Calgary Institute for the Humanities, held 24–27 March 1988. Bringing together scholars from several countries and a variety of disciplines, the conference addressed the question of the influence that Epicureanism and Stoicism – two philosophies of nature and human nature articulated during classical antiquity – exerted on the development of Western European thought from ancient times to the Enlightenment. Although the influence of these philosophies has often been noted in certain areas – for example, the influence of Stoicism on Christian thought, and the influence of Epicureanism on modern materialism – the essays in this volume make us newly aware of the degree to which these philosophies and their continued interaction informed European intellectual life, well into early modern times. Even then, many thinkers continued to perceive these two ancient schools as significant alternatives for understanding the human and the natural worlds.

Stoicism and Epicureanism arose more or less contemporaneously in Athens in the late fourth century B.C. Both Epicurus (341–271 B.C.) and Zeno of Citium (334–262 B.C.) developed their philosophies in reaction to the prevailing schools of Plato and

All the essays published here were presented at the conference on "Epicureanism and Stoicism," with the exception of Peter Barker's "Stoic Contributions to Early Modern Science." Louise Fothergill-Payne, J. J. MacIntosh, and Peter Barker all made important contributions to this introduction. Jane E. Jenkins and Stuart Hooper read drafts of it and helped remove infelicities of style.

Aristotle which dominated Greek thought at that time.[1] The Epicureans and the Stoics alike had Roman expositors whose writings facilitated the transmission of their thought into the medieval period and beyond. Cicero (106–43 B.C.) discussed Epicureanism and Stoicism in several influential works, such as the *Tusculans, De finibus, De natura deorum, De divinatione,* and *De fato.* His ultimate rejection of the moral teachings of Epicurus was countered by the Roman poet Lucretius (99–44 B.C.), whose lengthy philosophical poem *De rerum natura* is the most complete ancient exposition of the Epicurean philosophy of nature and human life.[2] The writings of Seneca (A.D. 4–65), especially his Letters to Lucilius, provided the readership of late antiquity, the Middle Ages, and the Renaissance with what came to be a widely diffused account of Stoic ethics, rivaled only by Cicero's *De officiis.*

Both the Epicureans and the Stoics endeavored to construct complete philosophies which would account for the natural world, human knowledge, and human action. Although the following words are attributed by Diogenes Laertius to the Stoics, they could apply equally well to the Epicureans:

Philosophy, they say, is like an animal, Logic corresponding to the bones and sinews, Ethics to the fleshy parts, Physics to the soul. Another simile they use is that of an egg: the shell is Logic, next comes the white, Ethics, and the yolk in the centre is Physics. Or, again, they liken Philosophy to a fertile field: Logic being the encircling fence, Ethics the crop, Physics the soil or the trees.[3]

The ultimate aim of both schools was ethical: They believed that a good life leads to the state of mental tranquillity which the Epicureans called *ataraxia* and the Stoics *apatheia.* For both, attainment of such tranquillity follows from the proper attitude toward

---

1 For a thorough treatment of both schools in antiquity, see A. A. Long, *Hellenistic Philosophy: Stoics, Epicureans, Sceptics,* 2nd ed. (Berkeley and Los Angeles: University of California Press, 1986). For an excellent introduction to the primary sources, see A. A. Long and D. N. Sedley, *The Hellenistic Philosophers,* 2 vols. (Cambridge: Cambridge University Press, 1987).

2 The writings of Epicurus himself are known primarily from the fragments of his letters preserved by Diogenes Laertius (c. A.D. 225–250) in Book X of his *Lives of the Eminent Philosophers.*

3 Diogenes Laertius, *Lives of Eminent Philosophers,* 2 vols. (Loeb Classical Library, 1925), VII, 40. See also Howard Jones, *The Epicurean Tradition* (London: Routledge, 1989).

the cosmos (physics), which, rightly understood (through logic) helps people to come to an understanding of themselves and their place in the world (ethics). According to the Epicureans, fear of the gods and of punishment in life after death is the chief cause of mental distress. In order to dispel this fear, and thereby attain tranquillity, they developed a philosophy of nature which explained all natural phenomena, as well as the human soul, in terms of chance collisions among atoms moving through void space. They relegated the gods to a beatified existence, far removed from the natural and human worlds, and denied that the soul survives the body at death. The Stoics claimed equal importance for logic, physics, and ethics. For them, tranquillity is to be attained only in a life of reason sufficient to itself and in harmony with the natural order. Unlike the Epicureans, however, their concept of the cosmos was one of divine and purposeful design (Logos) of which humanity is part. Both schools thought that questions about logic – reason and human knowledge – were closely related to physics and ethics. The Epicureans accounted for knowledge in terms of sensations, the causes of which were explained by their atomic theory and the consequences of which included the pleasure and pain that provided the foundation of their hedonistic ethics. The Stoics addressed questions in logic, and logic, for them, consisted of a theory of knowledge based on a precise succession of mental exercises. The proper discipline of one's reason protects the human mind from error, both in its understanding of the world and in its search for moral behavior.

Unlike Plato and Aristotle, whose writings have survived the centuries, nothing has come down to us of the philosophical works of the ancient Greek Stoics except a few fragments, and from Epicurus we have only three letters and some maxims. The recovery of a body of doctrine for these two schools has been gradual and piecemeal. Before the revival of classical learning in the fifteenth and sixteenth centuries, knowledge of them was largely confined to Latin sources, particularly ones approved of by early Christian apologists, fathers of the church, and medieval moralists, who pillaged pagan classics for their own purposes.

If one looks at this first, prehumanist stage, one finds that Epicureanism, which denied divine providence, reward and punishment after death, and the immortality of the soul, was regarded with horror. Epicurean ethics, with its goal of pleasure (*hēdone*), interpreted incorrectly as unrestrained sensuality, was deemed bestial: Our modern word "hedonism" still carries these negative connotations. There is virtually no evidence that Lucretius was read at all (only four ninth-century manuscripts survive),[4] and very little was gleaned from Cicero's discussions of both schools in philosophical dialogues such as his *Tusculans, De finibus, De natura deorum*, and, to a lesser extent, *De fato*. (An Academic skeptic, Cicero favored the Stoics, particularly on ethical issues, on which he attacked the Epicureans.) The Roman Stoic Seneca, although constantly railing against *hēdone*, nevertheless wove into his Letters to Lucilius many quotations from Epicurus, thereby contributing to a more favorable picture of Epicurus in later periods.

Stoicism, by contrast, with its emphasis on order (Logos) in the cosmos, divine providence working purposefully in nature, and the performance of moral duty according to the dictates of reason whatever the personal cost, exerted immense appeal. Cicero's account of divine providence (*De natura deorum*, Book III) found its way into the writings of Latin fathers of the church like Lactantius and Augustine and became a standard feature of Christian apologetics. In the Middle Ages and beyond, his *De officiis* was the most widely read treatise on Stoic ethics and political conduct.[5] Seneca, whose numerous writings condemning Epicurean *voluptas* (pleasure) and exalting the Stoic concept of virtue for its own sake, or *honestas*, assured him the admiration of medieval clerics, for he seemed to hint at the immortality of the soul.[6] In addition, "Seneca's fame grew fat on works which he had never written,"[7] such as an apocryphal correspondence between himself

---

4 See Ronald E. Latham, "Lucretius," in *The Encyclopedia of Philosophy*, ed. Paul Edwards, 8 vols. (New York: Free Press, 1967), 5:101.

5 See Michel Spanneut, *Le stoïcisme des pères de l'église* (Paris: Seuil, 1957), and *La permanence du stoïcisme de Zénon à Malraux* (Gembloux: Duculot, 1973). Also Harald Hagendahl, *Latin Fathers and the Classics*, Studia Graeca et Latina Gothoburgensia, no. 6 (Göteburg, 1958); J. Stelzenberger, *Die Beziehungen der frühchristlichen Sittenlehre zur Ethik der Stoa* (Munich: Huebner, 1933); Johannes Quasten, *Patrology* (Utrecht: Spectrum, 1953).

6 For detailed accounts of the availability of specific texts, see L. D. Reynolds, ed., *Texts and Transmission: A Survey of the Latin Classics* (Oxford: Clarendon Press, 1983), pp. 124–8 and 357–65, and Anthony Grafton, "The Availability of Ancient Works," in Charles B. Schmitt et al., eds., *The Cambridge History of Renaissance Philosophy* (Cambridge: Cambridge University Press, 1988), pp. 779 and 790.

7 L. D. Reynolds, *The Medieval Tradition of Seneca's Letters* (Oxford: Oxford University Press, 1965), p. 112.

and Saint Paul and a widely diffused little treatise on the four virtues. At one point "the saintly sage" was even thought to have ultimately become a closet Christian.[8]

Medieval Stoicism, it should be stressed, focused on ethics. The Renaissance recovery of classical texts unknown in the Middle Ages brought surprises for both Epicureanism and Stoicism and constituted a second stage in a fuller understanding of the two schools. In 1417, the Italian humanist Poggio Bracciolini brought to light a manuscript of Lucretius's magisterial poem *De rerum natura*, still the most important single source for our knowledge of Epicurean physics. Although not printed until 1473, from that time on *De rerum natura* served as a permanent and major source for Epicurean ideas.[9] Lucretius's difficult vocabulary and contentious stance ensured that the poem would take a long time to be properly understood, but it nevertheless became the model in the next century for philosophical and scientific poetry and, of course, an inexhaustible mine for Epicurean teaching. No comparable work of Latin Stoicism was recovered: All of Seneca's extant writings were already known by the early thirteenth century. Cicero's letters and rhetorical treatises, a find of early humanism, did not add to knowledge of Stoicism, but his philosophical dialogues, with their wealth of information about Stoic teachings, finally began to be studied carefully in the fifteenth century.[10]

A third major stage in shaping our modern knowledge of ancient Stoicism and Epicureanism came with the discovery and translation into Latin of Greek texts, also beginning in the fifteenth century: not the writings of ancient Greek Stoics and Epicureans themselves, but second- and third-century A.D. Greek summaries and compilations that were nevertheless more faithful

---

8 See especially G. G. Meersseman, "Seneca maestro di spiritualità nei suoi opuscoli apocrifi dal XII al XV secolo," *Italia Medioevale e Umanistica*, 16 (1973), 43–135, and Letizia Panizza, "Biography in Italy from the Middle Ages to the Renaissance: Seneca, Pagan or Christian?", *Nouvelles de la République des Lettres*, 2 (1984), 47–98.

9 For the important role that printing played in making the Renaissance revival of texts permanent, see Elizabeth Eisenstein, *The Printing Press as an Agent of Change*, 2 vols. (Cambridge: Cambridge University Press, 1979), pp. 181–224.

10 For the recovery of Seneca and other classical texts, see Reynolds, *Medieval Tradition*, and L. D. Reynolds and N. G. Wilson, *Scribes and Scholars* (Oxford: Oxford University Press, 1968). For Lucretius, see Anthony Grafton, "The Availability of Ancient Works," in Schmitt, *Cambridge History of Renaissance Philosophy*, pp. 784–5. For the slow reception of Cicero's skeptical works, see Charles B. Schmitt, *Cicero Scepticus* (The Hague: Nijhoff, 1972).

to the original teachings than Latin and Christian versions. *Lives of the Eminent Philosophers*, by Diogenes Laertius (third century A.D.), translated by the Camaldolese monk Ambrogio Traversari in 1431, has remained our most important single source for Greek philosophy from Thales to Epicurus. In his "Life of Zeno" and "Life of Chrysippus," Diogenes explained Stoic logic, physics, and ethics, and his "Life of Epicurus" preserved three letters and two collections of maxims from the master himself.[11] In addition to Diogenes, Epictetus (c. A.D. 55–135), the Greek slave who taught in Rome, began to be studied. His *Enchiridion*, a manual for healing the soul in deep distress, was translated twice in Italy, first by Perotti in 1450 and then by Poliziano, for Lorenzo de' Medici, in 1475. His fuller *Discourses*, recorded by Arrian, however, were not printed until 1535 and did not appear in Latin translation until the end of that century.[12] Plutarch (born before A.D. 50, died after A.D. 120), a Platonist, wrote several essays expounding and refuting the philosophy of the Stoics (see his *Moralia*, printed in two different Latin translations in 1570 and 1573); and Sextus Empiricus (second century A.D.), while also hostile to the Stoics, in *Adversus mathematicos*, which was first printed in Latin translation in 1569, nevertheless provided abundant information about their teaching, especially on logic.[13]

Greek Stoicism, in particular, proved itself to be far more intractable to Christian assimilation than its Latin derivatives. Stoic physics seemed at first to confirm the deepest suspicions one might have after reading the strictures of Augustine and Lactantius about the determinism and materialism of that school: Reason was in, and bound up with, matter; fate ruled the worlds of nature and history in an ineluctable and repeatable chain of causation; and there was no immortality. But with the lifelong efforts of two

11 Agostino Sottili, "Autografi et traduzioni di Ambrogio Traversari," *Rinascimento*, 5 (1965), 4–7; Lucien Braun, *Histoire de l'histoire de la philosophie* (Paris: Editions Ophrys, 1973), pp. 33–7, 52–9; and Luciano Malusa, "La premesse Rinascimentali all'attività storiografica in filosofia," in *Storia delle storie generali della filosofia*, ed. Giovanni Santinello, 2 vols. (Brescia: Editrice La Scuola, 1981), 1:7–13, 156–66.

12 R. P. Oliver, *Niccolò Perotti's Version of the Enchiridion* (Urbana: University of Illinois Press, 1954), and the introduction by W. A. Oldfather to the Greek text and his English translation of Epictetus (Loeb Classical Library, 1928).

13 D. Babut, *Plutarque et la Stoïcisme* (Paris: Presses Universitaires de France, 1969), and Richard H. Popkin, *The History of Scepticism from Erasmus to Descartes* (New York: Harper & Row, 1964).

historically oriented philosophers, Justus Lipsius (1547–1606) and Pierre Gassendi (1592–1655), the recovery and assimilation of ancient Stoicism and Epicureanism, respectively, reached a final, mature stage. Working in a humanist tradition, both men wove together classical Greek and Latin sources; in contrast to their medieval and humanist counterparts, however, Lipsius and Gassendi carefully made the logic and physics, as well as the ethics, of both schools compatible with Christian theology. Epicureanism, Gassendi's efforts assured, had never looked so respectable. Both Lipsius and Gassendi exerted considerable influence on seventeenth-century discussions of Epicurean and Stoic issues.[14]

The recovery of Diogenes Laertius's *Lives of the Eminent Philosophers* in the fifteenth century launched a new genre, the history of philosophy, which first expressed itself in the production of encyclopedic histories of ancient philosophy. Their authors, known as "polyhistorians," divided ancient philosophers into sects: the Academics, or Plato and his followers; Aristotle and the Aristotelians; the Stoics; the Epicureans; and finally the Skeptics, once the writings of Sextus Empiricus had been absorbed.[15] It is worth noting that in the first half of the seventeenth century, the renewed interest in Stoicism and Epicureanism brought about by Lipsius and Gassendi can be measured by the immense popularity of Thomas Stanley's *History of Philosophy* (1655), which devoted considerably more space to each of these two schools than to any other, even those of Plato and Aristotle.[16] The only other school to which Stanley devoted a comparable number of pages was the Pythagorean, an emphasis reflecting the growing importance of the mathematical sciences in the seventeenth century. Not only

14 The fundamental work in English on Lipsius is Jason Lewis Saunders, *Justus Lipsius: The Philosophy of Renaissance Stoicism* (New York: Liberal Arts Press, 1955). On Gassendi's role in restoring Epicureanism, see Bernard Rochot, *Les travaux de Gassendi sur Épicure et sur l'atomisme, 1619–1658* (Paris: Vrin, 1944); Lynn Sumida Joy, *Gassendi the Atomist: Advocate of History in an Age of Science* (Cambridge: Cambridge University Press, 1987); and Barry Brundell, *Pierre Gassendi: From Aristotelianism to a New Natural Philosophy* (Dordrecht: Reidel, 1987).

15 Santinello, ed., *Storia delle storie generali della filosofia*, 1:8, 68–83; Braun, *L'histoire de l'histoire de la philosophie*, pp. 77–84; Anthony Grafton, "The Availability of Ancient Works," in Schmitt, *Cambridge History of Renaissance Philosophy*, p. 790.

16 Of the 1,091 pages in this three-volume work, Stanley devoted 80 to the Stoics, 194 to the Epicureans, 96 to the Pythagoreans, 54 to Plato, and 52 to Aristotle. See Thomas Stanley, *The History of Philosophy: Containing the Lives, Opinions, Actions and Discourses of the Philosophers of every Sect*, 2nd ed., 3 vols. (London, 1687).

had both the Stoics and Epicureans been firmly incorporated into the philosophical canon, but they had also come to supplant the traditionally authoritative schools of philosophy. This important development continued into the eighteenth century, as reflected in Jakob Brucker's influential *Historia critica philosophiae* (1742–67).[17]

The essays in this volume attest to the enormous scope of the influence that Epicureanism and Stoicism exerted on European thought from the time of the Renaissance to the early Enlightenment, in literature, philosophy, political thought, and natural philosophy. During the Renaissance, the ideas associated with these schools appeared largely in literary and didactic contexts. Interest focused on ethics and political philosophy.[18] But during the seventeenth century, in the aftermath of the Reformation and the Copernican revolution, the ideas of the Epicurean and Stoic philosophies were frequently considered in the context of philosophy and natural philosophy as well, with particular concern about their religious and theological implications. Concepts articulated within these philosophies later reappeared in the form of theological presuppositions, scientific ideas, and philosophical problems.[19] For example, as Stoic and Epicurean ideas were incorporated into European thought, they precipitated a confrontation between philosophy and theology, science and religion. Stoic ideas about fate and necessity were seen either as expressions of divine providence or as impediments to divine will, and they brought about renewed consideration of the relationship between the natural order and Christian concepts of divinity.

17 Santinello, ed., *Storia delle storie generali della filosofia*: vol. 2, *Dall'età cartesiana a Brucker* (Brescia: Editrice La Scuola, 1979), pp. 564–603.
18 There is little work in English on the influence of Stoicism and Epicureanism on politics and literature. The only recent studies are Gerhard Oestreich's *Neostoicism and the Early Modern State*, trans. David McLintock (Cambridge: Cambridge University Press, 1982); Gilles Monsarrat, *Light from the Porch: Stoicism and English Renaissance Literature* (Paris: Didier-Erudition, 1984); and Louise Fothergill-Payne, *Seneca and Celestina* (Cambridge: Cambridge University Press, 1988). See also the important collection of papers edited by Ronald H. Epp, *Spindel Conference 1984: Recovering the Stoics, Supplement to Southern Journal of Philosophy*, 23 (1985). These recent studies put us in a position to present these influences, in some cases for the first time. A vital reference for all these issues is Schmitt, *Cambridge History of Renaissance Philosophy*.
19 For a useful discussion of the "transplantation" of ideas from one field of discourse to another, see Amos Funkenstein, *Theology and the Scientific Imagination from the Middle Ages to the Seventeenth Century* (Princeton: Princeton University Press, 1986).

Particular ideas from Stoic physics, such as the central role of the *pneuma* in the natural world and the construction of the heavens, were modified and incorporated into early modern science. Similarly, the Epicurean emphasis on chance as the causal agency in the natural world and on the reputed atheism of the ancient atomists brought to full consciousness the theological assumptions within which modern science was formed. In order for the atomic theory to be accepted as a conceptual framework for the new science, it had to be cleansed of its heterodox assumptions. Theological concerns lay behind Gassendi's baptism of Epicureanism, and similar problems continued to trouble the next generation of mechanical philosophers. Boyle's reluctance to embrace atomism, despite his endorsement of the "corpuscularian philosophy," was, at least in part, a reaction against the reputed atheism of ancient Epicureanism. Newton, consumed by his attempt to articulate a scientifically and theologically adequate philosophy of nature, drew on the ideas of both the Epicureans and the Stoics, tempering their ideas with his own idiosyncratic theological presuppositions.

Furthermore, British empiricism was deeply influenced by Gassendi's restoration of Epicurean thought.[20] Theories of knowledge, perception, and the soul – philosophical positions central to mainstream British philosophy in the late seventeenth and early eighteenth centuries – can be traced directly to Epicurean sources. Hume, in the mid-eighteenth century, still considered Stoicism an appropriate target for his skeptical and empiricist attacks.

Together with other philosophical schools, Epicureanism and Stoicism continued to exert identifiable influences on scientific and philosophical thought at least until the middle of the eighteenth century. Even after Aristotelian philosophy went into decline in the later seventeenth century, these alternative ancient philosophies continued to play an active role in intellectual life. The history of early modern thought, rather than simply being the triumph of the moderns, can perhaps be understood at least in part as the interplay of one set of ancient models with another.[21]

20 See also David Fate Norton, "The Myth of British Empiricism," *History of European Ideas*, 1 (1981), 331–44.
21 See Charles B. Schmitt, *Aristotle and the Renaissance* (Cambridge, Mass.: Harvard University Press, 1983), and Charles Lohr, "Metaphysics," in Schmitt, *Cambridge History of Renaissance Philosophy*, pp. 537–638.

1

Ethics and logic in Stoicism

GERARD VERBEKE

The Stoics were firmly convinced that philosophy possesses an organic unity, comparable to the coherence of a living being. They acknowledged that various disciplines may be distinguished within philosophy. Zeno of Citium, founder of the Stoic school, mentions three parts of philosophy – logic, physics, and ethics – all of which belong to the organic whole.[1] Logic is not an autonomous discipline, independent of physics and ethics, for it constantly refers to these other branches. Dealing with human thinking, logic is mainly concerned with protecting the mind from error.[2] This goal, however, could not be achieved without some ethical conditions. Only a wise man is free from error and falsity: His thinking is not jeopardized by irrational impulses, for he has suppressed and extirpated all irrational movements and inclinations. Thus he is guarded against emotional influences which might prevent him from attaining the truth.[3] The activity of the mind, therefore, is clearly related to moral behavior: A man who is dominated by irrational inclinations must inevitably go astray in the search for truth. Conversely, ethics is closely related to logic. Moral behavior is essentially a life in conformity with reason, Logos.[4] It is the proper function of logic to disclose the

1 Diogenes Laertius, *Lives of Eminent Philosophers*, trans. R. D. Hicks, 2 vols. (Cambridge, Mass., 1925), VII.39. H. von Arnim, *Stoicorum veterum fragmenta* (Stuttgart, 1903–5), 2:37. (Hereafter abbreviated *SVF*.)
2 Cicero, *De finibus*, trans. H. Rackham (Cambridge, Mass. 1914), III.21.72.
3 Stobaeus, *Ecl.* II.7.11g, p. 99, 3 W. (SVF, 1:216). Cicero, *Academica*, trans. H. Rackham (Cambridge, Mass., 1933), 2:144 (SVF, 1:66).
4 Stobaeus, *Ecl.*, II, p. 75.11 W. (SVF, 1:179).

true nature of reason and its activity. Logic tries to clarify fundamental notions such as knowledge, perception and understanding, judging and arguing, truth and error. Clarification of the notion of reason is important for understanding moral behavior.

Finally, physics is related to both logic and ethics. For Zeno, the three parts of philosophy may be distinguished but they cannot be separated, since they constantly refer to each other. Each part loses its meaning if separated from the others. Cicero stated that according to the Stoics logic is a virtue, a moral habit, because it protects us against error and enables us to withhold our assent from what is false. Physics is likewise a virtue, possessing a truly ethical dimension. The study of physics is necessary in order to distinguish between right and wrong, since moral conduct, for the Stoics, is characterized by its agreement with nature. Physical knowledge is needed in order to understand the meaning of some ethical precepts and, more generally, in order to grasp the role of nature in the practice of justice. Furthermore, physics is required in order to understand the unity of humanity and the true meaning of piety toward the gods.[5] The study of physics enables us to realize that all human beings are children of the same father, that they all belong to the same family. It follows that logic and physics are not independent of and separate from ethics: Both disciplines possess a truly moral relevance.

In Stoic thought the unity of philosophy belongs to the very essence of philosophical reflection, since it corresponds to the organic coherence of the cosmos. All parts of the universe are interdependent: They could not exist separated from each other. The cosmos is a living organism, animated and permeated by the creative Spirit or Reason. The soul of each individual is a particle of the divine substance that penetrates everywhere. Because of this divine animation, all parts of the cosmos are interrelated.[6] The Stoics called this cosmic coherence "universal sympathy." No

---

5 Cicero, *De finibus*, III.22.73.
6 Zeno was persuaded that the world is a living being, endowed with reason. He elaborated a number of arguments to support this view: The world is more perfect than anything else (Sextus Empiricus, *Adversus mathematicos*, trans. R. G. Bury (Cambridge, Mass., 1949), IX.104; Cicero, *De natura deorum*, trans. H. Rackham (Cambridge, Mass., 1939), II.21 (SVF, 1:111)); the cosmos produces living beings gifted with reason (Sextus Empiricus, *Adversus mathematicos*, IX.110; Cicero, *De natura deorum*, II.22 (SVF, 1:112–13)); finally, a part of the universe could not be more perfect than the whole (Sextus Empiricus, *Adversus mathematicos*, IX.85; Cicero, *De natura deorum*, II.22 (SVF, 1:114)).

part of the universe can be affected without exerting some impact on the whole: If one member of a living body is injured, the entire organism suffers. There was an ancient belief, in the Greek and Babylonian civilizations, that heavenly bodies exert some influence on what happens on earth, even on the course of human lives. Stoic thought confirmed and rationalized this opinion. Celestial bodies and their movements can affect terrestrial events, since they both belong to the same organic unity. Heavenly bodies are not in a privileged position: Human life, because it is included in an organic whole and develops in the course of time, is constantly influenced by the totality to which it belongs.[7]

Furthermore, the Stoics regarded universal sympathy as the basis of and justification for divination, the traditional practice of predicting some future events by observing the positions and changes of the heavenly bodies, the intestines of sacrificed animals, or the movements of birds in flight. Stoic teaching was able to encompass the possibility of divination of this kind; for, if all things in the universe are interrelated, it must be possible, starting from a particular observation, to draw conclusions about other parts of the universe. The method used in divination is comparable to that of medical diagnosis. Here again Stoicism clarified and consecrated an ancient popular belief.[8] Within the framework of Greek thought, the Stoic view on the organic unity of all things represented something new. Plato had thought that sensible reality depends on or participates in the transcendent forms but that there is no organic coherence between the two realms. Aristotle saw the whole world as striving toward the supreme act: Although there is some unity in the Aristotelian cosmos, it is not the unity of a living organism. Among previous philosophers it is Heraclitus who came closest to the Stoic viewpoint.

The organic unity of Stoic philosophy corresponds to the living coherence of the universe. The object of philosophy, as it was formulated by some later members of the school, is to embrace the

7 Although the doctrine of universal sympathy was emphasized and developed by Posidonius, the basic view on the unity of the cosmos was present from the beginning of Stoic philosophy. See K. Reinhardt, *Kosmos und Sympathie* (Munich, 1926), pp. 111–21.

8 Cicero, *De natura deorum*, II.19; Epictetus, *Dissertationes*, trans. W. A. Oldfather, 2 vols. (Cambridge, Mass., 1925), 1:14. By and large, Stoics did not oppose the mythological tradition, but they introduced a new interpretation of it, in conformity with their philosophical teaching.

whole of reality, all divine and human things.[9] If all things are organically related to each other, philosophy must present the same unitary structure: Whatever belongs to this all-encompassing organism must be studied in connection with the other parts of the universe. If some parts were isolated and studied on their own without taking the totality to which they belong into account, mistakes and false conclusions would inevitably follow. From this perspective, logic, physics, and ethics cannot be independent disciplines.

Some Stoics did not like to use the term "parts" – the term used by Zeno and Cleanthes – to refer to the various branches of philosophy.[10] Presumably this terminology was criticized because it does not reflect the fundamental unity of the various disciplines. The term "parts" may refer to a component which has some independent existence. Chrysippus prefers the term εἴδη, meaning "kind."[11] This terminology reflects the fact that philosophy does not include distinct parts but rather several kinds of reflection and inquiry, namely logic, ethics, and physics. In this view, each discipline represents a special orientation toward research, without losing its connection to the other branches of inquiry. The objects of logic, physics, and ethics are not identical. To deal with human thinking is not the same as to investigate the cosmos or moral behavior. Thus each discipline has its own characteristics and concerns, its own particular outlook. Chrysippus also modified the order in which the various philosophical disciplines should be studied. In his view, logic should be first, since it provides the methodology on which the other inquiries are based. He put ethics in the second place and physics in the third. He regarded the last discipline as the most fundamental.[12] He considered the investiga-

9 Aëtius, *Placita*, in H. Diels, *Doxographi Graeci* (Berlin, 1879), I. Proem. 2 (SVF, 2:35); Sextus Empiricus, *Adversus mathematicos*, IX.13; Seneca, *Epistulae morales*, trans. R. M. Grummere, 3 vols. (Cambridge, Mass., 1917), 89, 5; Cicero, *Tusculan Disputations*, trans. J. E. King (Cambridge, Mass., 1966) IV.26, 57; V.3, 7; *De officiis*, trans. Walter Miller (Cambridge, Mass., 1913), II.2, 5; Philo, *De congressu*, 79; *Quaestiones in Genesim*, I.6; III.43. This definition of philosophy had already been prefigured by Plato; see *Republic*, VI.486a.
10 Diogenes Laertius, *Lives*, VII.39, 41 (SVF, 1:482, 2:37); Cicero, *De finibus*, IV.4 (SVF, 1:45).
11 Diogenes Laertius, *Lives*, VII.39 (SVF, 2:37).
12 Plutarch, *De Stoicorum repugnantiis*, in *Moralia*, ed. M. Pohlenz, rev. R. Westman (Leipzig, 1959), Chap. 9, 1035a (SVF, 2:42); Epicureans also put logic in the first place.

tion of the cosmos and the discovery of its organic structure to be the ultimate ground of logic and ethics.

Chrysippus constructed elegant metaphors to illustrate his viewpoint. Philosophy is comparable to a living organism, logic corresponding to the bones and nerves, ethics to the flesh, and physics to the soul. Or, philosophy may be compared to an egg: The shell corresponds to logic, the white to ethics, and the yolk to physics. Finally, it might be compared to a garden: The fence represents logic, the crops ethics, and the soil and the trees physics.[13] The three metaphors, particularly the first two, unambiguously emphasize the unity of philosophical reflection: If philosophy is comparable to a living organism or an egg, it follows that its various branches must be interrelated. The third metaphor ought to be interpreted in the same perspective. The garden must be considered as a whole in which all the components refer to each other: Even the fence loses meaning, if there is nothing to protect. Moreover, all these images emphasize the view that physics is the very core, the heart of philosophy; as the study of the cosmos it is the most internal, while logic is more external. Logical argument and the analysis of moral conduct ultimately depend on the structure of the cosmos, including both human beings and God.[14]

Posidonius also compared philosophy to a living organism, but in his view physics should be studied first, because it corresponds to the flesh and blood, whereas logic refers to the bones and nerves and ethics to the soul.[15] Here again the organic unity of philosophy is emphasized, but the order has been changed. Although Posidonius, a very learned man, was interested in the study of physical matters, nevertheless he realized that the study of physical reality is on a lower level than inquiries about thinking and moral behavior. He regarded ethics as the soul of philosophical reflection: After all, the primary issue is to disclose how human beings must behave in life.

In the further development of Stoicism, philosophy increasingly concentrated on ethical matters: Human happiness and conduct

---

13 Sextus Empiricus, *Adversus mathematicos*, VII.16 (SVF, 2:38).
14 Plutarch, *De Stoicorum repugnantiis*, Chap. 9, 1035c–e.
15 Diogenes Laertius, *Lives*, VII.41. Sextus Empiricus, *Adversus mathematicos*, VII.16 (SVF, 2:38).

became the main topics of philosophical reflection.[16] Seneca, in his turn, accepted three philosophical disciplines: ethics, physics, and logic.[17] However, with respect to Posidonius, Seneca changed the order of the disciplines. He placed ethics first, because he regarded ethical issues as the most important. By contrast, logic is concerned with rhetoric and dialectics, the latter mainly studying words and their meanings.[18] Seneca complained bitterly of the evolution of philosophy in his time: Instead of being a study of wisdom, it had become a study of words, a *philologia*.[19] His own conviction is quite clear: Philosophy should not deal with words but with real things (*non in verbis sed in rebus est*).[20] Seneca constantly endeavored to help man in his search for true wisdom. Although there are only a few who attain this goal (maybe one in five centuries), there are many *proficientes*, people who make some progress toward it. It is the proper duty of philosophers to help those who are on the way.[21]

Epictetus increasingly reduced philosophy to an inquiry about ethical matters. Nevertheless, he recommended the study of dialectics, on the grounds that it may help us in our struggle against error.[22] He immediately added, however, that the study of dialectics may also become a source of vanity, in which case it becomes useless or even harmful.[23] Philosophers should investigate three essential topics. The first is the passions – a crucial issue, since they are the irrational impulses in human life, the main obstacle to moral conduct. The second concerns duties, referring to human behavior in the various circumstances of life: duties toward other people, duties in family life and society, and duties toward God, the father of all human beings. Finally, there is the most crucial of the issues, assent. Human choice is internal: Man is unable to change the course of events which is fixed by divine Reason, although he is able to agree or not to agree. A moral man is one who brings his own will into conformity with divine rule. Consequently, assent is the most fundamental moral attitude.[24]

16 According to Strabo, Posidonius had a special interest in Aristotle and in the discovery of causal relations. See Strabo, *Geography*, trans. H. L. Jones (Cambridge, Mass., 1917–33), II.3, 8.

17 Seneca, *Epistulae morales*, 78, 9.   18 Ibid., 89, 17.   19 Ibid., 108, 23.

20 Ibid., 16, 3.   21 Ibid., 72, 10.   22 Epictetus, *Dissertationes*, I.7.1–4.

23 Ibid., I.8, 4.

24 Ibid., III.2, 1–2. Epictetus criticized the philosophers of his time because they were only concerned with the third subject and disregarded the other two (III.2.6).

Marcus Aurelius was not concerned with questions of logic or physics. He stated that he did not want to study rhetoric, poetics, or other disciplines of that kind.[25] He was an internal man, eager for meditation and solitude. He wanted to concentrate on what he considered to be essential, and to him that certainly was human behavior and happiness. Studying them is the duty of a true philosopher. In this way, he reduced philosophy to ethics.[26] The question concerning the relationships among the various philosophical disciplines thus faded away.

From this discussion, we may conclude that all Stoics agreed on emphasizing the living, organic unity of philosophy. Logic also belongs to this unity, regardless of whether it is placed first, second, or third. What is essential is that logic be regarded as a truly philosophical inquiry. In this way the Stoic view differs from the one maintained by Aristotle, who thought that logic is very important and plays a decisive role in philosophical research but nevertheless denied that it is to be regarded as a part of philosophy in the same way that metaphysics, physics, psychology, and ethics are. What is the reason? Aristotle had included in philosophy all scientific research, not only the philosophical disciplines but also those which are now called the "sciences." Nevertheless, he had excluded logic, for logic deals essentially with the nature of science and scientific method but does not examine a particular area of reality. In this respect it is different from metaphysics, which is concerned with being as being; from physics, which is concerned with the sensible world and its becoming; from psychology, which is concerned with the human soul and its various activities; and from ethics, which is concerned with moral conduct and happiness. As for logic, since it studies science as such, it is preliminary and related to all kinds of scientific investigation, preparing us for the various philosophical disciplines without being part of any of them.[27]

In Stoic thought the situation is different. Logic is not merely a preliminary step but is regarded as truly philosophical reflection.

25 Marcus Aurelius, *Meditations*, I.7.
26 Ibid., V.9. It is a duty of philosophy to make man what he ought to be by nature: magnanimous, free, quiet, benevolent, and pious.
27 Aristotle, *De partibus animalium*, trans. W. Ogle, in *The Complete Works of Aristotle*, ed. Jonathan Barnes, 2 vols. (Princeton, 1984), I.1.639a4. See W. D. Ross, *Aristotle* 4th ed. (London, 1945), p. 20.

Logic is concerned with knowledge and truth and helps us avoid error and falsity. Thus it plays a role in the great struggle against irrational movements and impulses: It helps to promote and to guarantee the triumph of reason or Logos.

Stoic logic differs considerably from Aristotelian logic. Aristotelian logic is largely based on universal concepts that derive from sensible experience: The particular images resulting from perception are made intelligible; they are dematerialized and universalized under the influence of the active intellective principle.[28] Aristotle had used these universal concepts in his construction of syllogistic arguments. In such arguments, two extreme terms are linked by a middle term to which each of the extremes is related. At least one of these relationships must be understood to hold universally.[29] This way of arguing is impossible within the framework of the Stoic theory of knowledge, which is empiricist and according to which there are no universal concepts deriving from sense experience.

For the Stoics, the transition to rational understanding coincides with the passage from a particular image to its verbal expression. When a concrete image has been expressed in a language, the knowing subject reaches a higher level, that of rational insight. The object of this rational knowledge, however, is not a universal concept.[30] As a result, Aristotelian logic, based on universal notions, could not be adopted by the Stoics. Instead, they invented a new kind of logic, essentially grounded on relations among propositions. A simple form of this kind of arguing starts from a conditional proposition in which a particular antecedent entails a consequent. This manner of argument corresponds to the Stoic teaching on the organic coherence of the cosmos. The Stoic method of reasoning is, in fact, a constant confirmation of the

28 Aristotle, *De anima*, trans. J. A. Smith, in Barnes, ed., *Complete Works of Aristotle*, III.5. This text is very condensed: In the course of history there has been a debate about the question of whether all human beings participate in the same active intellect or whether each of them has his own active principle.
29 Aristotle, *Prior Analytics*, trans. A. J. Jenkinson, in Barnes, ed., *Complete Works of Aristotle*, I.4.25b32–5; cf. I.1.24b26–30.
30 Diogenes Laertius, *Lives*, VII.49 (SVF, 2:52). See W. Kneale and M. Kneale, *The Development of Logic* (Oxford, 1962), p. 153: "It seems likely that the Stoics were convinced in a similar way that *lekta* must be distinguished from the sentences which express them. . . . They did not in fact abstract from the attitude of speaker or writer in forming the concept of the *lekton*, but merely from the linguistic expression." This is true insofar as external language is concerned.

living unity of the universe. A new logic was created in light of an empiricist theory of knowledge and as a result of an organic interpretation of the cosmos.

In Aristotle's philosophy, the notion of cause plays an essential part. This notion was not invented by the Stagirite. It was used before him, but it was very extensive, rather undetermined and vague. Aristotle took a decisively progressive step when he introduced a precise distinction among four kinds of causes: material, formal, efficient, and final.[31] To these four kinds of causes, one may still add exemplary causes or patterns, which were at the center of Plato's metaphysics. Aristotle's whole approach to philosophical research manifests a characteristic way of proceeding: He repeatedly asks the same question, namely, Why does a particular phenomenon occur? Cause and effect are the main categories of his thought.[32] He considered the discovery of the ultimate causes of whatever exists to be the highest achievement of philosophy.[33]

Stoic philosophy did not pursue this pattern of thought. Instead of the categories of cause and effect, the Stoics used those of signs and what is signified. They replaced etiology with semiology. Indeed, Zeno, the founder of the Stoic school, is credited with having written a work entitled Περὶ σημείων (On signs).[34] Although we do not know the exact content of this treatise, it was probably connected with a basic shift in philosophical method. A Stoic philosopher proceeds like a physician, who starts from signs and symptoms in order to make a diagnosis. Divination is based on the same method.[35] The whole universe is permeated by divine Logos or Reason. All parts are interrelated, like those of a living body. The world is full of meaning. It must be possible, starting from the observation of one part, to draw conclusions regarding another.

According to Chrysippus, the first philosophical discipline to be

---

31 Aristotle, *Physics*, trans. R. P. Hardie and R. K. Gaye, in Barnes, ed., *Complete Works of Aristotle*, II.3.194b23–195a3.

32 Aristotle, *Metaphysics*, trans. W. D. Ross, in Barnes, ed., *Complete Works of Aristotle*, I.1.981a24–30.

33 Ibid., I.2.982b2–10.   34 Diogenes Laertius, *Lives*, VII.4 (SVF, 1:41).

35 The doctrine of signs is present in a treatise falsely attributed to Galen ("Οροι ἰατρικοί, in C. G. Kühn, ed., *Opera medicorum Graecorum*, 26 vols. (Leipzig, 1827–30), 19:394–6). According to Max Wellmann, it was written by a representative of the Pneumatic School no earlier than the third century. See his book *Die pneumatiche Schule bis auf Archigenes* (Berlin, 1895), p. 65.

studied is dialectics, which deals with two important subjects: that which signifies, and that which is signified. The first topic refers to the study of language, thus placing reflection on language at the beginning of philosophical inquiry. Chrysippus considered language to be the whole collection of signs that refer to something. Since semiology plays an important part in Stoic thought, it is quite understandable that the starting point of philosophy is an inquiry into language.[36] Within this context, a fundamental question arises: Is the meaning of words conventional or natural? The relative importance attributed to the study of language will largely depend upon the answer given to this question. If the meaning of words is merely arbitrary, language loses much of its value in philosophical reflection. According to the Stoics, however, the meaning of words is not merely arbitrary and conventional: The basic elements of language, the primordial sounds, resemble the things to which they indirectly refer. Thus the meaning of words is basically natural, not conventional. Accordingly, on the question of language the Stoics side with Heraclitus, Cratylus, and the Epicureans.[37] Aristotle did not share this opinion: In his *De interpretatione* he maintained that the meaning of language is conventional.[38] Because of their viewpoint on the nature of language, the Stoics had a strong argument for starting philosophy with the study of language. To speak is to utter words which refer directly to an intelligible object. In addition to external speech, there is also an internal speech, since all thinking activity is connected with words.

What is signified by words is called an "expressible" (*lekton*): something that can be expressed in words.[39] A further distinction is made between complete expressibles (judgments, including

36 Diogenes Laertius, *Lives*, VII.43.
37 Origen, *Contra Celsum*, in Alexander Roberts and James Donaldson, eds., *The Ante-Nicene Fathers*, trans. A. Cleveland Coxe, 10 vols. (Grand Rapids, Mich. 1989), vol. 4, I.24 (p. 341, Delarue); vol. 1, p. 74, 10 Kö. (SVF, 2:146). See A. A. Long, *Hellenistic Philosophy* (London, 1974), p. 134: "If words and what they signify have a natural relationship to each other, it is reasonable to suppose that there is some correspondence between the ways in which words have been formed and the formation of concepts." What the Stoics maintain is a resemblance between primordial elements of language and reality. See A. C. Lloyd, "Grammar and Metaphysics in the Stoa," in A. A. Long, ed., *Problems in Stoicism* (London, 1971), Chap. 4.
38 Aristotle, *De interpretatione*, trans. J. L. Ackrill, in Barnes, ed., *Complete Works of Aristotle*, 2.16a26–8; cf. 1.16a5–8.
39 Diogenes Laertius, *Lives*, VII.63; Sextus Empiricus, *Adversus mathematicos*, VIII.70.

propositions and arguments) and incomplete expressibles (various kinds of predicates).[40] An expressible is not a universal concept; it is an object that subsists as a rational image. This representation is rational because the sensible object has been translated into language. In this way, the sensible image has been rationalized, while at the same time the Stoics could remain faithful to their empiricism. If a complete expressible corresponds to reality, it may be called true.[41] Indeed, words do not directly refer to reality but rather to the rational image in the mind. Language is a kind of sign, directly related to an intelligible content and indirectly related to reality. Consequently, the study of language is a way of getting access to the world: Words are natural but indirect signs of real things.

But what exactly is a "sign" in Stoic thought? According to Chrysippus, it may be described as the antecedent of a conditional proposition. Let us take the most simple form of a conditional proposition: if *A*, then *B* (εἰ τσδε, τσδε). This proposition does not affirm that *A* is the case; rather it declares that there is a connection between *A* and *B*: If *A* is realized, then *B* also must follow.[42] The question arises as to the nature of this link. As we have already explained, the Stoics did not adopt the Aristotelian categories of cause and effect; hence, they did not regard the antecedent as the cause of the consequent. In Chrysippus's view, the antecedent is incompatible with the contradictory proposition of the consequent. Therefore, the antecedent always and necessarily refers to the consequent.[43]

The Stoics maintained that all parts of the universe are linked to each other like the parts of a living organism. Although they did not question the precise nature of these connections, they emphasized the point that one factor may reveal the presence of another. A philosopher is thus comparable to a physician or a diviner: Starting from the observation of some particular aspect, he is able to draw conclusions about what is hidden. The human being is endowed with a discursive and synthesizing mind. Consequently he is capable of inferring what is hidden on the basis of what he

40 Diogenes Laërtius, *Lives*, VII.63–4.
41 Sextus Empiricus, *Adversus mathematicos*, VIII.12.   42 Ibid., VIII.276.
43 Sextus Empiricus, *Outlines of Pyrrhonism*, trans. R. G. Bury (Cambridge, Mass., 1933), II.111. See Benson Mates, *Stoic Logic* (Berkeley and Los Angeles, 1961), p. 48; J. B. Gould, *The Philosophy of Chrysippus* (Leiden, 1970), pp. 75–82.

perceives. In this way, each part of the world is what it is in itself, but in a sense it is also what it is not, since it refers to other parts of the cosmos. In dealing with the sensible world, man is able to comprehend its coherence. On the basis of this universal sympathy, he endeavors to know what is not given in perception and even that which is not perceptible. For example, from the movements of the body, man can understand the soul, which is its animating principle.

When the whole of reality is taken into consideration, four categories of beings may be distinguished. First, there are things which are directly and immediately evident. They belong to the world in which we live and are part of our environment. We are familiar with them and can perceive them whenever we want. We do not need the help of a sign to discover the presence of such beings, since they are always perceptible. Second, there are things which by nature are perfectly knowable but are temporarily hidden from a particular individual. When somebody is far from Athens, this city will not be knowable to him. It will be hidden, even though it is perceptible. The fact that the city is hidden is a consequence of where the particular subject happens to be. In this case, Athens may either be known directly, if the subject moves to the place where it is located, or it may be known indirectly, by means of signs that refer to it. Third, there are beings which are nonevident by nature. They are not and never will be perceptible. In his account, Sextus Empiricus provided two examples: empty space beyond the world and the pores of the skin. They could not be perceived by any of the senses, and yet they are knowable by signs which indicate their existence. Sextus Empiricus did not agree with this Stoic teaching. If some things are hidden by nature, he argued, they could never be known, either directly or indirectly. How could one ever be sure what a sign referred to, if the object could never be perceived? Fourth, there are things which are nonevident in an absolute way. They will never be the object of a cataleptic knowledge. It will never be possible to know whether the number of stars is odd or even, just as it will never be possible to know the number of grains of sand on the Libyan shore.[44] In both cases, the objects are perceptible, and it is theoretically possible to calculate their number, but nobody will be able

44 Sextus Empiricus, *Adversus mathematicos*, VIII.145–7.

to bring this calculation to an end. In this case, the use of signs could not be helpful. Consequently, signs are useful only to discover objects of the second and third categories.

Furthermore, the Stoics made a distinction between "commemorative" and "indicative" signs. The former refer to events of the past. Since history does not deal with present phenomena that can be immediately observed, and since things will never return within the same cycle of historical development, such events cannot be directly observed. Hence they must be studied on the basis of signs.[45] Indicative signs mainly refer to things that are hidden by nature, such as the human soul and God. Such things will never become perceptible; they will never be discovered by any of the senses. They are nevertheless knowable, because they manifest themselves in what is perceptible. The soul reveals itself in the movements of the body. God constantly reveals himself in the orderly arrangement and evolution of the cosmos.[46] The divine Logos is present everywhere, permeating all parts of the universe. The human soul participates in divine reason; in fact it is a part of this all-embracing principle.

History is also an achievement of the divine Logos. According to the Stoics, the development of history is cyclical. Each cycle offers the same content. There is no change or progress from one cycle to the next. Looking at history as a whole, the Stoics maintained that each cycle is a masterpiece of the divine Reason: Progress from one cycle to the next is impossible, since each cycle is perfect.[47] In order to understand this viewpoint, one has to take into account the fact that human freedom is very limited: It is a merely internal freedom, unable to change anything in the actual course of events. Man can only give his assent or refuse it to the development of history, but, whatever his internal attitude, he cannot change the course of events.[48] Judging from the relative rarity of human wisdom, one might be tempted to underestimate the achievements of the divine Logos and to consider them imperfect. However, according to Seneca, there are many people

45 Ibid., VIII.151–3.   46 Ibid., VIII.154–5.
47 Gerard Verbeke, "Les stoïciens et le progrès de l'histoire," *Revue philosophique de Louvain*, 62 (1964), 12–14.
48 Epictetus, *Man*, Chap. 53; Seneca, *Epistulae morales*, 107, 10 (SVF, 1:527). The last sentence in Seneca's text is a condensed expression of Cleanthes' viewpoint: "Ducunt volentem fata, nolentem trahunt" (Fate guides those who accept it and compels those who are reluctant).

who do make moral progress. In return for their efforts, they are less and less subject to the power of their passions. Nonetheless, it remains extremely difficult to suppress the influence of irrational impulses. So it remains possible for the Stoics to claim that each cycle of world history is a masterpiece. If many individuals refuse to give their assent, they are morally responsible, but the development of history remains the same. Because they restricted freedom to an internal attitude, the Stoics were able to maintain their optimistic philosophy in the face of their negative evaluation of concrete moral behavior. Even if many people are not faithful to the ethical ideal as it was conceived by the Stoics, there is no reason to propose a pessimistic view of the world and its history. The work of the divine Logos cannot be abolished by human weaknesses and failures.[49]

In the Stoic perspective, the universe and its history are full of meaning. They are manifestations of hidden powers working within. What we observe in the structure and development of the world reveals the soul that is at the origin of this orderly arrangement and harmonious evolution. Among the many factors which are significant in this whole, moral life holds a privileged position. A moral person is one who has suppressed irrational movements and who lives in perfect conformity with the divine Logos. He tries to change neither the world nor the course of events. On the contrary, he recognizes the work of the creative Reason and attempts to live in harmony with it.[50] Moral life, more than anything else, discloses the true nature of the Logos, dismissing whatever is irrational. In this way, it reveals the deepest ground of the universe. A moral person constantly manifests Reason in its purest and finest essence. If divine Reason is the animating principle of the universe, then it is most clearly revealed by moral behavior. Furthermore, ethical conduct is an acknowledgment of the organic coherence of the cosmos. A moral individual gives his assent to the structure and development of the world. In this way moral life has a logical and physical dimension: It demonstrates that Logos is truth, harmony, and freedom from passions.

A wise person is a sign, a symbol that elucidates the deepest roots of the universe and its history.

---

49 The Stoics also emphasized that their viewpoint deals with the cosmos as a whole rather than only particular aspects.
50 Diogenes Laertius, *Lives*, VII.87 (SVF, 3:4); Cicero, *De finibus*, II.34 (SVF, 3:14).

2

Medieval connectives, Hellenistic connections: the strange case of propositional logic

CALVIN G. NORMORE

One of the more striking differences between medieval and early modern philosophy is the status accorded the study of the theory of valid arguments. The study of this theory dominated the first two years of a medieval arts curriculum, and its influence permeated university life in the Middle Ages. There is nothing shocking about a medieval theologian who is discussing predestination or the creation of the world in time stopping his argument to spend a folio or two outlining some part of a logical system which he intends to use in resolving the question.[1] It is partly this interpenetration of logic and substantive philosophical discussion which makes much medieval philosophy seem so familiar to someone brought up in the twentieth-century Anglo-American tradition.

Within early modern philosophy the attitude is very different, an attitude exemplified by Descartes's discussion of the role of deduction in his *Regulae*:

> Let us now review all the actions of the intellect by means of which we are able to arrive at a knowledge of things with no fear of being mistaken. We recognize only two: Intuition and deduction.
>
> By intuition I . . . mean . . . the conception of a clear and attentive mind which is so easy and distinct that there can be no room for doubt about what we are understanding. Alternatively, and this comes to the same

I would like to thank especially Margaret J. Osler for saving me from a number of blunders and omissions.

1 For examples cf. Robert Holkot's discussion in II Sent. Q. II in Robertus Holkot, *In quatuor libros sententiarum quaestiones* (London, 1518), reprinted by Minerva (Frankfurt, 1967).

thing, intuition is the indubitable conception of a clear and attentive mind which proceeds solely from the light of reason. Because it is simpler it is more certain than deduction, though deduction, as we noted above, is not something a man can perform wrongly. Thus everyone can mentally intuit that he exists, that he is thinking, that a triangle is bounded by three lines and a sphere by a single surface and the like....

There may be some doubt here about our reason for suggesting another mode of knowing in addition to intuition viz. deduction, by which we mean the inference of something as following necessarily from some other propositions which are known with certainty. But this distinction had to be made, since very many facts which are not self-evident are known with certainty provided that they are inferred from true and known principles through a continuous and uninterrupted movement of thought in which each individual proposition is clearly intuited. [Rule 3][2]

What seems to be at work here is a rather different conception of deduction from that familiar in the Middle Ages. Whereas medieval thinkers hoped to algorithmize deduction, at least partially, and devoted much effort to working out rules of valid inference, Descartes seems to see deduction as a process governed by intuition rather than by rule. On this view, we *see* logical relations (presumably relations among ideas) in (by?) our natural light. Rules could presumably record our practice, but they could never add to it. Hence formal logic can bring us no new knowledge.[3]

With the rich development of logic in this century has come a new appreciation of what medieval logicians (particularly those working in the early fourteenth century) were trying to do and an increasing scorn for the early modern dismissal of logic as sterile. There is much about this attitude that is merely Whiggish. As we are able to render more of fourteenth-century logic into our own notations and interpret it within our own philosophical programmes, we think the better of it. As our philosophical pro-

2 René Descartes, *Regulae ad directionem ingenii*, in *Oeuvres*, ed. Charles Adam and Paul Tannery, 11 vols. (Paris: Vrin, 1974), 10:369–70; trans. in *The Philosophical Works of Descartes*, ed. John Cottingham, Robert Stoothoff, and Dugald Murdoch, 2 vols. (Cambridge: Cambridge University Press, 1985), 1:14–15.
3 An interesting discussion of Descartes's attitude to deduction can be found in Ian Hacking, "Proof and Eternal Truths: Descartes and Leibniz," *Proceedings of the British Academy*, 59 (1973), 1–16.

grammes move farther from the intuitionism of figures like Descartes, so our opinion of their work declines. What is needed, I think, is to keep both the transition from ancient to medieval logic and the transition from medieval to modern logic in view. To do this for even a significant fragment of logic would be a mammoth task, well beyond the scope of a single essay. The task attempted here is the more modest one of trying, by focusing on a very narrow logical issue, to illustrate some of the factors involved in the transition from medieval to early modern logical theory.

The early modern rejection of Scholastic logic involves a nearly simultaneous shift in several areas of logic, and a full discussion of the factors involved will be a very large project. In this essay I would like to concentrate on just one corner of the puzzle. It is widely recognized that Hellenistic logic did not accord either sentential logic or the theory of truth-functions the central role it has in twentieth-century logic. Among the ancient schools only the Stoics seem to have had a clear appreciation of a logic of sentential or propositional relations, and only they took much interest in truth-functions. Yet during the Middle Ages a truth-functional sentential logic was developed and elaborated in a quite remarkable way. What I would like to do in this essay is to trace the *fortuna* of one connective, conjunction, from ancient debates, through the Hellenistic period into the Middle Ages. I hope to show that the debates about the nature and role of conjunction, isolated though they may seem, illuminate some of the thornier problems of Hellenistic, medieval, and perhaps early modern semantics and suggest avenues by which the influence of the Hellenistic schools, especially Stoicism, in this case, was felt both in the Middle Ages and in the early modern period.

Conjunction is perhaps the least controversial connective in modern propositional logic. Getting students to believe the palpable falsehood that there is a significant truth-functional conditional in English takes art.[4] Getting students to believe that all negation is external negation takes only a little less. Even disjunction seems often to be regarded as somehow intensional. But

---

4 A good example of this art can be found in W. van O. Quine, *Methods of Logic*, rev. ed. (New York: Holt, 1959).

conjunction is a clear case of an almost uncontroversially truth-functional connective. If p, q, and r are sentences, then (p and q and r) is true if and only if p is true and q is true and r is true; otherwise it is false. Few claims are more evident. Moreover, it seems clear to practically everyone that a sentence like "Joanna is untenured and harried" is equivalent to "Joanna is untenured and Joanna is harried." So here, if anywhere, there seems no tension between term-forming and sentence-forming operators. Among connectives, conjunction is a model citizen.

In his article "Aristotle on Conjunctive Propositions," Peter Geach summarizes what he calls the "classical doctrine" of the conjunctive proposition as consisting of four claims:

 (I) For any set of propositions there is a single proposition that is their *conjunction*, they being its *conjuncts*.
 (II) The conjunction of a set of propositions is true if and only if each one of them is true and false if and only if some one of them is false. It therefore always has a truth-value, given that its conjuncts all have truth-values.
(III) Like any other proposition, a conjunction may occur as an unasserted part, e.g. as an *if* or *then* clause, of a longer proposition.
(IV) Conjunctive propositions frequently occur in ordinary discourse: "kai" and other words in Greek, "et" and other words in Latin, and "and" in English, are connectives that serve as signs of conjunction in very many ordinary sentences.[5]

Geach adds that "this doctrine is sometimes disputed, but only, I think, through confusion."[6] It may then come as a bit of a surprise to find that among the Greek schools only the Stoics took a view of conjunction much like this and that their view was loudly and bitterly contested throughout antiquity.

According to Diogenes Laertius, the Stoics defined a conjunction as "a statement [*axioma*] composed by certain coupling conjunctions, e.g. 'It is day and it is light.'"[7] And Galen too says that "the followers of Chrysippus . . . use the term 'conjunction'

5 Peter Geach, "Aristotle on Conjunctive Propositions," in *Logic Matters* (Berkeley and Los Angeles: University of California Press, 1980), pp. 13–27. The passage quoted is on pp. 13–14.
6 Ibid., p. 14.
7 Diogenes Laertius, *Lives of Eminent Philosophers* (Loeb Classical Library, 1925), VII.72.

for all propositions compounded by means of the conjunctive connectives."[8]

The truth-conditions for conjunctions are as *we* would expect. According to both Sextus and Epictetus, a Stoic conjunction is true just in case all of its conjuncts are true. It is false if even one conjunct is false. Since the Stoics accepted bivalence, this exhausts the possibilities.[9]

Although the Stoic accounts both of what a conjunction is and of its truth-conditions are just what Geach and, for that matter, contemporary linguists and logicians accept, neither was uncontroversial in antiquity. The Stoic account of what a conjunction is ran directly counter to the view Aristotle had proposed in *De interpretatione*. There he argues,

But if one name is given to two things which do not make up one thing there is not a single affirmation. Suppose, for example, that one gave the name "cloak" to horse and man; "a cloak is white" would not be a single affirmation. For to say this is no different from saying "a horse and a man is white," and this is no different from saying "a horse is white and a man is white." So if this last signifies more than one thing and is more than one affirmation, clearly the first also signifies either more than one thing or nothing (because no man is a horse).[10]

Now it is possible to interpret this passage in the light of Aristotle's remark in *De interpretatione* (17a8) that "the first statement-making sentence is the affirmation, next is the negation. The others are single in virtue of a connective." The interpretation would be that conjunctions are single statements but affirm more than one thing. This is Ackrill's interpretation and seems the common twentieth-century reading, but it is, I think, a reading heavily colored by modern logical assumptions.

Another way of understanding it, perhaps the most important so far as its influence on the early Middle Ages is concerned, was expressed in Boethius' account of the *vis* of a sentence. As C. J. Martin has argued, Boethius develops (or more likely adapts) a

8 Galen, *Institutio logica*, ed. Karl Kalbfleisch (Leipzig: 1896), Chap. 4, par. 6, p. 11.
9 Sextus Empiricus, *Adversus mathematicos* (Loeb Classical Library, 1949), VIII.125; Epictetus, *Dissertationes* (Loeb Classical Library, 1925), II.9.8.
10 Aristotle, *De interpretatione*, trans. J. L. Ackrill (Oxford: Oxford University Press, 1963), 18a18–26.

theory which is designed to account for the difference between a sentence and a mere list of words. Part of that difference lies in the force (*vis*) of a sentence, in something like the sense in which "force" is used in twentieth-century speech-act theory. Declarative sentences of the sort the Stoics called *axiomata* assert or deny something, and that they do so is their *vis*. In a categorical sentence it is the copula which indicates the *vis*. In a noncategorical sentence things are more complicated. If we consider the kind of noncategorical sentence in which Boethius is most interested – the conditional – we see that neither its protasis nor its apodosis is asserted. When I claim "If it is day it is light," I assert neither that it is day nor that it is light. What I assert is a certain connection between the two embedded categorical sentences. Again, when I assert "Either Socrates is sick or he is well," I do not assert either disjunct. What I assert, rather, is that a certain condition holds. Boethius himself thought that this condition was that one and only one of the disjuncts could be true. But there seems to be no analogous *vis* for the conjunctive particle in a conjunction to have. If the conjunction really is, as we moderns think, truth-functional, then, in asserting it, I do assert each of the conjuncts and nothing more. But if I assert nothing more, then the conjunctive particle adds nothing over and above what is asserted by the conjuncts. It has no *vis*, then, and so cannot bind the conjuncts together.

Not all of the Hellenistic debate about conjunction revolved around its status as a connective. Among those who granted that expressions of the form *kai . . . kai* had *vis*, there was considerable debate over just what the truth-conditions for such expressions were. At the center of the debate were the Stoics, who claimed, according to both Sextus Empiricus and Aulus Gellius, that a conjunction (*sumpeplegmenon*) was true just in case each and every one of its conjuncts was true. This condition, which was, as we shall see, the butt of a number of attacks, was obviously also the object of some wonder in antiquity, for when Aulus Gellius reports it he remarks,

That which among them [is called] *sumpeplegmenon*, we call "conjunction" or "copulative" which is of this sort: "P. Scipio, was son of Paulus and twice consul, and he had a triumph and held the office of censor and L. Mummius was his colleague as censor." Moreover in every conjunction if one [conjunct] is a falsehood then even if all the rest are true, the whole is said to be a falsehood. For if to all those truths which I have said about Scipio I were to add "and he conquered Hannibal in Africa" which is false, that whole which the things said conjoined are, on account of this one falsehood which would have been added, because they are said together, would not be true.[11]

Gellius's surprise suggests that the Stoic truth-conditions would have seemed strange to the ordinary Roman bilingual. Perhaps this is because conjunctive particles are much used in Latin and Greek as ways of smoothing transitions. It would not be impossible for most of a speech to be knit together by such particles, and it would be odd to regard a speech as false because one claim made in it was false.

This is one of the lines of attack taken up by Sextus Empiricus.[12] Sextus complained that a conjunction with one true conjunct and one false conjunct should be "no more true than false." Since it is true if both its conjuncts are true and false if they are both false, then the intermediate state should really be intermediate. He extended this criticism by arguing that a *sumpeplegmenon* with very many conjuncts almost all of which are true should be much closer to being true than to being false.

The inspiration from Sextus' attack seems connected with, if not exactly derived from, a picture he has of conjunction as a kind of mixing. This appears more directly in his second attack, which focuses specifically on how different conjunction on the Stoic model is from other kinds of mixing, "for just as what is compounded of white and black is no more white than black ... so also the true is in fact only true and the false only false, and the compound of the two must be described as no more true than false."[13] But, Sextus pointed out, on the Stoic account if you mix sentences of different truth-value by conjoining them you always get a falsehood.

Sextus knew the Stoic response – which is that just as we call a cloak torn if it has one tear, so we call a conjunction false if it has one false conjunct – but he thinks this merely to reflect an inappropriate willingness to carry the inexactness of ordinary discourse into philosophy. In a remarkable paper, "Le modèle conjunctif," Jacques Braunschweig argues that Sextus failed to

11 Aulus Gellius, *Noctes Atticae* (Loeb Classical Library, 1946), XVI.8.
12 Sextus Empiricus, *Adversus mathematicos*, VIII.125–9.   13 Ibid.

take into account the full resources of the Stoic theory of mixtures.[14] Braunschweig suggests that on the Stoic view the conjunction has a *function* which it fulfills only if all its conjuncts are true – just as a cloak has a function which it can fulfill only if it has no holes.

Unlike Aristotle, Sextus did seem to suppose that the *sumpepleg-menon* is a kind of sentence, and to suppose that the *kai* or other conjunctive particle does have a force of its own. That he did not think its force is the force the Stoics had proposed is clear, but it is less clear what the presuppositions of his own attack may be. One option is that he shares these presuppositions with Galen, who devoted part of Chapter 4 of his *Institutio logica* to an attack on the Stoic position.

Galen classified connectives according to the kind of semantic relationship that they expressed. He recognized two basic kinds of semantic connection: conflict and consequence.

Although Braunschweig may be right in thinking that conjunction presented for the Stoa a kind of model for how a system hangs together, there is a more mundane reason why it is important. The third Chrysippean indemonstrable is an argument form involving a negated conjunction. It is

$$
\frac{\begin{array}{c} -(P \ \& \ Q) \\ P \end{array}}{-Q}
$$

This indemonstrable played a rather important role in Chrysippean thinking. Because of Chrysippus' view that a conditional is true only if there is some necessary connection between antecedent and consequent, he seems to have held that simple predications about the future should not be expressed by conditionals like "If he is born under the Dog Star he will die at sea" but by negated conjunctions like "Not both: he is born under the Dog Star and he does not die at sea." The third indemonstrable provides a way of reasoning using such expressions.

To play the role which Chrysippus intended for them, negated conjunctions – and so conjunctions – have to be able to express the merely accidental semantic relationships on which astrology re-

---

14 Jacques Braunschweig, "Le modèle conjunctif," in J. Braunschweig, ed., *Les Stoiciens et leur logique* (Paris: Vrin, 1978).

lies. Hence it was important for Chrysippus both that conjunction be a real connective and that there be no special relevance conditions on what it may connect.

This is the part of the doctrine which Galen attacked. In Chapter 4 of *Institutio logica*, he insisted that in order for a negated conjunction to be true there must be opposition (*mache*) between the conjoined elements. The full display of Galen's doctrine of *mache* and his associated doctrine of *akolouthia* (consequence) would take us far afield; what is important here is just that a negated conjunction expresses what Galen called incomplete opposition, that is, the conjuncts cannot both be true but could both be false. In Chapter 14 Galen claimed that it is only where this incomplete opposition is found between the two sentences joined by a negated conjunction that we can apply the third indemonstrable. Thus whereas Chrysippus wanted to use the third indemonstrable to capture the claims of sciences such as astrology, which make empirical predictions, Galen claimed it can be used only where there is semantic opposition between the terms involved.

Although it seems very difficult to trace the roots of Galen's view, Posidonius seems to have played an important role. In his *De conjunctione* (214.1.4–20), Apollonius Dyscolus attributed to Posidonius the view that all grammatical connectives (including *kai*, one presumes) have a semantic force which requires that they not merely continue the flow of speech but modify it in some way – that is, they have to express some semantic relationship between the things they connect.[15] This seems to be Galen's view. He suggests that the appropriate relationship for *sumpeplegmenon* is incomplete opposition.

Some such view may well lie at the back of Sextus' attack as well. Sextus, of course, did not present a positive doctrine, but the tenor of his attack suggests that it is precisely the combination of the claim that any two sentences can be joined by conjunction that his attack is directed against. If Sextus had been operating out of a position which insisted that conjunction requires nontrivial semantic relations between the conjuncts, then this is the attack we would expect.

Despite the ubiquity in the ancient world of the opinion that

15 Cf. L. Edelstein and I. G. Kidd, eds., *Posidonius*, 2 vols. (Cambridge: Cambridge University Press, 1972): vol. 1, *The Fragments*, frag. 45, pp. 59–60.

what Geach calls the "classical" view of conjunction is false, it is the classical view that we find not only expressed but taken for granted in works at the beginning of the fourteenth century. Ockham, for example, wrote,

The copulative is that which is composed from several categorical sentences by means of this conjunction "and" or by means of some part equivalent to such a conjunction, just as in this copulative "Socrates runs and Plato disputes." . . . Moreover for the truth of a copulative it is required that both parts be true and so if any conjoined part is false the copulative itself is false. . . . It should be known immediately that the contradictory opposite of a copulative is a disjunction composed from the contradictories of the parts of the copulative.[16]

With the exception of the claim about the opposite of a conjunction being a certain kind of disjunction – something the Stoics, committed as they were to exclusive disjunction, would have rejected – Ockham's position was exactly Chrysippus'. Moreover Ockham merely repeated an account we can find in Burleigh, Pseudo-Scotus, Buridan, and practically any other fourteenth-century logician. Is this a case of rediscovery or of transmission?

We are not yet in a position to answer this question definitively. But we can say that if it is a case of transmission, it is not through the endorsement of any of the usual channels. The most authoritative source of early medieval logic – Boethius – was a firm opponent of the Stoic view. As C. J. Martin has argued, Boethius thought that a conjunction is just a list and is not properly speaking a proposition at all. Boethius here reflected the consensus of the ancient commentators on Aristotle available to him. The fact is that there does not seem to have been a single ancient source available to the medievals that reported the Stoic view of conjunction with approval.

Moreover, if we look at logical works up to and including the

16 William Ockham, *Summa logicae*, II.32. In *William Ockham opera philosophica, et theologica*, 17 vols. *Opera philosophica*, vol. 1, ed. P. Boehner et al. (St. Bonaventure, N.Y.: Franciscan Institute, 1974), pp. 347–8 (my translation): "Copulativa est illa quae componitur ex pluribus categoricis mediante hoc coniunctione 'et' vel mediante aliqua parte aequivalente tali coniunctione. Sicut ista est copulativa 'Sortes currit et Plato disputat' . . . . Ad veritatem autem copulativae requiritur quod utraque pars sit vera et ideo quaecumque pars copulativa sit falsa, ipsa copulativa est falsa . . . . Sciendum est statim quod opposita contradictorie copulativae est una disiuntiva composita ex contradictoriis partium copulativae."

time of Abelard, we do not find conjunction listed or treated among the acceptable types of sentence connection. Until the second half of the twelfth century, the classification of hypothetical sentences standardly followed appears to be that of Boethius in his *De syllogismis hypotheticis*. Boethius divided hypothetical sentences into conjunctions and disjunctions, and he meant by disjunction something like classical exclusive disjunction, but what he meant by conjunction is a conditional with a particle such as *si* or *cum*. He definitely did not mean sentences formed with *et* or analogous particles. This is the division we find in works like the *Dialectica* of Garland, and it is the division we find in Abelard's *Dialectica*.

Suddenly however, somewhere around the middle of the twelfth century we find a new classification of types of hypothetical sentence beginning to appear in logical treatises. A good example is the mid-twelfth century *Ars emmerana*, whose anonymous author wrote, "There are moreover seven species of hypothetical sentence: conditional, local, causal, temporal, copulative or conjunctive, disjunctive, adjunctive."[17] The author went on to say, "The copulative or conjunctive is that in which through a joining [*copulativa*] conjunction the consequent is joined to the antecedent as in 'Socrates is a human and Brunellus is an ass.'" Here we have the grammatical part of the "classical" account. The truth-conditions come next. In a treatise clearly closely related to the *Ars Emmerana*, namely the equally anonymous *Ars Burana*, we find the same seven types of hypothetical sentence, with almost the same example for conjunction, but we also find an account of their truth-conditions. In that account the author wrote, "Every copulative is false of which either part is false."[18]

The last piece of the classical doctrine appears in another roughly contemporary work, the *Fallacie parvipontani*. In this work, apparently from the school of the Paris-based but English-born dialectician Adam of Balsham, we find,

---

17 In L. De Rijk, ed., *Logica modernorum*, 2 vols. (Assen: Van Gorcum, 1967), II.II, p.158. "Sunt autem ipotheticorum propositionum septem species: conditionalis, localis, causalis, temporalis, copulativa vel coniuncta, disiuncta, adiuncta . . . .

"Copulativa est, vel coniuncta, in qua per copulativam coniunctionem consequens copulatur antecedenti ut 'Socrates est homo ⟨et⟩ Brunellus est asinus.'"

18 Cf. *Logica modernorum*, II.II, p. 191, 1.4–6. "Omnis copulativa falsa cuius altera pars falsa. Item omnis disiunctiva vera, cuius altera pars vera" [Munich 4652, 104r–116r].

For granted that this sentence "Socrates and Plato run" does not thus contain several sentences, yet it has several parts and the force [*vis*] of several. Moreover when several sentences are put forward conjunctively they are sometimes so put forward that one sentence is made from them as when they are predicated copulatively, and sometimes [they are put forward] so that there is not any one sentence made from them, as when they are predicated aggregatively, as in the syllogism.[19]

Here we have at least some form of the doctrine of *vis* but with the understanding that a conjunction which joins sentences has a *vis* and that sentences with conjoined subjects are equivalent to such conjunctions.

It seems, then, that a single generation in the second half of the twelfth century witnesses the shift from the picture of conjunction which Boethius had built on Aristotelian foundations to the Stoic view, and that with this goes a new way of conceiving hypothetical sentences.

I have already suggested earlier that we do not yet know exactly where the twelfth-century logicians got their knowledge of Stoic conjunction. It is not impossible, given what we do know, that they simply discovered it independently, but this hypothesis is by no means forced upon us. Not only were (admittedly unfavorable) reports of the Stoic position available through writers like Gellius, but there was pressure at least to count conjunction as a real connective from the late Latin grammarians, particularly from Priscian. Priscian devotes a short book of his *Institutes* to hypothetical sentences and their associated connectives. The book contains lists of types of hypotheticals, and those lists are supersets of the lists we find in the "modern" twelfth-century texts. Comparison suggests very strongly that the twelfth-century logic books borrowed from the grammarians the distinctions they found important.

A second question – perhaps more important – is why the later twelfth century abandoned the Aristotelian picture of conjunction for the Stoic view. Again the state of current research permits only

19 Cf. *Logica modernorum* I, p. 607, 1.27. "Licet enim hic propositio 'Socrates et Plato currunt' non sic plures propositiones continet, tamen partes plurium habet et vim plurium. Quando autem proponuntur plures propositiones coniuncte, quandoque ita proponuntur ut ex eis fiat una propositio, ut quando copulative predicantur, quandoque ita quod ex eis non est una aliqua propositio, ut cum aggregative praedicantur, ut in syllogismo."

very tentative suggestions. Perhaps the most promising of these, however, is that precisely because conjunction is the purest case of purely truth-functional sentential composition, interest in it fluctuated with interest in truth-functional composition as such. Up until the middle of the twelfth century there was little interest in truth-functional logic. In particular, the accounts of validity current neither required nor encouraged special attention to the truth-values of the premises or conclusion of an argument. These accounts emphasized, rather, the semantic relations between them and were, as Christopher J. Martin has suggested, akin in spirit to twentieth-century relevance logic.[20] But for a variety of reasons these accounts of validity came to seem more problematic, and when Abelard, who had been championing the older picture of validity, was shown by Alberic of Paris to be defending an inconsistent view, there seems to have been a general flight to an account of validity much like that we use today, one where an argument is regarded as valid just in case it is impossible that its premises be true and its conclusion false. Such an account natural-ly focused interest both on modal concepts and on questions about how the truth-value of one sentence is related to that of another. Within such a context, Stoic concepts were more useful than Aristotelian conceptions, and the twelfth century looks like a period in which logic moved steadily closer to the form in which the Stoa had practiced it.

But the tides of logical fashion change, and while accounts of validity which emphasized truth-values and truth-functionality dominated the thirteenth and early fourteenth centuries, by the late fourteenth century there was already beginning a move to a more intensional account, which required that a valid argument be one in which the conclusion was somehow contained in the premises.[21] This shift was slow, and never complete. As E. J.

---

20 Cf. Christopher J. Martin, "William's Machine," *Journal of Philosophy*, 83 (1986), 564–72.

21 We find this account as early as Thomas Bradwardine, who, in his *Insolubilia*, requires that the conclusion of a valid consequence be "understood" (*intelligitur*) in the antece-dent. Cf. M.-L. Roure, "La problematique des propositions insolubles au XIIIe siècle et au debut du XIVe, suivi de l'edition des traités de W. Shryeswood, W. Burleigh et Th. Bradwardine," *Archives d'Histoire Doctrinale et Littéraire du Moyen Age*, 37 (1970), 205–326; cf. p. 299, no. 6.053. In suggesting this shift I diverge from the views expressed by E. J. Ashworth in her *Language and Logic in the Post-Medieval Period* (Dordrecht: Reidel, 1974).

Ashworth has emphasized, there were authors who adhered to the modal account of validity throughout the postmedieval period. Moreover, the change in accounts of validity was confounded by a different shift from interest in medieval theories of *consequentia* back to an emphasis on syllogistic. This shift refocused logic on the theory of terms and focused attention on proof-theoretic, rather than semantic, accounts of validity. What is remarkable, however, is that these dramatic changes were not accompanied by abandonment of the truth-functional character and the proposition-forming force of conjunction. Thus the notion of a propositional logic was kept alive and was available for its renaissance in the nineteenth century. We have here, I suggest, one fragment of Stoic logic whose influence was uninterrupted from the twelfth century to our own.

3

Stoic psychotherapy in the Middle Ages and Renaissance: Petrarch's *De remediis*

LETIZIA A. PANIZZA

The course of Stoicism in Italy in the Middle Ages and the Renaissance, from about 1350 to 1550, has not by any means been fully charted. It is generally agreed that a renewal of Latin Stoicism there gathered momentum with the humanist Francesco Petrarca (1304–1374) and was bound up with his desire to write philosophy in an elegant Latin that educated laymen, as well as professional scholars, could understand. The kind of letters Petrarch wrote, in fact, are close in spirit to the urbane letters of moral and spiritual guidance written by the Roman Stoic Seneca to his younger friend and disciple Lucilius – intimate and moving, yet instructive. Petrarch's renewal of Stoicism moved along two main paths. The first was his presentation, in *De vita solitaria* and *De otio religioso*, of a Stoic / Christian way of life in which cultivation of the scholarly life and ethical perfection are one.[1] Petrarch's model looks remarkably like Seneca's *sapiens* – that

This article and earlier ones are contributions to a volume in preparation on Latin Stoicism in Renaissance Italy. See L. Panizza, "Gasparino Barzizza's Commentaries on Seneca's Letters," *Traditio*, 33 (1977), 297–358; idem, "Textual Interpretation in Italy, 1350–1450: Seneca's Letter I to Lucilius," *Journal of the Warburg and Courtauld Institutes*, 46 (1983), 40–62; and "Biography in Italy from the Middle Ages to the Renaissance: Seneca, Pagan or Christian?", *Nouvelles de la République des Lettres*, 2 (1984), 47–98.

1 See edition of the former with facing Italian translation by G. Martellotti in Petrarch, *Prose* (Milan: Ricciardi, 1955); reprinted separately (Turin: Einaudi, 1977). English translation by J. Zeitlin, *"The Life of Solitude" by Francis Petrarch* (Urbana: University of Illinois Press, 1927). Critical edition of *De otio* by G. Rotondi (Rome: Vatican Press, 1958); Latin and facing Italian translation by A. Bufano in Petrarch, *Opere Latine* (Turin: UTET, 1975). See also H. Cochin, *Le frère de Pétrarque et le livre du "Repos des religieux"* (Macon: Protat Frères, 1902).

self-sufficient sage, as rare to find as the phoenix, delineated in the *Dialogues*. The second was Petrarch's elaboration of Stoic psychotherapy in his huge compendium, *De remediis utriusque fortunae*, based on a mutilated work of similar name attributed to Seneca. The principles of psychotherapy that Petrarch advocates, he tells us explicitly in *De remediis* and in his more intimate *Secretum*, are derived mainly from Cicero's *Tusculans*, the richest Latin source on the subject, and also from Seneca's *De tranquillitate animi* and the Letters to Lucilius. Although considered for at least two centuries after Petrarch's death a major work of moral philosophy, and printed more than any other of his Latin works, *De remediis* is nowadays his most neglected and least-liked writing.[2]

Since *De remediis* has been classified for centuries as moral philosophy, let me explain why I seem to be changing the label. I am using the word "psychotherapy" in its etymological sense of "healing the soul," or, more precisely for this context, "healing the passions," understood as disturbed emotions. All the major ancient philosophical schools had opinions about psychotherapy and used analogies from bodily illness and health to discuss the soul. The Stoics, too, provided remedies or cures meant to restore the soul overwhelmed by the passions to inner *tranquillitas*, or equanimity. Just as philosophy used to be understood in a very wide sense, embracing the whole of learning, moral philosophy, in turn, included ethics proper, politics, and economics, as well as practical advice about how to live and face death and how to temper the emotions. If one wants nowadays to examine the history of psychotherapy, of the analysis of various emotional states and their treatment, one has to do some digging in the variegated mines of moral philosophy.[3]

How were emotional illnesses defined? What was the treatment proposed? How was the therapy supposed to work? Was it actually effective? In trying to give answers to these questions

2 The tide is beginning to turn; see note 23, this chapter, for modern translations and studies. The little work attributed to Seneca, very popular in the Middle Ages, entitled *De remediis fortuitorum liber ad Gallionem*, can be found in Seneca, *Opera quae supersunt. Supplementum*, ed. Fr. Haase (Leipzig: Teubner, 1902), pp. 44–55.

3 See P. Laín Entralgo, *La curación por la palabra en la Antigüedad clásica* (Madrid: Revista de Occidente, 1958). English translation: P. Laín Entralgo, *The Therapy of the Word in Classical Antiquity*, ed. and trans. L. J. Rather and J. M. Sharp (New Haven: Yale University Press, 1970).

about Stoic psychotherapy and Petrarch's *De remediis*, we need to look first at Cicero and Seneca, glance briefly at medieval Christian writings different in spirit but also imbued to some extent with Stoicism – monastic treatises on the spiritual life, for example, and the *De contemptu mundi* genre – and compare the *De remediis* with Petrarch's earlier attempt at self-therapy, the *Secretum*.

I

Petrarch certainly exploited Cicero's *Tusculans* more fully than anybody we know before. Cicero, however, as he tells us frequently, was not himself a Stoic but an Academic or Skeptic. He actually attacks Stoic determinism, using the Skeptics' weapons, in *De fato* and *De divinatione*.[4] Nevertheless, in *De natura deorum* he has the Stoic spokesman Balbus provide a full picture of a benign divine providence purposefully ordering the universe, an account frequently borrowed by early Christian writers like Lactantius and Augustine. In a dialogue about the main ethical schools, *De finibus*, Cicero clearly favors the Stoics and despises the Epicureans; and in *De officiis*, the most widely diffused Latin moral treatise of the Middle Ages and the Renaissance, his express aim is to adapt the teachings of the Greek Stoic Panaetius to Roman tastes. In *Tusculans* he also addresses contemporary Romans, conscious of writing at a turbulent moment of their history, the extinction of the Roman republic. Cicero's aim was practical: how to overcome crushing anxieties about suffering and sudden death. He surveys philosophical opinions about the immortality of the soul and, in Books III and IV, turns from physical suffering to emotional distress and Stoic psychotherapy.[5]

Underlying Cicero's interpretation is the basic Stoic belief that although one cannot alter the outside world – under the control of providential fate, anyway – one can alter one's inner emotions. Hence the importance of psychotherapy. We are all afflicted by emotional disturbances, *perturbationes*, caused by errors of judgment. The errors, in turn, arise from discrepancies between our

4 For Cicero's skepticism and the Renaissance, see C. B. Schmitt, *Cicero Scepticus* (The Hague: Nijhoff, 1972).
5 All references are to book and section of the Loeb Classical Library edition of *Tusculan Disputations* (1966).

subjective mental impressions, to which the imagination contributes, of something desirable or harmful – *opinio* – and the same object viewed by sound judgment or right reason, *ratio*. Our emotions are continually affected by the imagination's power to exaggerate or underestimate the desirability or undesirability of an object. Philosophy offers the cure, *medicina*, by pointing out the discrepancies and thus changing the *opiniones*, which should in turn change the emotions. To be cured means to be free from unwanted emotions, fully in control of ourselves and our feelings. This is the state of the self-sufficient Stoic sage, or *sapiens*. As Cicero sums up, "The whole purpose is for you to be master of yourself."[6]

The *perturbationes*, which render the soul unstable and ill, *insanus*, are to be distinguished on the one hand from madness, or *furor*, and on the other from well-regulated reasonable feelings, *affecti*. If one is mentally ill, *insanus*, one can still discharge the duties of ordinary life, but *furor* takes away one's power of judgment. It is blindness of the mind (III.11), and psychotherapy cannot deal with it. The Stoics recognized four main *perturbationes*: two stirred up by subjective impressions of something good or evil in the present, *voluptas* and *aegritudo* respectively; and the other two stirred up by *opiniones* of future good or evil, *cupiditas* and *metus*. Cicero's definitions emphasize the intensity, the abnormality, and the uncontrolled quality of these feelings. Thus *voluptas* is not an ordinary pleasure but an exhilarating, exuberant kind, an elated feeling of joy that is excessive ("voluptas gestiens... praeter modum elata laetitia"). The modern notion of mania might be an appropriate equivalent. *Aegritudo* is a subjective impression of a serious present calamity so great that it rightly appears as a cause of anguish ("opinio magni mali praesentis... ut in eo rectum videatur esse angi"). The term "depression" and its older relative "melancholia" may come closest to the substance of this untranslatable word. *Cupiditas* Cicero defines as an excessive appetite or desire, uncontrolled and immoderate craving, ambition, or lust ("immoderata appetitio... non obtemperans... vel libido"); and *metus*, or dread, is a subjective impression of a great threatening evil ("opinio magni mali impendentis"), which

---

6 "Ut tute tibi imperes," in II.47, repeated shortly afterward in II.53: "Totum... est, ut tibi imperes."

sounds like "anxiety." The two most severe disturbances are *aegritudo* and *cupiditas*, and *aegritudo* is simply the worst of all (III.23–5).

Although the Stoics were accused, especially by the Christians, of suppressing all emotion, Cicero does not seem to share such a view. In contrast to the disturbed emotions just listed, he reports that the Stoics allow good emotions in accord with reason, called *constantiae*. So instead of *voluptas* there can be *gaudium*, or calm enjoyment; instead of *cupiditas*, *voluntas*, or wishing; and instead of dread, *cautio*, or mild apprehension. Interestingly, there is no reasonable emotional state for *aegritudo* (IV.12–14).

Cicero is not always sure about applying medical terminology to do with bodily illness to the soul – an issue besetting later writers as well. While the Greeks, he says, use *pathe* for disease of both body *and* soul, Cicero prefers to distinguish in Latin between *morbus*, for bodily illness, and *perturbatio*, for emotional disorder (III.7–8). He accepts Zeno's definition of the latter as "an agitation of the soul alien from right reason and contrary to nature" ("aversa a ratione contra naturam animi commotio," IV.47), but on the other hand also declares that illnesses of the soul – now called *morbi* – are more numerous and harmful than bodily ones (III.5). On another occasion, he refuses to go as far as Chrysippus in drawing analogies between physical and mental illnesses (IV.23). Then he also disagrees with the Aristotelians, who consider emotional disturbances natural, to be merely restrained rather than extirpated, for this would suggest they did not need to be cured like any other illness (III.22). Nevertheless, Cicero accepts that there is *medicina* for the soul as well as the body and compares health and illness in the body to wisdom and folly in the mind, using the same words for both, *sanitas* and *insanitas*. "It follows," says Cicero, "that wisdom is a sound condition – *sanitas* – of the soul, unwisdom ... a sort of unhealthiness – *insanitas* – which is unsoundness and also aberration of mind" (III.10). Cicero is in less doubt about the origin of illness: Physical disease befalls one involuntarily, but in emotional disturbance there is an element of the voluntary, at least in succumbing to those *opiniones*. Otherwise, one could not cure oneself. Cicero never resolved to what extent mental illness and its cure were matters of the will, however, and the same issue vexed later thinkers.

Of all the *perturbationes*, *aegritudo* resembles bodily sickness the most; its very name, Cicero clarifies, is linked with pain (III.23). It is synonomous with wretchedness, *miseria* (III.22,27), and includes worry, anxiety, and anguish, a lowering, sinking, and breaking down of one's spirits. Indeed, while all disturbance is wretchedness, *aegritudo* is actually being put on the rack. Cicero's discussion of remedies is mainly in connection with this distress, just like Petrarch's in the *Secretum*. There is no question for Cicero of bearing pain, physical or mental, for its own sake; it must be unavoidable. Being in control at all times, the "stiff upper lip" attitude, is admired, while moaning and shrieking are condemned, and weeping and sobbing utterly despised (II.55–6). Even with physical pain, however, Cicero recognizes the imagination's power to make pain worse and also to lessen it (II.42).

Cicero's discussion of remedies for *aegritudo* is loose and rambling, often repetitive, and hence difficult to pin down. Much time is spent disagreeing with Peripatetics, Cyrenaics, and above all Epicureans in order to establish a qualified preference for the Stoics. Epicurus' therapy of compensation, which Cicero rejects, sounds very sensible. It required diverting the mind from its present anguish by, first, the recollection of past pleasant and happy memories, and second, the enjoyment of sense pleasures such as fine music, sweet-smelling flowers, and good food (III.43). The Cyrenaics put the blame for *aegritudo* on sudden, unexpected calamities; the remedy lay in *praemeditatio futurorum*, that is, diligent reflection on all the vicissitudes that fall to the human lot, before they actually happen – a therapy of prevention (III.29–30) which found great favor in the Middle Ages and certainly with Petrarch. While unable to accept that sudden blows of fortune are responsible for all *aegritudo*, Cicero finds this kind of meditation admirable, as long as we remember that our subjective impression of the disasters, and not the disasters themselves, cause the emotional disturbance. If evil resided in things or events themselves, he asks pointedly, why would meditation make them easier to bear? ("Si enim in re esset [*malum*] cur fierent provisa leviora?", III.32). In contrast, the Epicureans advise against examining future possible evils, because they are difficult enough to bear when they arrive. Cicero briefly touches on the usefulness of interior dialogues in which we imagine the personified virtues Fortitude, Temperance, Justice, and Prudence exhorting our infirm soul not to succumb passively to Fortune (III.36–7).

The cures Cicero most approves of in treating emotional disturbances boil down to two. First, we examine at length the particular circumstances that bring on the disturbance, and then we persuade ourselves, as we would do in a rhetorical exercise, that we exaggerate or underestimate its desirability or undesirability. The second, related cure has to do with finding *exempla*: deeds and sayings of others who have gone through and endured similar experiences. Both methods involve meditation and reflection – Cicero uses the words *cogitari* and *meditari* (III.56–8). He says there already exist prepared discourses on specific calamities, or *casus*, such as poverty, exile, slavery, invasion, infirmity, and blindness, as well as *consolationes* on losing a loved one (III.81–2). He is confident that this therapy of changing one's point of view will at least lessen the *perturbationes*. In Book IV, he sums up his therapeutic goal thus: "Just as where evil is expected, the prospect must be met with endurance, so where good is expected, the objects held to be momentous and delightful must be regarded in a calmer spirit."[7] We have here in a nutshell the basic principle of Petrarch's *De remediis*.

Regarding the relationship of Stoic psychology to ethics, it may come as a surprise to realize that emotional balance and imbalance are synonymous with virtue and vice. Thus virtue is "adfectio animi constans conveniensque" – an even-tempered and harmonious disposition of the soul (IV.34). From such a soul spring all good wishes, thoughts, desires, and deeds that make up right reason. A person who has reached that state is truly happy, *beatus*, like the Stoic sage. The disordered soul, on the other hand, is full of defects and faults, *vitia* and *peccata*, and under each of the four main disturbances is a long list of these defects. *Aegritudo* embraces envy, rivalry, jealousy, anxiety, grief, mourning, torment, lamenting, brooding, worrying, and despondency (IV.16).

While Cicero's approach in *Tusculans* is expository, Seneca's is more personal and dramatic. Not only was he a practicing Stoic, albeit an eclectic one who was forever quoting his rival Epicurus, but he presents himself as a healer, addressing Serenus in *De tranquillitate animi*, and Lucilius in the letters called in Latin *Epistulae morales*. Both were younger and less experienced in the

7 "Ut in malis opinatis tolerabilia, sic enim in bonis sedatoria sunt efficienda ea quae magna et laetabilia ducuntur" (IV.65).

Stoic way of life than Seneca himself and turned to him for counsel.[8]

Serenus is baffled by his malady; he is not physically ill, but neither is he well. Let us listen to him carefully, for the similarities between Serenus and, later, Petrarch describing their *aegritudo* are remarkable. "I am in all things attended by this weakness of good intention" ("bonae mentis infirmitas"), Serenus confides, "In fact I fear that I am gradually fading away, or, what is even more worrying, that I am hanging like one always on the verge of falling" (1.15). Lacking determination, restless, pulled in opposing directions, Serenus begs Seneca for any remedy "by which you could stop this inner fluctuation of mine, and deem me worthy of being indebted to you for tranquillity" (1.17). With this illness, Seneca sees no need for the customary harsh Stoic measures such as "the necessity of opposing yourself at this point, of being angry with yourself at that, of sternly urging yourself on at another" (2.2). Rather than these inner dialogues, Seneca recommends self-confidence and the belief that one is on the right track: mild treatment, and not so far from what the Epicureans recommended.

While for Cicero all four *perturbationes* and their numerous offspring take away one's inner equilibrium, for Seneca instability – elation one minute, despondency the next – is the main problem. "What you desire," he tells Serenus, "is something great and supreme and very near to being a god – to be unshaken," an abiding stability of mind ("stabilem animi sedem"), which the Greeks call *euthymia* and he *tranquillitas* (2.3). But Serenus' illness is just the opposite, and its symptoms include fickleness, disgust with life, and continual shifting of purpose ("levitate ... taedio adsiduaque mutatione propositi," 2.6). As a result, one is always starting afresh and never finishing anything, living forever in suspense and hoping always for something new. Seneca sums it up with the phrase "to be always dissatisfied with oneself," *sibi displicere* (2.7) – Petrarch's description of his own malady. The advice continues to be mild: One must distinguish between what one would like to do and what one is really able to achieve. Feverish ambitions lead quickly to despondency and need to be

---

8 See Loeb Classical Library editions of *De tranquillitate animi* (1965), and *Epistulae morales* (1962–7). References are to chapter or letter, and section.

curbed: "You ought not to approach a task from which you are not free to retreat. . . . Leave untouched those that grow bigger as you progress and do not cease at the point you intended" (6.4). Seneca recommends a judicious balancing of work and retirement, of solitude and dedication to a common good, especially by communicating the fruits of one's study in writing – a program he followed himself and which Petrarch found congenial too. In his personifications of *Fortuna* and in his metaphors about the storms and struggles of the emotions, blown this way and that by the accidents of life and reason, Seneca visually dramatizes the process of psychotherapy more than Cicero. Fortune battles unsuccessfully with the sage, whose desires are completely restrained and who therefore sets little value on what she has to offer: "Of my own free will I am ready for you to take what you gave me before I was conscious – away with it" (11.4). Fortune plays with mankind, but if one is prepared for all her blows – sickness, captivity, disaster, conflagration – *tranquillitas* can be maintained (11.6–7). The message is similar to Cicero's, yet with Seneca the soul itself becomes the only stage of significant moral activity, the site of human conflicts, defeats, and triumphs.

Letter 28 to Lucilius further illustrates this dramatic aspect of self-therapy. Lucilius has been afflicted by a bout of low spirits, *tristitia*, and thought a change of scenery might help. For Seneca, distraction is of no use, for wherever Lucilius goes, his problems will go with him. A better remedy would be to act out within himself the roles of prosecutor, judge, and defense, for unless he can see clearly that he is ill, there will be no possibility of a cure. In Letter 11, Seneca advises Lucilius to choose a person he both likes and admires and have him present at all times in his imagination. Lucilius should then turn to this imaginary healer for comfort and advice. And in 13, dealing with anxieties, Seneca declares that we suffer more through the *opiniones* than through things themselves; "We are all in the habit of exaggerating, imagining or anticipating painful things." One remedy is to conduct an inner dialogue, with questions, accusations, retorts, and denials; another, to put before the distressed mind pleasant future events and even drive out one fault by another, tempering dread and anxiety by hope ("vitio vitium repelle, spe metum tempera," 13.12). Petrarch appears to have taken such recommendations very much to heart in his *Secretum*.

In the last analysis, however, there is no halfway measure with the *perturbationes*. They must be extirpated, says Seneca, not tempered as the Peripatetics teach (Letter 116). Nevertheless, unlike the Cynic – Cicero had made the same point – the Stoic does have feelings. Seneca explains that the wise man, emotionally self-sufficient, always serene, rejects unwanted, distressful feelings that would undermine the supreme goal of *tranquillitas* (Letter 100). Seneca puts it another way, in *De vita beata*: "The highest good is the inflexibility of an unyielding mind, its foresight, its sublimity, its soundness, its freedom, its harmony, its beauty" (9.4). Little wonder that Petrarch often places Seneca side by side with Saint Augustine as one of the two masters of introspective literature.

II

Petrarch's interest in Stoic psychotherapy was not just part of his general fascination with classical thought and literature. In his *Secretum* or, more precisely, *De secreto conflictu curarum mearum* – "On the hidden conflict of my own worries" – Petrarch revealed the conflict of his own passions that had brought about what he calls both *accidia* and *aegritudo*. He also makes known remedies he has tried, some of which were more successful than others. There are obvious connections, therefore, with the later and larger *De remediis*. Drafts of the *Secretum* were written in 1347 and 1349,[9] and a final version in 1353, with a few finishing touches in 1358. The first record of *De remediis* goes back to 1354; it was completed in 1366.

The *Secretum* is a psychological and allegorical dialogue between himself and Augustine, with Truth present as a silent judge.

---

9 All references are to the edition by E. Carrara with facing Italian translation, in *Prose*. English translation by W. H. Draper, *Petrarch's "Secret, or The Soul's Conflict with Passion"* (London: Chatto & Windus, 1911). Translations in this article my own. For the *Secretum's* sources, composition, and secondary bibliography, see F. Rico's magisterial *Vida u obra de Petrarca*: vol. 1, *Lectura del "Secretum"* (Padua: Editrice Antenore, 1974). I follow Rico's redating of the *Secretum* (pp. 7–16). See also K. Heitmann, *Fortuna und Virtus: Eine Studie zu Petrarcas Lebensweisheit* (Cologne: Bohlan Verlag, 1957); F. Tateo, *Dialogo interiore e polemica ideologica nel "Secretum"* (Florence: Le Monnier, 1965); C. Trinkaus, *The Poet as Philosopher: Petrarch and the Formation of Renaissance Consciousness* (New Haven: Yale University Press, 1979); and N. Mann, *Petrarch* (Oxford: Oxford University Press, 1984).

Petrarch presents himself as the sick patient, *morbus*, and Augustine as both voice of conscience and rational healer. In other words, we have an inner dialogue of the sort prescribed by Cicero and Seneca as therapeutic, with the self split into opposing parts. There were numerous later examples for Petrarch to turn to as well, notably Augustine's own *Soliloquies* and Boethius's *Consolation of Philosophy*.[10]

Petrarch brings together in the *Secretum* two not always concordant approaches to healing: the ancient and pagan one we have just looked at, and the Christian, found in patristic and monastic ascetical literature about the spiritual and moral life. In fact, the *Secretum* has as a second title *De contemptu mundi*, in recognition of its similarity to works in that genre like Pope Innocent III's noted *De miseria humane conditionis*. The pope had set out to illustrate graphically the wretchedness of the human lot "in order to bring down pride" ("ad deprimendam superbiam"), and also had in mind a companion volume to raise up despair, never completed. In this plan we can recognize faintly a Stoic theory put to Christian ends.[11] Petrarch practiced the meditation on death recommended by the pope, imagining, in every disgusting detail, the horrible suffering of the dying body and the torments to follow in hell. If carried out with sufficient intensity, this "cogitatio vehemens" was meant to change one's attachments to the pleasures and satisfactions of this life into loathing and dread – the "contemptus mundi." Significantly, Petrarch found that this therapy by aversion, far from producing the desired effect of strengthening his will, made his anxiety and depression worse. He never returned to it.[12]

Petrarch has an even more exaggerated sense of the role of the will in curing emotional disturbances than his pagan models. He has Augustine explain his own famous conversion as an effective example of meditation and decision: "As soon as I fully willed to change, I was at once able to; and with marvellous and most

10 On Petrarch's fondness for Augustine, see P. P. Gerosa, *Umanesimo cristiano del Petrarca* (Turin: Bottega d'Erasmo, 1966), and studies listed in note 9, this chapter.

11 Lotharii Cardinalis (Innocent III), *De miseria humane conditionis* (Padua: Editrice Antenore, 1953). C. Trinkaus sees Petrarch reviving the genre; see his "Human Condition in Humanist Thought: Man's Dignity and His Misery," in *In Our Image and Likeness*, 2 vols. (London: Constable, 1970), 1:171–99.

12 *Secretum*, pp. 29–45. The meditation is discussed by Rico, *Vida u obra*, pp. 41–102, with ample documentation of earlier examples.

fortunate swiftness, I was transformed into another Augustine."[13] Critics have noticed how mistaken Petrarch is about Augustine, who saw his conversion in Pauline terms of divine grace freeing him from the slavery of error and lust, rather than as an exercise of the will putting down the *perturbationes*.[14]

But Petrarch is not unusual in blurring boundaries between Christian and Stoic; throughout the Middle Ages, Christian writers adapted and modified Roman Stoic teachings, especially on ethics. There are three issues relevant for literature on the passions. First, the *opiniones* become less and less exaggerated subjective impressions or imaginings of an outside reality verifiable by another mental faculty – that is, our own judgment or somebody else's – and overlap with errors of religious belief. These are to be corrected by illumination from above or religious instruction, sources of truth outside oneself. Second, the terms *peccata* and *vitia* come to acquire connotations of sin. Now sin does not make one merely unhappy or less perfect but deserving of penance and divine punishment. At the same time, the Christian scheme of the seven capital sins retains something of the Stoic understanding of *peccatum* as faulty inner disposition. (In Dante's *Purgatorio*, all the souls have to undergo a long and painful purgation of the capital sins before they can move on to paradise.) Third, there is increasing ambivalence about the will's role in causing and curing "moralized" emotional illness. Cicero and Seneca made an important distinction between a general will to be cured and the various kinds of medicine or remedies for healing the passions. Both were necessary, and the former was certainly not a substitute for the latter. But Christians are in a dilemma: On the one hand, they should *a fortiori* have more effective wills than pagan Stoics in willing away disturbed passions; and on the other hand, God's grace should be at work in mysterious ways to bring about sudden changes of the will, as in the case of conversion.[15] Petrarch himself

---

13 "Postquam plene volui, ilicet et potui, miraque et felicissima celeritate transformatus sum in alterum Augustinum" (*Secretum*, p. 20). Compare Augustine, *Confessions*, VIII.10–12.

14 See P. Courcelle, *Les confessions de Saint Augustin dans la tradition littéraire* (Paris: Études Augustiniennes, 1963), esp. pp. 329–43, and Rico, *Vida u obra*, pp. 70–9.

15 For the diffusion in the Middle Ages of Stoic moral values among monastic orders, see M. Spanneut, *Permanence du Stoïcisme de Zénon à Malraux* (Gembloux: Duculot, 1973); K.-D. Nothdurft, *Studien zum Einfluss Senecas auf die Philosophie und Theologie des zwölften Jahrhunderts* (Leiden: Brill, 1963); L. D. Reynolds, *The Medieval Tradition of*

is led by his own experience to confess that neither a great amount of sheer willing nor, apparently, God's grace has in fact delivered him.

Significantly, Petrarch turns explicitly to Cicero's *Tusculans*, Book III, and Seneca's *De tranquillitate animi* for advice about that most intractable disturbance, *aegritudo*. Petrarch, it seems to me, has explicitly identified these Latin sources of Stoic psychotherapy, recognized their similarity to Augustine and later, more derivative Christian works, exploited their fuller description of mental illness, and found their remedies more helpful. "You have a deadly kind of disease of the soul," he has Augustine inform him during an examination of conscience on the seven capital sins, "which the moderns call *accidia* and the ancients *aegritudo*." Reinforcing this union of classical and ascetic Christian thought on the matter, Petrarch adds other synonyms: *tristitia* – Augustine's favorite word – and *taedium*, *dolor*, and *morbus*.[16] Petrarch describes the illness as one in which all things are "et aspera et misera et horrenda" ("both harsh and vile as well as frightening"). He is tormented day and night and cannot shake it off; he imagines his soul is assailed by *Fortuna*, leading an army of life's woes which weaken and then crush him with anxiety. As with Seneca's Serenus, nothing in his own life or others' pleases him; everyday occurrences, even city noise, are a source of distress.

Nevertheless, where will and grace have not succeeded, Petrarch has found two cures. The first is to commit to memory helpful pieces of advice, *sententiae*, culled from his reading. Just as good doctors always carry medicines with them at all times to treat illnesses without delay, so he will have such arguments ready, especially for sudden bouts of anger. Petrarch quotes a

---

Seneca's Letters (Oxford: Oxford University Press, 1965); P. Faider, "La lecture de Sénèque dans une abbaye du Hainaut," in *Mélanges Paul Thomas* (Brussels, 1930), pp. 208–47; J.-M. Déchanet, "Seneca noster: Des Lettres à Lucilius à la lettre aux Frères du Mont-Dieu," in *Mélanges Joseph de Ghellinck*, 2 vols. (Gembloux: Duculot, 1951), 2:753–66; G. G. Meersseman, "Seneca maestro di spiritualità nei suoi opusculi apocrifi dal XII al XV secolo," *Italia Medioevale e Umanistica*, 16 (1973), 43–135.

16 "Habet te funesta quaedam pestis animi, quam accidiam moderni, veteri egritudinem dixerunt" (*Secretum*, p. 86). The literature on *accidia* is vast; for Petrarch, see studies in note 9, this chapter, esp. Rico, *Vida u obra*, pp. 197–220, and also S. Wenzel, *The Sin of Sloth: Acedia in Medieval Thought and Literature* (Chapel Hill: University of North Carolina Press, 1967), esp. pp. 155–63, 185–6. Augustine discusses *tristitia* in *City of God*, XIV.8. Understood as sorrow for sin, he sees it as a good emotion.

passage from Virgil (*Aeneid*, I.52–9) about destructive, stormy winds, which, interpreted as an extended metaphor to do with the passions, he apparently found apposite. Augustine advises him also to make marginal notes or signs to help him recall particular passages. The suggestion anticipates the *De remediis*, especially as Petrarch states that philosophers as well as poets are to be included in his collection.[17] The second is about tempering ambition, a form of *cupiditas* that Seneca advised Serenus about. Petrarch should take comfort and rejoice in his many gifts of good fortune and, at the same time, consider the example of others before him who have sustained similar misfortunes. Most of all, like a survivor on the shore contemplating a shipwreck, he should consider how much luckier he is than the drowning crew. Although a remarkably unchristian sentiment, Petrarch is actually following Seneca's advice in Letter 15: "When you see many ahead of you reflect how many are behind. If you would be grateful to God [Seneca actually has "gods"] and satisfied about your past life, reflect on how many men you have surpassed."[18] Although the *Secretum* has been judged inconclusive, the Roman Stoics did help Petrarch on cures for *aegritudo*. They encouraged him to consider the *perturbationes* in terms other than guilt and punishment, as something that could be dealt with at least partially by practical rational means, especially internal dialogues between himself and revered authorities. The will cannot do everything, but neither, by implication, can grace.

III

De remediis utriusque fortunae takes its name from a pamphlet attributed to Seneca, *De remediis fortuitorum*, mentioned first by the church father Tertullian, reduced to *flosculi* (excepts) for Vincent of Beauvais' encyclopedia, and merged into Innocent III's *De contemptu mundi*.[19] This probably apocryphal work is a short

17 *Secretum*, pp. 102–4.
18 Ibid., p. 92. Seneca's words are, "Cum aspexeris quot te antecedant, cogita quot sequantur. Si vis gratus esse adversus deos et adversus vitam tuam, cogita quam multos antecesseris."
19 See Haase, *Opera*, for *Testimonia*; for the work's *fortuna*, see R. G. Palmer, *Seneca's "De remediis fortuitorum" and the Elizabethans* (Chicago: Institute of Elizabethan Studies, 1953); and P. Faider, *Études sur Sénèque* (Gand: Van Rysselberghe & Rombaut, 1921), p. 111.

collection of Reason's often one-line answers to emotional laments
expressing all kinds of irrational fears; as manuscripts sometimes
put it, "ratio confortans" ("a consoling reason") dialogues with
"sensum conquerens" ("a whining emotion").[20] Minimal remedies are provided for dread of death, bitter and violent death,
dying young, lack of burial, illness, defamation, exile, pain,
poverty, hunger and nakedness, loss of money, blindness, deaf-
ness and dumbness, death of one's children, shipwreck, enemies,
loss of friends, and loss of a wife. As there is no corresponding
remedy for loss of a husband, the remedies would seem to be
intended for men: "I am without a wife," is the last complaint,
to which follows the misogynistic rejoinder, "And without an
adversary – you can now begin to be master of yourself and your
possessions."[21]

The leaflet's remedies address only one Stoic perturbation, *timor*
or *metus*, treating it crudely by denying, as we just saw, that there
are true reasons for grief. Interestingly, the kinds of suffering
listed have on the whole to do with domestic and family problems, a plainly ripe area for therapy: "Happy is the man who does
not seem so to others but to himself; you see, however, how
rarely this happiness is found at home."[22]

Petrarch's addition of "utriusque" and his change of Seneca's
original "fortuitorum" to "fortunae" are not without significance.
In the dedicatory letter (dated 1354) preceding Book I, addressed
to Azzo da Correggio, lord of Parma, he boldly asserts that he will
improve on Seneca's popular little tract and expand it. (Petrarch
thought it was genuine.) While denying that he has ever stolen
from or pillaged others' works, Petrarch also affirms that virtue
and truth are public property and so there is room for him to add
his own contribution to the ancients'. What Seneca did for Gallio,

20 See rubrics in, for example, Milan, Ambrosiana, C293 (saec. XIV), and London, British
   Library, Harley 3436 (saec. XV), which also indicate how highly the booklet was
   valued: "Fecit [Seneca] illum sub dyalogo ut sit sensus conquerens et ratio confortans.
   Liber autem iste et sensuum maiestate et eloquii claritate et sententiarum brevitate
   refulget" ("[Seneca] composed this dialogue with a consoling reason and complaining
   emotion. This book shines on account of the sublimity of its thought, the brilliance of
   its style, and the brevity of its maxims.")
21 To the complaint "Uxorem bonam amisi. . . . Sine uxore sum," Reason answers, "Et
   sine adversario, iam tui rerumque dominus tuarum esse incipis" (Haase, *Opera*, p. 54).
22 "Felix est non qui aliis videtur, sed qui sibi: vides autem, quam rara domi sit ista
   felicitas" (ibid., p. 55).

in *De remediis fortuitorum*, Petrarch will do for Azzo, but in addition he will deal with the other side of fortune that Seneca had left out – hence both kinds, or "utriusque fortunae." Petrarch explains, "I believe it is far more difficult to behave under the rule of favorable than adverse fortune, and I believe myself that fair fortune is more dangerous and deceptive than foul."[23] What led him to this conviction was not specious reasoning ("sophismatum") but his life's experience ("rerum experimenta vitae"). How many have borne the most dreadful calamities with equanimity, only to lose their heads when showered with wealth, honors and power! A popular proverb about what an uphill struggle it is to handle success proves the point: "Magni laboris esse, ferre prosperitatem."[24] Petrarch absorbs all the subjects listed in Seneca's booklet and adds many many more. Book I contains 122 dialogues combating "prospera fortuna"; Book II contains 132 dialogues combating "adversa," making a grand total of 254. The dialogues are all examples of familiar psychological allegory: The soul is torn apart by conflict between Reason, on the one hand, and the four Stoic *perturbationes* – elation and craving, sadness and dread – on the other. *Ratio* puts *Gaudium* and *Spes* into place in Book I and resists and controls *Dolor* and *Metus* in Book II.

---

23 "Difficilius prosperae fortunae regimen existimo, quam adversae: aliquantoque fateor apud me formidilosior et quod constat insidiosior est fortuna blanda quam minax" (*Opera omnia* [Basel, 1581], fol. †††4r). Translations my own. There is as yet no modern critical edition of *De remediis*, but A. Sottili is preparing one. C. H. Rawski has completed an English translation with notes that awaits publication. Two early English translations in print are F. N. M. Diekstra, ed., *A Dialogue between Reason and Adversity: A Late Middle English Version of Petrarch's "De Remediis"* (Assen: Van Gorcum, 1968), with very complete introduction on the work's history; and *"Physicke against Fortune", 1579, First English Translation by Thomas Twyne*, facsimile reproduction with introduction by B. G. Kohl (Delmar: Scholars' Facsimiles and Reprints, 1980). See also selection in *Four Dialogues for Scholars*, ed. and trans. C. H. Rawski (Cleveland: Western Reserve University Press, 1979). In Italian, see *"De' rimedii dell'una e dell'altra fortuna," volgarizzati per D. Giovanni Dassaminiato* (1427), ed. Don C. Stolfi, 2 vols. (Bologna, 1867), and selections in Petrarch, *Prose*, ed. Martellotti; and in German, see selections translated and annotated by R. Schottlaender, with up-to-date bibliography by E. Kessler (Munich: Fink, 1975). Discussions of *De remediis* can be found in Heitman, *Fortuna und Virtus*; Rico, *Vida u obra*; and Trinkaus, *In Our Image*. Azzo and Petrarch were close friends, dating from their Avignon days in the 1330s. When Azzo came to power in Parma, Petrarch became a favored adviser and received benefices; Azzo, in turn, accompanied Petrarch to Rome and Naples in 1341, when he was crowned poet. Azzo was then visited relentlessly by adversities, which he apparently bore in exemplary fashion. He died in 1362, before the completion of *De remediis*. See ample notes in G. Fracassetti, *Lettere di F. Petrarca: Varie* (Florence, 1892), pp. 525–33.

24 *Opera omnia*, fol. †††4r.

Petrarch redefines *fortuna*, drawing attention away from chance events – *fortuita* – to the Stoic emotional disturbances themselves. The preface to Book II makes clear that the greatest struggle of the human condition is not with events befalling us but with the *perturbationes*, forever in conflict with one another and with reason. Not only are there so many appetites to divide and devour the psyche, but we also oscillate relentlessly from one feeling to its contrary. Petrarch supplies a long list of the psyche's opposing fluctuations: willing / not willing, loving / hating, flattering / threatening, mourning / having compassion, showing mercy / becoming angry, being downcast / being elated, forgetting / remembering, and so on – all of which prevent us from attaining our inner tranquillity. Strongest of all are the four Stoic passions, whose winds buffet the fragile soul without respite, those old friends "Sperare seu Cupere et Gaudere, Metuere et Dolere," known to Virgil and Augustine, Petrarch points out, drawing them once again into the Stoic circle.[25]

Writing about everyday life for ordinary people, he begs pardon from educated Christian readers not only for putting down the chapters as they came to mind but also for using the pagan word *fortuna*. "Do not take offense," he warns Azzo, in case of misunderstanding, "for you have heard very often what I think of it." But addressing people of average education, "I saw it would be necessary to use a word in common usage, knowing well what others have said." Saint Jerome, for instance, stated, "There exists neither fate nor fortune."[26] *Fortuna*, in other words, is a colloquial expression for chance events and the passions, in this two-part work, and at one point Petrarch interchanges the two, calling his subject matter "passionum ac fortunae." The stress is on natural, psychological phenomena that are caused (directly) neither by God nor the devil, nor stellar influences. Following Petrarch's explanation, one could recast the title thus: "Treatment for various extreme and irrational forms of elation and depression, often brought on by numerous illusions, events, and circumstances to which everyone is subject."

25 Ibid., p. 106; and *Aeneid*, VI.733: "Hinc metuunt cupiuntque, dolent gaudentque," quoted also by Augustine in *City of God*, XIV.3.

26 "Neque vero te moveat fortunae nomen . . . saepe quidem ex me, quid de fortuna sentiam audisti, sed . . . haec necessaria praeviderem, noto illis, et commui vocabulo usus sum, non inscius, quid de hac relate alii, brevissimeque Hieronymus ubi ait: Nec fatum, nec fortuna" (*Opera omnia*, p. 106).

The *De remediis utriusque fortunae* is thus remarkable because it is the first example since classical times of popular Stoic psychotherapy addressed to lay people. Although written in Latin and meant for Christians, it is distinctly unclerical; it is not about the religious or even the spiritual life. Most important of all, it breaks treatment for disturbed emotions away from strictly ethical and theological contexts, so developing a trend started in the *Secretum*. Petrarch thus reforms the "De contemptu mundi" tradition by restoring it to its Stoic matrix. He adds, as he says, the other side of *fortuna*. He greatly increases the number of remedies. He turns each dialogue into a literary piece, crammed with quotations from ancient philosophers and poets and with *exempla*, some of which amount to anecdotes. True to their Stoic nature, the perturbations repeat their cries of elation or woe with little variation, while Reason, also true to Stoic nature, cajoles and persuades and always has the last word.

At the same time, it must be said, Petrarch does not liberate himself entirely from "De contemptu mundi" attitudes. The dedicatory letter to Azzo da Correggio, prefaced to Book I, explaining why Petrarch composed remedies for inner distress and how he meant Azzo and other readers to use them, also reveals a deep pessimism about the universe and human life and a troubling uncertainty about whether tranquillity can ever be attained. And the leitmotiv of the preface to Book II is Heraclitus's cosmic law that all things come into being through strife: "Omnia secundum litem fieri." (This pessimism is at odds with classical Latin Stoicism, and also with some strands of Christian doctrine, as Lorenzo Valla would later pick up.) Writing just after the Black Death, in war-torn northern Italy, Petrarch felt there was nothing more fragile and unstable than the affairs of mankind. Our finest natural gifts – memory, intellect, and foresight – are turned against us, and we are continually tormented by anxiety about the past, present, and future: "Thus perhaps we seem to dread more than anything else not being wretched at all."[27] In phrases reminiscent of Innocent III, Petrarch laments that at life's beginning there is blindless and oblivion; in its course, distress; at its end, pain; and throughout, error and illusion.[28] What day of our life has not

---

27 "Ut nil magis metuere videamur, quam ne quando forte parum miseri simus" (*Opera omnia*, fol. †††3v).

28 "Vitam . . . miserandum ac triste negocium efficimus, cuius initium caecitas et oblivio possidet, progressum labor, dolor exitum, error omnia" (ibid.). The titles of *De*

turned out to be fuller of burdens and troubles than peace and tranquillity? Petrarch asks. Endowed with uncanny malevolence, *Fortuna* – in the double sense mentioned earlier – tosses us about hither and thither, like playthings. It would almost be better to lack reason; at least then, like other animals, we would not be tortured by mental distress. The best safeguard against life's incessant flux lies in frequent conversation with wise men and continual reading of texts speaking to our condition – the remedy sketched previously in the *Secretum*. Petrarch is moved particularly by writers who lived so long ago, yet whose precepts guide us now to the harbor of inner tranquillity "like so many brilliant stars stuck in the firmament of truth, so many gentle and pleasant breezes, so many expert pilots at the tiller."[29] Petrarch, fond of nautical metaphors likening the soul to a frail bark blown about by the storm of the passions, adopted similar imagery from Virgil in his *Secretum* to describe *aegritudo* and composed an entire sonnet on the theme.[30]

In the preface to Book II, where the theme of strife is predominant, *Natura*, not just *Fortuna*, is imagined as bitterly hostile to mankind, a cruel stepmother rather than a provident mother. All creatures war against one another to their mutual destruction, from the humblest insect upward, and, as if that were not enough, all nature – from earthquakes and fire to snakebites and fleas – harms mankind. The inner war of the passions is still more relentless and destructive; it would sometimes appear that peace of soul itself is as much an illusion as the other *opiniones*.

So what must Azzo do? Petrarch takes his role as Azzo's adviser seriously. He approves of the ruler, burdened by responsibilities, nevertheless setting aside some hours every day for reading and reflection. Petrarch recommends short and practical books that can easily be committed to memory, like the dialogues he has

---

miseria's three books are "De miserabili humane conditionis ingressu," "De culpabili humane conditionis progressu," and "De dampnabili humane conditionis egressu." Throughout, the pope sees life as nothing but strife – the hostility of nature against mankind and of all men against each other. See the introduction by Maccarrone to the edition cited in note 11, this chapter.

29 "Interque perpetuos animorum fluctus, ceu totidem lucida sydera, et firmamento veritatis affixa, ceu totidem suaves ac foelices aurae, totidem industrii ac experti nautae, et portum nobis quietis ostendunt" (fol. †††3v).

30 *Aeneid*, I.52–7, 8–59; VI.730; *Secretum*, pp. 102–4; *Canzoniere*, 189, beginning "Passa la nave mia colma d'oblio / per aspro mare a mezza notte il verno / enfra Scilla et Caribdi . . .".

prepared. "By means of these brief, succinct, and authoritative sayings – *sententiae* – you will be able to protect yourself against all the attacks and sudden changes of *fortuna*, just as if you were always armed with handy weapons."[31] Petrarch is most concerned that Azzo should not have to "empty out the entire armory at each and every sighting or rattling of the enemy"; rather, Azzo should "always and everywhere have ready against a twofold illness a brief remedy prepared by a friend, like a proven antidote kept in a little box for every evil and harmful good proceeding from both kinds of Fortune."[32] Thus the entire book is like a pharmacopoeia, and each dialogue a specific medicine for a specific malady. There is nothing to stop Azzo from reading several dialogues from both sections, for the volume is also meant to prepare him in advance for each and every vicissitude, so he will be "paratus ad omnia, promptus ad singula" ("ready for everything, in general and particular").[33] Petrarch urges Azzo to practice this "praemeditatio futurorum," discussed by Cicero, quoting in further support Aeneas's words to the Sybil: "No kind of strange or unexpected hardship can befall me; I have meditated on and embraced all."[34]

Petrarch shares Cicero's worry that medical terminology may not be entirely appropriate for healing the *perturbationes*. He admits that to many people verbal *medicamenta* will seem to have no real effect. On the other hand, "just as the sicknesses of the soul are invisible, so are the cures." And if the illnesses are brought about by mistaken subjective impressions, "falsis opinionibus," they can likewise be cured by true and authoritative sayings, "veris sententiis."[35] Petrarch's gift to Azzo is thus a complete manual of

31 "His ... brevibus ac praecisis sententiis, quasi quibusdam expeditis atque continuis armis, contra omnes insultus, omnemque repentinum impetum, hinc illinc iugiter sis instructus" (*Opera omnia*, fol. †††4r).

32 "Ad id maxime respexi ne armarium evolvere ad omnem hostis suspitionem ac strepitum sit necesse, quin mali omnis et nocentis boni atque utriusque Fortunae remedium breve sed amica confectum manu quasi duplicis morbi, ut non inefficax antidotum in exigua pixide, omnibus locis atque temporibus ad manum ... et in promptu habeas" (ibid.).

33 Ibid.

34 For Cicero, see this chapter, section I. For Virgil, see *Aeneid*, VI.103–5: " ... non ulla laborum, / O virgo, nova mi facies inopinave surgit: / Omnia praecepi, atque animo mecum ante peregi."

35 "Nec me fallit ... medicamenta verborum multis inefficacia visum iri, sed nec illud quoque me praeterit, ut invisibiles animorum morbos, sic invisibilia esse remedia" (*Opera omnia*, fol. †††4v). For Cicero, see this chapter, section I.

Stoic psychotherapy in which Azzo's own reason will use Petrarch's words to quell "those four famous and related passions of the soul . . . which two sisters, prosperity and adversity, gave birth to at the same time."[36] The treatment, as Cicero too said, works on the principle of opposition: "Both attacks of Fortune are to be dreaded, just as both are to be resisted; one lacks a bridle; the other, solace; in one place the soul's elation must be brought down, and in another its labors encouraged and raised up."[37]

Following the loose order and ordinary style announced by Petrarch, the remedies in Book I for excessive *gaudium* and *spes* cover various family, social, political, and cultural circumstances and events, as well as various inner states (we are reminded of Cicero's remedies for both the disturbances themselves and occasions giving rise to them). Personal qualities include youthful vigor, health, agility, and beauty; intelligence and memory; even virtue and piety, if one boasts overconfidently about them. Family and social qualities include noble lineage, an eloquent and fertile wife with a generous dowry, pleasant relatives, valiant sons and chaste daughters, banquets, hunting, various games and sports, gems and precious objects, even paintings and statues. Some of the dialogues could apply to the lowly as well as the high and mighty; others are meant only for aristocrats, like posts at court, being a good courtier or lord, having a good army, winning a victory and establishing peace. Only one is specifically about the religious: on becoming pope. There are none specifically for women, but there are several that speak ill of them to husbands, fathers, and lovers. A pair of dialogues attacks foolish belief in the predictions of astrologers and diviners, and in dreams.

There are limits to what Petrarch will allow Stoic psychotherapy to do. "De tranquillo statu" (I.90), for example, has Reason denying that there can be any such thing in *this* life; to rejoice in peace of mind is a grave illusion, *opinio*, for it is precisely then that Fortune will strike. There are similar warnings about trusting or enjoying feelings of happiness (I.108), or indulging in good hopes (I.109). Petrarch is at his most ascetic in Book I, Chapter 122,

---

36 "Quatuor ille famosiores et consanguineae passiones animi . . . quas duae sorores aequis partibus prosperitas et adversitas peperere" (*Opera omnia*, fol. †††4v).

37 "Utraque fortunae acies metuenda, veruntamen utraque tolleranda est, et haec quidem freno indiget, illa solatio, hic animi elatio reprimenda, illic refovenda ac sublevanda fatigatio" (ibid.).

closely following Augustine, for whom the only possible earthly joy is hope in the life to come.[38]

In Book II, the subjects are associated with arousing extremes of emotional distress, the contrary of Book I. Personal qualities like illness, blindness, chronic infirmity, stupidity, a humble lineage, illegitimacy, and poverty are dealt with. So are family matters like being married to a shrew and losing a wife or relatives especially children. There is a long list of calamities, from shipwreck and imprisonment to loss of one's patrimony and torture. But there is another category: The same circumstance or event – such as having many children – that was the occasion of false elation and illusion can also be the occasion of the opposing perturbation. In themselves, the qualities or events take on a morally indifferent hue.

Petrarch's main technique for altering the *opiniones* is to attack whatever the irrational emotion is craving or dreading, regardless of whether there is good reason to feel elated or downcast. It is this aspect of the *De remediis* which modern readers, I believe, find most unattractive. This aspect is also where Petrarch differs most from Augustine, who points out that since you cannot really desire a good object too much, not all the *perturbationes* are bad (*City of God*, XIV.9). Cicero's treatment of craving and sexual passion, *cupiditas* and *libido*, is the model for Petrarch. Cicero recommends, first, blackening the object of desire – therapy by aversion – so that women appear loathsome and sex degrading. Then the desire is diverted to other occupations, and a change of environment prescribed. In addition, Cicero warns about the criminal acts that are the effects of passion (*Tusculans* IV.68–76). The successful suppression of *libido* is the end to be achieved (IV.62). Petrarch, who rejected vivid meditation on death as an effective remedy for *aegritudo*, reintroduces the "cogitatio vehemens et assidua" on the degrading effects of lust, following Cicero carefully (I.69).

Perhaps Petrarch's most novel move in Book II is his arrangement there of the so-called seven capital sins, with the banishment, however, of the names *aegritudo* and *acidia* and the substitution of the far more neutral term, *torpor animi* (II.109), or sloth pure and simple. Why? Was it to avoid the connotations of

38 *City of God*, XIX.20.

acidia – its association with monastic and religious life – in a lay work? Was it simply to use a term understood by lay people? In the *Secretum*, Petrarch had noted that *aegritudo* was a word used by the ancients for what moderns called *acidia*.[39] Now, in a dialogue near to but separated from the group on the seven capital sins, "De tristitia et miseria" (II.93), Petrarch takes up in quite different terms the same problem. Only at the very end of the dialogue, for instance, does Petrarch name his old illness and the Stoic philosophers, as if to play down their importance: "With respect to casting out 'aegritudo animi' – for that is what the philosophers call it – and restoring tranquillity, it is useful to know what Cicero discussed in Book III of his *Tusculans*, and Seneca in his *De tranquillitate animi*."[40]

Another change has taken place. In the *Secretum*, Petrarch had found Cicero and Seneca more helpful than the "De contemptu mundi" tradition about *aegritudo*. Now he apparently defers to Augustine and a religious solution. Petrarch uses Augustine's word, *tristitia*, for the worst emotional disturbance. But to the outbursts of *Dolor* about the wretchedness of this life, Reason recommends happy thoughts about the dignity of man and his myriad inventions – about the beauty and marvels of nature, from tiny plants and insects to the order of the stars – just the contrary of the prefatory epistles. The person who suffers *tristitia* or *aegritudo* should get out of himself, away from introspection, and look to God. Petrarch has taken to heart Augustine's *City of God*, Book XXII, where Augustine gives a rapid sketch of man's miseries (Chaps. 22–3) and goes on to contrast them with his outstanding achievements and the many pleasant satisfactions in life given by God (Chap. 24). "All these," Petrarch says, "are solaces for wretched mankind."[41] We need to be aware, however, that while seeming to give way to Augustine, Petrarch is perhaps more genuinely Stoic than in other dialogues where dour Reason will not allow room for either joy or "tranquillitas animi." Augustine's homily is derived from the Stoic account of divine

39 See this chapter, section II.
40 "Ad aegritudinem vero animi (ita enim hanc philosophi appellant) et depellendam et tranquillitatem revehendam, proderit nosse quid de primo Cicero in Tusculano suo tertia luce disseruit; quid de secundo, Seneca in eo libro quem de animi tranquillitate composuit."
41 "Haec omnia miserorum sunt . . . solacia" (XXII.24).

providence and the excellence of man in Book III of Cicero's *De natura deorum*, known to earlier church fathers like Minucius Felix, Arnobius, and above all Lactantius, whose *De opificio Dei* was probably Augustine's immediate source. Petrarch would thus be reviving yet another strand of Latin Stoicism in this "dignity of man" dialogue and associating Augustine with it.[42] We must not forget, either, that Seneca, in *De tranquillitate animi*, also advises Serenus that there are times when the afflicted soul must be drawn into rejoicing, and gloomy sobriety banished for a while.

By including the dialogues on the seven capital sins under the title of "fortuna adversa," Petrarch finally is making even clearer than in the *Secretum* that emotional disturbances, like everything else to do with this unstable goddess, befall us. They are not a product of our will, and still less of grace. We have some power to bring them under the control of Stoic Reason. But despite Petrarch's improvements on earlier medieval prescriptions for wayward emotions, despite his keener understanding of Cicero and Seneca, *De remediis* remains for the modern reader an unappealing book, albeit a fascinating one for its recognition of emotional illness. Its main fault, I think, lies in Petrarch's failure to carry out the proposed symmetrical structure of treatment by opposite emotions – raising sadness and curbing elation. Nearly all the dialogues attack, denigrate, vituperate, and devalue the objects arousing desire or dread. There is not enough praise to balance the blame, not enough liking to balance the loathing. There is hardly any room for joy, especially for joy in striving or achievement. What is lacking above all is motivation, a desirable and attractive goal that is the reason for undertaking the painful process of tempering the *perturbationes* in the first place.

IV

In conclusion, I would like to give a very brief account of the history of Petrarch's *De remediis* in the fifteenth and sixteenth centuries. According to D. W. Fiske, "Until after the days of Erasmus, no secular production in modern Latin literature was

---

42 See G. Gawlick, "Cicero in der Patristik," *Studia Patristica*, 9 (1966), 57–62; C. Trinkaus, in "Human Condition," credits Petrarch with reviving the genre but does not mention Augustine's contribution.

familiar to so large a public."[43] More recently, N. Mann has confirmed Fiske's findings, supplying detailed information about manuscripts, early printed editions, translations, and abridgments throughout Europe.[44] In 1366, Petrarch's favorite scribe, Giovanni Malpaghini da Ravenna, completed the final version. It found immediate favor in Italy, but after about 1425 it does not seem to have been read much, except for – or perhaps because of – two translations into Italian. The first, dated 1427, by the Florentine Camaldolese monk Fra Giovanni Dassaminiato, was printed only in the last century; the second, by another friar, Remigio Nannini, also Florentine, was printed in Venice several times: 1549, 1584, 1589, and 1607.[45] Outside Italy, the work's diffusion was far greater, beginning with France, where it was translated as early as 1378 by Jean Daudin, and extending to the Low Countries, Germany, England, the Spanish peninsula, and eastern Europe. Although *De remediis* was intended by Petrarch for lay people, evidence about ownership points to religious orders and a predominantly clerical audience, who absorbed *De remediis* back into devotional and ascetic literature on the vanity of this life and the world. Latin abridgments by Adrian the Carthusian, Arnold Geilhoven, and Albrect von Eybe mutilated Petrarch's text, eliminating the prefaces, drastically curtailing the number and length of the dialogues, and even removing the names of the four Stoic passions. These compilations, too, found their way into the libraries of religious orders, including the Brethren of the Common Life.[46] The astonishing popularity of the work in Germany is

43 "Francis Petrarch's Treatise *De remediis*: Texts and Versions," *Bibliographical Notices III* (Florence, 1887), p. 2.
44 See esp. "The Manuscripts of Petrarch's *De remediis*: A Checklist," *Italia Medioevale e Umanistica*, 14 (1971), 57–90. The treatise, he reports, is preserved entire in at least 150 manuscripts and is abridged and translated in another 94. A further 70 manuscripts are no longer extant.
45 *De' rimedii* (see n. 23, this chapter); I have seen Nannini's 1589 edition. In the dedicatory letter (no pagination), Polidoro Rolli gives the impression that he knew of no earlier translation and that he was surrendering his own copy of Nannini's translation to the press by popular demand for "quella bella dotta e santa opera" ("that beautiful learned and holy work"). Rolli highly recommends the dialogues for their usefulness: "gran consolatione sia leggerli a chi di presente ne ha bisogno, et utilissimo preparamento a quelli che potessero incorrere in alcuna disgratia" ("To those in present need, reading them brings great consolation; and those to whom some misfortune might befall, they bring the most helpful preparation").
46 In general, see Diekstra's introduction to *A Dialogue betwen Reason and Adversity*,

attested most of all by Spalatin's German translation, accompanied by magnificent woodcuts for each dialogue by the anonymous Master of Petrarch. Printed first in Augsburg, in 1532, it was frequently reprinted. Enthusiasm for *De remediis* outside Italy began to wane only in the seventeenth century.

In Italy, the dwindling interest in Petrarchan Stoicism has both a negative and positive side. I have suggested elsewhere that a distorted understanding of Latin Stoicism, and especially Seneca – to which not only Petrarch but also other humanists like Coluccio Salutati and Gasparino Barzizza subscribed – began to come under criticism in the second quarter of the fifteenth century.[47] Lorenzo Valla's *De vero falsoque bono*, whose first version can be dated around 1430, was aimed at exposing the contradictions of Latin Stoic ethics in general and Petrarchan Stoicism in particular. The Stoic spokesman's speech in Valla's dialogue is demonstrably a parody of Petrarch's two prefaces.[48] On the positive side, there is continued interest in Cicero and Seneca on healing the passions. Guarino Veronese, a contemporary of Valla, and Cristofero Landino, member of Marsilio Ficino's Neoplatonic Florentine circle at the end of the fifteenth century, wrote commentaries on Cicero's *Tusculans*, and Leon Battista Alberti composed two treatises which draw directly from Cicero's *Tusculans* and Seneca's *De tranquillitate*. *Teogenio*, dedicated in 1442 to Leonello d'Este, ruler of Ferrara, on the death of his father, provides advice for the

---

pp. 24–32, and N. Mann, "Recherches sur l'influence et la diffusion du *De remediis* de Pétrarque aux Pays Bas," Mediaevalia Lovaniensia, I, no. 1 (Louvain: Louvain University Press, 1972), 78–88. For England, see Palmer, *Seneca's "De remediis fortuitorum".* Adrien Monnet, prior of the Charterhouse in Liège and professor of theology before his death in 1411, does not even name Petrarch in his compendium: *Incipit liber de Remediis utriusque fortune prospere et adverse. Compilatus per quendam Adrianum Cartusienum et sacre theologie professorem* (Cologne, 1470). The theologian addresses "piis animis," urging them to rise above the miseries of this life to the joys of the next. He admits he has drawn freely from others and mixed opinions from philosophers and poets with Scripture. Albrect von Eybe (1420–1475), a canon lawyer who spent some time at the Roman Curia during the papacy of Pius II, reduces some of the dialogues to a single sentence in which "Respondeo" briskly puts down "Doleo." About toothache we find, "Doleo. Dentibus eger sum. Respondeo: Invalidum atque caducum animal es cui etiam que prevalida videbantur infirma sunt" (*Opusculum Remediorum adverse fortune ex Francisco Petrarca oratore et poeta sane clarissimo* [Leipzig, 1504], fol. Aiiiv).

47 See my "Gasparino Barzizza's Commentaries," pp. 332–41.

48 Critical edition by M. Lorch (Bari: Adriatica Editrice, 1970); see also her study *A Defense of Life: Lorenzo Valla's Theory of Pleasure* (Munich: Fink, 1985). On Valla and Petrarch's *De remediis*, see my "Lorenzo Valla's *De vero falsoque bono*, Lactantius and Oratorical Scepticism," *Journal of the Warburg and Courtauld Institutes*, 41 (1978), 95–9.

"animo perturbato" ("distressed spirits") overwhelmed by "casi avversi" ("bad luck") which also "perturbino la quiete e tranquillo stato" ("disturbs the peace and calm disposition") of most people. The speaker, like Seneca and Petrarch, believes that prosperity has a worse effect on one's emotional stability than adversity. In *Profugiorum ab aerumna*, known better by its Italian title *Della tranquillità dell'anima*, Alberti paraphrases Seneca's advice to Serenus about treating the psyche mildly and gently in times of emotional distress.[49] In the second half of the fifteenth century, the revival of the Greek Epictetus's writings on practical Stoic healing of the passions put Petrarch even farther into the background. Niccolò Perotti translated the *Enchiridion* in 1450. The title means both "dagger" and "manual"; one was supposed to have Epictetus's advice to hand, like a trusty weapon, for times of severe emotional distress. Perotti explicitly compared the manual to Book III of Cicero's *Tusculans*.[50] In 1475, Poliziano translated the *Enchiridion* anew, dedicating it to Lorenzo de' Medici after the shattering Pazzi conspiracy, in which Lorenzo's younger brother Giuliano was murdered. Poliziano's Latin translation was published along with his other works from 1497 onward, but Epictetus's much fuller and richer *Discourses* were not in print until 1535, and not in more accessible Latin translation until the end of the sixteenth century.[51] By that time, the center for Stoicism was France and the Low Countries, and Cicero and Seneca's psychotherapy flourished vigorously in the writings of the late Renaissance Stoic, Justus Lipsius. His edition of Seneca's *De constantia libri duo*, of 1584, alone ran into more than eighty editions.[52] I have left the question, Did Stoic psychotherapy work? to the end, because it is not possible to answer directly with a simple yes or no. But if one looks at the demand for and diffusion of Petrarch's *De remediis* and other related works, ancient and modern, one can conclude that writers and readers certainly thought it did work – or at least hoped that it would.

49 L. B. Alberti, *Opere volgari*, ed. C. Grayson, 3 vols. (Bari: Laterza, 1966), 2:53–104 (*Teogenio*); and 107–83 (*Profugiorum*).

50 R. P. Oliver, *Niccolo Perotti's Version of the Enchiridion* (Urbana: University of Illinois Press, 1954). Perotti's translation was not printed during the Renaissance.

51 See G. Capitolo, *La filosofia stoica nel secolo XVI in Francia* (Naples: Perrella, 1931), pp. 17–18.

52 Ibid., pp. 39–49, and J. L. Saunders, *Justus Lipsius* (New York: Liberal Arts Press, 1955).

4

Alonso de Cartagena and John Calvin as interpreters of Seneca's *De clementia*

NICHOLAS G. ROUND

Nobody who has read Karl Alfred Blüher's admirable history of Seneca's reception in Spain can be in any doubt that the reign of Juan II of Castile (1406–1454) marks a distinctive epoch in that history.[1] It was then that the misleading but endlessly seductive *topos* of Seneca as Spaniard first began to be widely current, largely as a result of a series of vernacular translations. A couple of relatively late items – a partial revision of a thirteenth-century *De ira* and a wholly new version of the *Apocolocyntosis*, both associated with Nuño de Guzmán[2] – were relatively restricted in their immediate influence. These apart, we are concerned with four main instances. There were two translations of Seneca's *Tragedies*, the earlier and less complete of which was apparently produced under the patronage of the marquis of Santillana. There were, again, two versions of the *Letters to Lucilius*; here, too, there are links with one of the minority of lay patrons in mid-fifteenth-century Castile, Fernán Pérez de Guzman. In both these cases, translations into other Romance vernaculars appear to have been the immediate source for the Castilian renderings; the textual histories, complex for both *Tragedies* and *Letters*, have been studied

I thank the British Academy, and the Calgary Institute for the Humanities, for grants supporting this project. I also have to thank Ms Patricia Odber de Baubeta for kindly obtaining for me the microfilm of MS Escorial T III 6, from which I have quoted here.

1 See Karl Alfred Blüher, *Séneca en España: Investigaciones sobre la recepción de Séneca en España desde el siglo XIII hasta el siglo XVII* (Madrid: Gredos, 1983), pp. 113–55. Mario Schiff, *La bibliothèque du Marquis de Santillane*, Bibliothèque de l'École des Hautes Études, no. 153 (Paris: École des Hautes Études, 1905), is still of value, esp. pp. 124–31.

2 Blüher, *Séneca en España*, pp. 153–4; also Louise Fothergill-Payne, *Seneca and Celestina* (Cambridge: Cambridge University Press, 1988), pp. 10, 13.

in articles by myself and Mario Eusebi respectively.[3] The other two major examples of Senecan translation in our period were executed directly out of Latin, and their patron was King Juan II himself. In the early 1430s Alonso de Cartagena translated a series of extracts and treatises, some authentic and some not. About ten years later, again by royal command, the judge of appeal Pero Díaz de Toledo made Castilian versions of the pseudo-Senecan *Proverbia* (which he glossed at length) and of *De moribus*. The immediate scholarly background to these two undertakings is, in great part, the same.

Not all the items translated by Cartagena belong to the mainstream of Senecan (or even pseudo-Senecan) tradition.[4] The socalled *Libro de la cavallería*, apparently added to the series at a late date, turns out to consist of sentences taken from Vegetius. The *Libro de amonestamientos e doctrinas*, a *florilegium* from pseudo-Senecan and other sources, is a rarity in Latin collections of Seneca's works. Blüher identifies it with the *De legalibus institutis* which appears at the end of one such manuscript, the fourteenth-century MS. 10238 of the Biblioteca Nacional in Madrid. (This is now cataloged with Santillana's books, though, in fact, it may well not have belonged to him.) The work which Cartagena says he translated first is a selection of extracts, extensively glossed, from many books of Seneca. This, and a similar collection from the *Declamations* of Seneca the Elder, both come from the much longer *Compilatio secundum alphabetum*, made by Luca Manelli, bishop of Citò, in the early fourteenth century. A copy of Manelli's Senecan *Compilatio* had been sent, in 1406, to King Martin of Aragon by the antipope Benedict XIII (Pedro de Luna); it was later translated into Catalan. There is a Latin text in two volumes

---

3 Nicholas G. Round, "Las traducciones medievales catalanas y castellanas de las *Tragedias* de Séneca," *Anuario de Estudios Medievales*, 9 (1974–9), 187–227; Mario Eusebi, "La più anticha traduzione francese delle Lettere Morali di Seneca e i suoi derivati," *Romania*, 91 (1970), 1–47.

4 On Cartagena and Seneca, see Blüher, *Séneca en España*, pp. 133–48; on *De legalibus institutis*, ibid., pp. 68n and 140; Schiff, *Bibliothèque*, pp. 102–3; Antonio Fontán, "Algunos códices de Séneca en bibliotecas españolas y su lugar en la tradición de los diálogos," *Emérita*, 17 (1949), 9–41, at 36–7; on Manelli, Blüher, *Séneca en España*, pp. 131–2, 135–6; also Guillermo Antolín, *Catálogo de los códices latinos de la Real Biblioteca de El Escorial* (Madrid: Imprenta Helénica, 1910–23), 5:421; Nicholas G. Round, "The Mediaeval Reputation of the *Proverbia Senecae*: A Partial Survey Based on Recorded MSS," *Proceedings of the Royal Irish Academy*, 72, C, 5 (1972), 103–51, at 134.

in the Escorial library. Juan II's copy may have reached him either from the Aragonese court or – perhaps more probably – from Benedict's library at Peñíscola, after the latter's death in 1423. Our initial clue to the provenance of the other pieces translated by Cartagena comes from the king's subsequent commission to Pero Díaz.

He had asked for versions, so Pero Díaz recalls, of "los proverbios de Séneca e el libro que conpuso que se intitula *De las costunbres* e ansí mesmo ciertas actoridades notables de la philosophía moral de Aristóteles que fueron sacadas de la traslación arábica en latín" ("the proverbs of Seneca and the book which he wrote entitled *On Customs* and also certain noteworthy authorities of Aristotle's moral philosophy, which were taken from the Arabic translation into Latin").[5] This prompts two questions. How did Pero Díaz come to know the translation history of this last item? And how did Juan II come to ask for it to be translated, along with two pseudo-Senecan works? The answer to the first question is that the detail came from a rubric title; the description was, in fact, a traditional part of the title given to Hermann the German's version of the *Summa Alexandrinorum*, compiled from Aristotle's *Ethics*. And it occurs in a Senecan context in MS. Q I 8 of the Escorial library, a fourteenth-century Italian "collected works" of Seneca. The one avowedly non-Senecan item in this manuscript is headed "Excerpta de libro Aristotelis ethicorum secundum translationem de arabico in latinum" – which points the way to a solution for our second problem.

The text of the *Proverbia Senecae* in MS. Q I 8 differs significantly in its detail from the one translated by Pero Díaz, and the Escorial copy itself was almost certainly still in Italy in the last years of the fifteenth century. So this was not Juan's own manuscript. But we can reasonably conclude that he did possess a copy of Seneca's works of much the same tradition, that this copy – like Q I 8 – also contained the *Excerpta Aristotelis*, and that he asked Pero Díaz to make his three translations from this. It is a more conjectural step, though a likely one, to suggest that Cartagena,

---

5 Pero Díaz de Toledo, *Proverbios de Séneca*, MS Escorial S II 10, fol. 2v; see Jaime Ferreiro Alemparte, "Hermann el Alemán, traductor del siglo XIII en Toledo," *Hispania Sacra*, 35 (1983), 9–56, at 19, 21–2. On MS Escorial Q I 8 see Fontán, "Algunos códices," pp. 32–5; Antolin, *Catálogo*, 3:356–8; marginalia on fols. 99r and 143v establish its provenance and late fifteenth-century location as Italian.

ten years earlier, had used the same royal volume. Escorial Q I 8, after all, includes every Senecan item translated by him, except three: *De la cavallería* (an afterthought in the series); the Manelli extracts (whose separate provenance we know); and *Amonesta-mientos e doctrinas* (which does appear, however, in the not too dissimilar BN 10238). We might think it likely, too, that the king's collected Seneca included this last item. If it did, then virtually everything that Juan II ordered to be translated as the work of Seneca came from one or other of two manuscripts in his possession: a copy of Manelli (probably in two volumes), and a one-volume "collected works" closely resembling Q I 8.

The two texts may even have reached Juan from the same source. Manuscripts of the Q I 8 type, as Antonio Fontán has observed, tend to be associated with the papal court of fourteenth-century Avignon.[6] The obvious route by which such material was likely to reach Spanish readers was via the library of Benedict XIII, with which, as we have seen, the Manelli compilation had yet clearer links. After Benedict's death, so Fontán again informs us, most of his books were taken over by the papal legate, Cardinal Foix. The cardinal, as it happened, had spent an anxious couple of days in Juan II's company in July 1429, pleading unsuc-cessfully for an end to armed hostilities with Aragon. That was hardly an occasion for literary topics to arise, but by January 1431 Juan's military attention had begun to turn toward the prospect of war with Muslim Granada, and serious peace negotiations with the eastern neighbor were in train. A tactful present, in line with the young king's known bookish tastes, would have made much sense at about this time. And it is within a few months of this that

6 Fontán, "Algunos códices," pp. 33–5; also p. 37 (the very similar MS Madrid BN 10238 is perhaps a direct import from Italy); see also L. D. Reynolds, "The Medieval Tradition of Seneca's *Dialogues*," *Classical Quarterly*, 62 (1968), 355–72. On Cardinal Foix in 1429, see *Crónica de Juan II*, ed. Cayetano Rosell, in *Crónicas de los Reyes de Castilla*, II, Biblioteca de Autores Españolas, no. 68 (Madrid: Rivadeneira, 1877), pp. 460–1; Pero Carrillo de Huete, *Crónica del Halconero de Juan II*, ed. Juan de Mata Carriazo, Colección de Crónicas Españolas, no. 8 (Madrid: Espasa-Calpe, 1946), pp. 38, 40; on peace talks in 1431, Manuel de Bofarull y de Sartorio, *Guerra entre Castilla, Aragón y Navarra: compromiso para terminarla*, Collección de Documentos Inéditos del Archivo General de la Corona de Aragón, no. 37 (Barcelona: Imprenta del Archivo, 1869), p. 17. Cordova Cathedral described in a late gloss to Manelli's *Compilatio*, MS Madrid BN 6765, fol. 200v; Juan II there in May and July 1431, Carrillo de Huete, *Crónica*, pp. 93, 108–9; Cartagena with him, Luciano Serrano, *Los conversos D. Pablo de Santa María y D. Alfonso de Cartagena* (Madrid: CSIC, 1942), p. 130.

we find our earliest datable evidence of Cartagena translating Seneca at Juan's command – first a selection from Manelli and then the treatises. It is a colorable suggestion that the texts had come to hand only recently and that they came – by way of Cardinal Foix – from what remained of the library of Peñíscola.

The translator was then in his midforties.[7] His father, Pablo de Santa María, had once been rabbi of the Jewish community in Burgos. Converted in 1390, he had become a great personage in Christian Castile, rising to be chancellor of the kingdom, for a time, and bishop of his native city. Don Pablo's son, Alonso de Santa María, or de Cartagena, had made good use of his advantageous start in life. He was dean of Santiago de Compostela, a judge of appeal, a member of the king's council, active in its legal business. He had been several times an ambassador in Portugal, where the scholarly King Duarte had made much of him. He had translated a number of classical Latin texts and had recently written a critique of the new translation of Aristotle's *Ethics*, by the Florentine humanist Leonardo Bruni. This was risky, for Cartagena knew no Greek, though he knew what he liked about Aristotle. But it was further evidence of a powerful and original mind. In the mid-1430s he was to make an international name for himself as a Castilian delegate at the Council of Basel, defending national prestige and papal prerogatives with much zeal and much skill. In July 1435, he succeeded his father in the see of Burgos.

On the evidence of rubrics which describe him still as "dean of Santiago," most of his Senecan versions are earlier than this. Probably they belong to the period before May 1434 when he set out for Basel, for their glosses and prologues addressed to the king strongly suggest that the translator was in regular contact with him. They also contain cross-references and allusions to contemporary events, which among them enable us to date the series in a more detailed way.[8] Thus, the two earliest items are the Manelli compilation (apparently nearing completion in the summer of

7 Biography in Serrano, *Los conversos*, pp. 119–235; Francisco Cantera Burgos, *Alvar García de Santa María y su familia de conversos* (Madrid: CSIC, 1952), pp. 416–64; on his writings, see Karl Kohut, "Der Beitrag der Theologie zum Litteraturbegriff in der Zeit Juans II von Kastilien. Alonso de Cartagena (1384–1456) und Alonso de Madrigal, gennant El Tostado (1400?–1455)," *Romanische Forschungen*, 89 (1977), 183–226.
8 For the date of the Manelli version, see n. 6, this chapter; other datings in Blüher, *Séneca en España*, pp. 142–3n.

1431), and the first book *De providentia*; the second book of that
title (more correctly *De constantia sapientis*) presumably followed.
The *Artes liberales* (Letter 88) cannot be much earlier than 1434; the
De vita beata is later again. The translation of the two books *De
clementia* is probably best located between the *De providentia* pair
and the *Artes liberales*, for its prologue can still refer to the wars
of 1429 and 1431 as taking place "poco tienpo ha" ("a short time
ago"). The same phrase is used in one of the glosses to Book I of
De clementia, in relation to "un cavallero deste regno . . . que dezía
en latín unas palabras que dixo Vejecio, que quieren en nuestro
lenguaje dezir 'Quando toda la esperança cesa, el miedo usa de
armas,' e cuidólo poner en plática, e fallóse mal dello" ("a gentleman of this kingdom . . . who used to quote in Latin some words
from Vegetius, which in our language mean 'When all hope is at
an end, fear has recourse to arms,' and he thought to put it into
practice and found himself the worse thereby").[9] Few Castilian
rebels were capable of such scholarly quotation, but two of those
arrested by Juan II on 7 February 1432, on suspicion of plotting
against him, might well have been exceptions. Both the count of
Haro and Fernán Pérez de Guzmán were later to have Latin works
dedicated to them by Cartagena himself. The likeliest view is that
the *cavallero* in question was Fernán Pérez, which would make the
version of *De clementia* I later than the date of his arrest. Book II
was actually the earlier of the two to be translated, on the ground
that Book I was a mere exhortation to the virtue of clemency; the
definition, which ought logically to precede it, comes in Seneca's
second book.[10] This reversal of order, explained in Cartagena's
introduction to Book I, is reflected in cross-references in several of
its glosses. A date for both books in 1432 – or at most, a very little
later – would fit our evidence best.

9 Gloss to *De clementia* I, MS Escorial T III 6, fol. 126r. All references here to Cartagena's
   version are to this copy (fols. 91r–169r), which dates from 1447 (see Julián Zarco
   Cuevas, *Catálogo de los MSS castellanos de la Real Biblioteca de El Escorial* (Madrid:
   Imprenta Helénica, 1924–6), 2:395–7; unlike Zarco Cuevas, I have used the foliation
   given in the manuscript). See also *Crónica de Juan II*, ed. Rosell, p. 504; Carrillo de
   Huete, *Crónica*, p. 58; Fernán Pérez de Guzman, *Generaciones y semblanzas*, ed. Robert
   Brian Tate (London: Tamesis, 1965), p. x. For works dedicated to Haro and Fernán
   Pérez, respectively, see Jeremy N. H. Lawrence, *Un tratado de Alonso de Cartagena sobre
   la educación y los estudios literarios* (Barcelona: Universidad Autónoma, 1979), and the
   presently unedited *Duodenarium* (this chapter, n. 31).
10 Fols. 94v–95v; cf. glosses on fols. 97r, 126r.

Cartegena's presentation of the *De clementia* texts is in line with his handling of similar treatises in the Senecan series. A general prologue explains to Juan II the relevance which this work of Seneca ought to have for him: that is, the "external cause" of the translation. A separate introduction to each book considers the "internal cause": the ethical subject matter.[11] Both the introductory material and the translated texts are supplied with glosses: forty-five in all, ranging in length from a dozen words to several hundred. Taken together, the glosses run for just over one-quarter of the length of the text: slightly less, in the much longer Book I, but fully one-third in the brief Book II. Though the Book II glosses are, in fact, much shorter on average than those in Book I, they are much more frequent. The contrast is rooted in their subject matter. Overwhelmingly, the glosses in the second book are designed to explain Seneca's use of particular words in their context, or stages in the development of his argument. In Book I, though such linguistic explanations are by no means absent, the commonest function of the glosses is to supply the background of various references to Roman history or customs. This, in turn, happens because in Book II Seneca is explaining in the abstract what clemency is, while in Book I he is telling the Emperor Nero what he ought to do about it.

In neither case does Cartagena, as commentator, wear his learning at all obtrusively. He refers by name to less than a score of sources, few of them more than once. There are a handful of references to other writings by Seneca (mostly to Cartagena's other translations), and a couple to Cicero on the teachings of the Stoics; a few of Seneca's *exempla* are amplified, out of Valerius Maximus. For the rest, Aristotle, Vegetius, and Virgil make up the classical tally; single references to Augustine, Boethius, Isidore, and Jerome are the only patristic items; among more recent authorities named are Aquinas, the *Romuleon*, and a standard Latin glossary – John of Genoa's *Catholicon*. The handful of legal references includes the *Decretum*, Ulpian on the *Digest*, and the *Partidas* of Alfonso the Wise. But Cartagena seems almost to prefer a vague style of allusion – to "historians," "philosophers," "the

---

11 On internal and external causes and the construction of medieval prologues, see Alastair J. Minnis, *Medieval Theory of Authorship: Scholastic Literary Attitudes in the Later Middle Ages* (London: Scolar Press, 1984), p. 28.

Civil Law," and so forth. Against the Stoic belief that all lapses from virtue are equally grave, he cites, characteristically, "Santo Tomás donde esta materia tracta" ("Saint Thomas, where he deals with this matter") (fol. 162v). This paucity of reference is no reliable index of his own scholarship, which his other works reveal as solid and wide-ranging for its day. It had more to do with the audience for which he was writing: the king and his lay retainers. As he explained to the count of Haro in the 1440s, Cartagena did not believe that laymen ought to dabble in the properly professional studies of theologians or philosophers: "Ne altiores quam comode possumus aprehendere scienciarum investigationes querentes in crassam ignorantiam inopes sciencie incidâmus" ("lest, pursuing our researches into knowledge higher than we conveniently can, we fall, for lack of knowledge, into crass ignorance").[12] After completing their basic grammatical studies, he argued, they should move directly to a wholesome diet of simple moral instruction: "ad moralia documenta, saltem sub vulgari lingua et groso tradendi modo" ("to moral examples, if only in the vulgar tongue and roughly conveyed"). That was what he sought to supply here.

Nonetheless, the glosses had, for Cartagena, a quite specific function in enhancing the impact of the translation. In his discussion of Bruni's Aristotle, he had compared the different idioms required in a text and in its gloss with those appropriate to a king and his ambassador, or to a judge and an advocate: "nam breviter textus nos docet, glossule vero quid textus senserit aperire solent" ("for the text teaches us succinctly, but the glosses open up the meaning of the text").[13] The role of the gloss (as of the introductory matter) was communicative; it was there to open up the text, to render it available and convincing to readers whose formative cultural and other experience had not fully prepared them for it. The information, especially of a philological kind, which Cartagena had at his disposal for this task was not uniformly reliable. Augustus's honorific title, for example, was probably not due to his having "greatly increased" the Roman domains (fol. 116v), and it is doubtful whether the Greek root underlying

12 Lawrance, *Un tratado*, p. 56; see also p. 57.
13 Aleksander Birkenmaier, "Der Streit des Alonso von Cartagena mit Leonardo Bruni Aretino," in *Vermischte Untersuchungen zu Geschichte der mittelälterlichen Philosophie*, Beiträge zur Geschichte der Philosophie des Mittelalters, 20, 5 (Münster: Aschendorf, 1922), pp. 129–236, at 167.

triumphus (fol. 153r) really did mean " a song for three voices." It
was natural enough to turn to the *Catholicon* for a definition of
muraenas, but lampreys are not, as it happens, boneless (fol. 137v).
Yet Cartagena could be both shrewd and scrupulous in his choice
and deployment of information. His expansion of an image drawn
from hunting (fol. 126v) shows that as dean of Santiago he had
observed the methods used by Galician huntsmen closely and
intelligently. Writing about Dionysius of Syracuse, he is careful to
refer to "Curaguça de Cicilia" (fol. 125r), presumably lest anyone
should suppose him to have meant Ragusa. His down-to-earth
paraphrases of Seneca's incidental metaphors – "ut quodammodo
speculi vice fungerer"; "Parcetur ubique manibus tuis"; " e ludo in
carcerem ire" (glossed respectively on fols. 97r, 152r, and 157v) –
create the sense of a tactful, commonsense presence, efficiently
shepherding readers through the text. He is equally willing to
clarify, as this becomes necessary, aspects of his own work and
wording, emphasizing, for example, that *sabidor*, in this text, does
not mean "omne sotil en ciencia" but "omne virtuoso e perfecto"
(fol. 167v). More generally, as he explains in introducing Book II,
he has avoided as inexact the usual translations of *clementia* –
misericordia and *piedat* – preserving instead the less familiar Latin
term (fol. 155r). This, he says, is what translators from Greek into
Latin did when no single word would meet their needs: "dexáron-
lo griego como yazía, declarando su propiedat por otras palabras"
("they left it in Greek as it was, explaining its true meaning by
other words") (fol. 155v). (The same observation, applicable both
to Ciceronian and to medieval translation practice, forms part of
his argument against Bruni.[14]) But here again the gloss has an
essential part to play: "aquí llamaremos clemencia como la llama
el latín, e la significación suya entenderá quien quisiere, por las
declarasciones que della en este tratado se fazen mención" ("here
we will give clemency the name which it is given in Latin, and
anyone who wishes to understand its meaning can do so from the
explanations of it which are given in this treatise"). It was a
principle which he formulated in more general terms in his version
of Cicero's *De inventione*, completed at very much the same date as
these Senecan treatises.[15]

14 Ibid., pp. 167–9; cf. again fol. 155v.
15 Alonso de Cartagena, *La Rethórica de M. Tullio Cicerón*, ed. Rosalba Mascagna,
   Romanica Neapolitana, no. 2 (Naples: Liguori, 1969), pp. 31–2.

One reason why Cartagena felt that his readers needed guidance not provided by the Senecan text was his belief that Seneca, though an inspiring teacher of virtue and a man whose moral life, if it matched his writings, could well be regarded as saintly, had been alarmingly muddled in his treatment of moral issues. It was not simply that certain Stoic teachings clashed with Christian moral theology: the equal status of all virtues; the equal gravity of every sin; the dismissal of pity as a mere passion, which the man of perfect wisdom must overcome. These points could readily be corrected, and Cartagena's marginal glosses supply the corrections as necessary, although rather laconically. Pagans, after all, were bound to have got some things wrong. The problem was that, even in those matters where Seneca was in the right, his teaching about the virtues was not easily assimilated to those mnemonic and didactic schemes according to which medieval Christians organized their ethical theory. "Sed in hac inquisitione virtutum" ("But in this inquiry into the virtues"), wrote Cartagena, again in the course of his argument with Bruni, "et illorum scientifica discussione quam summarie se habuerit, quam improprie discus-serit facillimum est videre" ("and in his attempt to discriminate between them in a scholarly way, it is all too easily seen how hasty were his procedures, how inappropriate the distinctions which he drew").[16] The impression was reinforced by the pseudo-Senecan *De quatuor virtutibus*, which Cartagena took to be genuine. But even *De clementia* posed difficulties. Clemency found no place in Aristotle's *Ethics*, "aquel antiguo solar de las virtudes" ("that ancestral home of the virtues"), though it was perhaps implied there (fol. 96v). But Cartagena found it necessary to make no less than three attempts, in his ancillary material, to show readers how this material might relate to traditional schemes. His prologue puts forward a variant, for the use of princes, on the conventional list of four cardinal virtues: Juan II is to show justice, generosity, fortitude, and clemency (fol. 91r). (The gloss, at this point, gives the series as "justice, *truth*, fortitude, and clemency," though for no very obvious reason.) The introduction to Book I affiliates clemency within the family structure of the virtues. It is related to temperance (because it involves restraint of anger); to *epiqueya* (equity), and thus to justice; and to charity (because it derives from

16 Birkenmaier, "Der Streit," p. 174 (reference to *De quatuor virtutibus*, p. 175).

the mutual love between rulers and their subjects).[17] Finally, the introduction to the second book arrives at a definition by distinguishing between clemency and either *piedat* or *misericordia*. The insertion of Seneca's ideas into a context of familiar orthodoxy is here accomplished with a notable thoroughness.

On matters of ancient history, by contrast, the immediate priority was to tell readers things that they did not presently know: the identity of exemplary figures, like Mucius Scaevola or Phalaris; the history of Augustus's wars; the family relationships of the Caesars (fols. 105v, 161v, 116v,121v). There were also matters of Roman custom to be explained: what a triumph was like; why soothsayers might be expected to take an interest in dead bodies; who was liable to be thrown to wild beasts (fols. 153r, 112v, 165v). With these matters Cartagena deals competently – at least within certain limits. There are topics on which he evidently finds it better not to dwell at length: the scandalous conduct of Augustus's daughter Julia gets very summary mention (fol. 122v); the notion of religious sanctuary, when slaves took refuge at the statues of emperors, is deliberately played down (fol. 137v). Sources are seldom specified, and some of those which are – the *Romuleon*, for example – inspire little confidence. A gloss which is eloquent both of Cartagena's limitations and of his real strengths tackles the central problem of Seneca's fulsome praise of Nero. Everyone agrees that Nero was the worst of emperors. So did Seneca not realize this – which would be incredible in one so intelligent? Or did he know it and choose to flatter Nero – surely a great fault in a philosopher whose reputation stood so high that Jerome thought him a saint? Cartagena's answer is that for the first few years of his reign Nero was indeed virtuous but that he became wicked thereafter. He notes that Seneca gives the emperor's age as twenty-two; at this stage, then, praise was in order. Later, Seneca did not praise Nero, as is shown by the veiled reference, in one of his letters to Saint Paul, to an unnamed "público malfechor" ("public evildoer").[18] The argument is boldly stated, the evidence well marshaled. But a crucial item,

---

17 Fols. 95v–96v. The background on equity is in Aristotle, *Ethics*, V,10; a few years later Leonardo Bruni was to comment unfavorably on Cartagena's preference for the Greek term *epichea*, as opposed to the standard lawyers' paraphrase *ex bono et aequo* (Birkenmaier, "Der Streit," p. 208).
18 Fol. 99v. Though not translated in Spain, the Paul–Seneca correspondence was accepted

the correspondence with Saint Paul – still unquestioned among fifteenth-century Spaniards – is simply not authentic. Cartagena, for his part, does not think of questioning its authenticity.

What he does try to do, on several occasions, is to link information about the ancient world with elements of contemporary experience. References to Roman law were one obvious and natural way of doing so. In the legal tradition, the Castilian *Partidas* – a text which Juan II quite certainly knew and handled – stood in clear line of descent from the *Digest* and the classical codes.[19] There were other analogies, too. When Cartagena, commenting on the civil wars of Rome, observed that the Castilian word for such rebellious episodes was *asonadas*, he gave the example of the twelfth-century Count Manrique de Lara:[20] That was safe enough. But Juan II was well aware of far more recent civil strife. As another gloss reminded him, the preceding century's struggle between Pedro the Cruel and his half-brother Enrique, founder of Juan's dynasty, had left grudges as lasting as the wars of Caesar and Pompey. But this modern instance, Cartagena continues, was providential, for the descendants of Pedro and Enrique intermarried "de que salistes vos, como vuestra gloriosa planta, en que segunt la escriptura a propósito dize, se fizieron amas partes una, e cesó todo escrúpulo e diversidat" ("from which you sprang, and your glorious lineage, in whom, as the scripture aptly says, both were made one, and all scruple and all difference was at an end") (fol. 122r). Juan II, then, is a God-given restorer of concord and the bringer of an Augustan peace. A few years in the early 1430s were the only part of his reign in which it was actually possible to believe this.[21] But the rhetoric persisted: "El César novelo," Juan de Mena was to call him, in the less propitious circumstances of the following decade.

there in the fifteenth century (as elsewhere in Europe); cf. the translated preamble to vernacular versions of Seneca's letters (see Fothergill–Payne, *Seneca and Celestina*, pp. 2–4, 145, and references given there).

19 See Cartagena's gloss referring to the *Lex Cornelia* (fol. 133v): "segunt podedes ver en la ley duodécima del título de la Partida" ("as you may see in the twelfth law in the section of the *Partida*"); also Nicholas G. Round, *The Greatest Man Uncrowned: A Study of the Fall of Don Alvaro de Luna* (London: Tamesis, 1986), pp. 123–4.

20 Fol. 116v. Manrique Pérez de Lara, count of Lara, fought in the civil wars of Alfonso VIII's minority and was killed in the battle of Huete in 1164; see *Diccionario de historia de España*, ed. Germán Bleiberg, 3 vols. (Madrid: Revista de Occidente, 1979), 2:874.

21 Round, *Greatest Man Uncrowned*, pp. 7–8; Juan de Mena, *Laberinto de fortuna*, ed. Louise Vasvari Fainberg (Madrid: Alhambra, 1976), p. 77.

Sometimes, too, Cartagena evokes an aspect of contemporary religious life. A propos of the tyrannical maxim "When I am dead, let fire consume the earth," he adds that "algunos ay agora que dizen que querrían que quando ellos moriesen, que luego fuese la fin universal del mundo" ("there are some nowadays who say that when they die they would have it be the universal end of the world"), and he criticizes the selfishness of this outlook.[22] It seems very possible that those who were reported as voicing it may have been millenarian heretics of some sort. Another gloss, on Seneca's distinction between *religio* and *superstitio*, makes clear Cartagena's distaste for other forms of unofficial zeal: "algunos toman algunas vanas imaginaciones, cuidando servir a Dios, e non son aprovadas por la iglesia" ("some people adopt certain vain imaginings, believing that they are serving God, and they are not approved by the Church") (fol. 162v).

Undisciplined Christian zealots, we might relevantly reflect, did much to make the world unsafe for people of Jewish descent. But the principal factor linking together Cartagena's religious and political reflections was surely his deep attachment to a principle of authority. That was a crucial theme in the Castilian politics of the early 1430s, when it seemed that Juan II and his chief minister Alvaro de Luna had at long last made royal authority truly effective.[23] The achievement was still flawed by hidden tensions with the magnates, by Luna's arbitrary security measures (such as the arrests of February 1432), and by too lavish a flow of grants from tax revenues. But it was a real achievement, celebrated at length by the royal chronicler, Cartagena's uncle, Alvar García de Santa María. His account makes plain the key role which was played by the legally trained career bureaucracy. Both Alvar García and his nephew were closely associated with this group; their *converso* background was shared by many of its members.

22 Fol. 159v; cf. his view of the Bohemian Hussites in a later work, the *Defensorium unitatis Christianae*, ed. Manuel Alonso (Madrid: CSIC, 1943), p. 286. On links between millenarian heresy and anti-*converso* movements, see Nicholas G. Round, "La rebelión toledana de 1449: Aspectos ideológicos," *Archivum* (Oviedo), 16 (1966), 385–446.

23 Alvar García de Santa María, *Crónica de Juan II*, Colección de documentos inéditos para la historia de España, no. 100, pp. 302–11, esp. 308–11; discussion in Round, *Greatest Man Uncrowned*, pp. 12–29; on "princeps legibus solutus" see ibid., pp. 115–16, 121–2. See also Alan Deyermond, "Historia universal e ideología nacional en Pablo de Santa María," in *Homenaje a Alvaro Galmés de Fuentes*, 2 vols. (Oviedo: Universidad, 1985), 2:313–24; M. Jean Sconza, "A Reevaluation of the *Siete edades del mundo*," *La Corónica*, 16 (1987–8), 94–112.

Such men had a deep personal stake in peace and patronage and thus a more rooted attachment to royal supremacy than any of its merely political partisans – even, perhaps, than Luna himself. They helped to develop a conscious ideology in which Juan II appeared as the heir in Spain to the universal monarchy of Romans and Goths – a notion decisively furthered in the writings of Cartagena's own father. Their training also led them to see the Castilian king as the *princeps* of the Roman legal code, unassailable by any subject and unfettered by positive law. To much of this the *De clementia* version seems, in places, very closely addressed.

The Senecan original, indeed, was one of the founding texts in the tradition of absolutist legal theory. "Animus reipublicae tu es" ("You are the soul of the state"), Seneca had told Nero, "illa corpus tuum" ("it is your body"); still more significantly, he had praised the emperor as "in terris deorum vice" ("the gods' deputy on earth").[24] These views had been echoed and developed by medieval jurists from the thirteenth century on. The echoes had not gone unheeded in Spain and may well themselves have contributed to Juan II's choice of *De clementia* as a Senecan text for early translation. Cartagena, though, has no specific comment to offer on either passage. Rather, he takes up a more general theme, still closely related to the notion of absolute royal power: the idea of clemency as the pursuit of greater equity, overruling the rigors of positive law. A long gloss on the introduction to Book I illustrates this from the experience (shared by Cartagena himself) of Juan II's council.[25] Sometimes the nature of the case leads that body to abandon "la vía ordinaria del derecho escripto" ("the normal course of written law") and follow a procedure known as *espediente*. If suitably applied, with good intention and "razonable igualdat" ("reasonable evenhandedness"), and if it leads to a mitigation of penalties, this will be an act of *epiqueya*. The *princeps* must have observed due process at all earlier stages and must obey natural reason now. The case must be such that the original

24 See Gaines Post, *Studies in Medieval Legal Thought: Public Law and the State, 1100–1322* (Princeton: Princeton University Press, 1964), p. 355; E. H. Kantorowicz, *The King's Two Bodies: A Study in Medieval Political Theology* (Princeton: Princeton University Press, 1957), p. 92; also Round, *Greatest Man Uncrowned*, pp. 106, 111, 128.

25 Fol. 96r; for Cartagena's activity in the *consejo de justicia*, see Rosell, ed., *Crónica de Juan II*, p. 461; on legal derogation and absolutism, see Round, *Greatest Man Uncrowned*, pp. 87–129, and references there.

lawgiver himself would no longer wish to apply the law in question. And there must be no taint of favor or vested interest. Failing these conditions, the result will not be equity but injustice. Yet if it is exercised correctly, this virtue belongs uniquely to the "soberano poderío" ("sovereign power") of the *princeps*, for he is above positive laws, while other judges are beneath them. The lesson in absolutism could hardly be more explicit. In the renewed civil strife of the midcentury, the derogation of laws was to be put to intensively political use, for it could readily be made to cover breaches of positive law which served the interest of the executive power. Yet Cartagena's insistence on the proper limits of the exercise suggests a more principled approach than would have commended itself to a thoroughgoing practitioner of power politics like Alvaro de Luna.

Such an approach seems implicit, too, in the prologue, with its review of Juan II's performance in terms of the "four virtues." The praise of Juan's fortitude in the Aragonese and Granadine wars was only to be expected. So was the greater praise reserved for the latter conflict as an "antigua e entera enemistad" ("ancient and authentic enmity") (fol. 93v), fulfilling both human and divine law. (It was also, though Cartagena does not say so, much likelier to unite the Castilian nobility, many of whom had close ties with Aragon.) But there are clear ideological resonances behind the further comment that Juan's victory over the Moors would qualify him for a Roman triumph, and the description of Granada – never hitherto a part of Castile – as "la rebelde cibdat" ("the rebel city").[26] The latter follows the same logic as had made Cartagena declare, in his prologue to *De providentia*, that Seneca, born in Cordova, owed Juan allegiance as his "natural" vassal. The Castilian kings had inherited the Roman and Gothic title to the whole territory of Hispania and the loyalties that went with it. On justice, Cartagena's verdict is oddly hedged with reservations: "con grant voluntad la mandades guardar. E si en algunas partes a las vezes fallesce, algo trae la calidat de los tienpos, algo por

---

26 Fol. 93v; cf. *Cinco libros de Séneca* (Seville: Meinard Ungut and Stanislaus Polonus, 1491), fol. 53r; see also Alan Deyermond, "The Death and Rebirth of Visigothic Spain in the *Estoria de España*," *Revista Canadiense de Estudios Hispánicos*, 9 (Spring 1985), 345–67; also Deyermond, "La ideología del Estado moderno en la literatura española del siglo XV," in *Realidad e imágenes del poder: España a fines de la Edad Media*, ed. Adeline Rucquoi (Valladolid: Ambito, 1988), pp. 171–93, and references there.

ventura viene de otras razones, pero todos con grant confiança en Dios esperamos que en vuestros días irá de bien en mejor" ("most earnestly you give orders for it to be maintained. And if it is, at times, lacking in certain places, some of that is due to the temper of the times and some, perhaps, to other causes, but all of us hope, with great trust in God, that in your days things will go from good to better") (fol. 93v). This seems little better than giving Juan the benefit of the doubt – a doubt which may have arisen either because his writ still did not run in all areas of the kingdom or because of Alvaro de Luna's legally dubious security methods. Cartagena's comment on Juan's generosity is laconic to the point of being ambiguous. There is no need to write of this, he says, "ca muchos son los que en sus faziendas lo sienten" ("for there are many who feel its effects on their property")[27] – and he has already observed that taxpayers do not always approve of an openhanded ruler. As for clemency, he opts to illustrate this with the example of the Pardon of Segovia – the general amnesty of all pending cases with which, in 1428, Juan II had prepared the ground for Luna's return to power after temporary eclipse.[28] It was, then, an episode whose political tendency favored Don Alvaro, but it was first and foremost an irenic gesture, designed to unite the kingdom. Like Seneca addressing Nero, though perhaps with a better hope, Cartagena sought not merely to praise Juan II but to give him matter for thought.

That aim informs the translation as a whole. It is, overall, longer than its original: in terms of word count, something like twice as long. Both Seneca's peculiarly concise prose style and the contrast between the two languages tend to magnify this disparity. Deliberate expansions contribute little more than one-fifth of the increased length. Yet it is clear that, both consciously and unconsciously, Cartagena favored the analytic expansion of Seneca's often condensed and gnomic meanings. Cartagena's practice varied a great deal, however, from one part of the text to another. Book II, for instance, with its more theoretical subject matter, is rendered at relatively greater length than Book I. In general, those

---

27 Fol. 94r; cf. fol. 91v: "Los que pechan non querrían en el señor muy larga franqueza" ("Those who pay tax would not want their lord to be a man of very great liberality"). But "those who pay tax" were, by definition, the *estado llano*; Cartagena and most of his potential readers belonged to the tax-exempt group whom royal generosity chiefly benefited.

28 Fol. 94r; for the *perdón de Segovia*, see Rosell, ed., *Crónica de Juan II*, p. 444.

passages where Seneca's meanings are most clearly expressed –
mainly in the form of continuous direct or reported speech – are
the ones most concisely translated. Elsewhere, two elements
account for most of the deliberately added material. There are
"discourse markers" which help to place particular clauses and
sentences within the larger argument. And there are expansions –
some informative, others merely pleonastic – of this or that word
or phrase. In vocabulary, Cartagena sometimes has recourse to the
practice which he cites with reference to the term *clementia* itself,
introducing and explaining, as necessary, a calque on some Latin
term. Normally, however, he adheres to the principle, set out in
his *De inventione* prologue, of respecting the "modo de fablar que a
la lengua en que se pasa conviene" ("mode of expression appropriate to the target-language").[29] Robustly vernacular equivalents
appear, even for Seneca's specifically Roman lexis of titles, institutions, and customs. Nor is Cartagena afraid to paraphrase, even at
some length, items for which no single Castilian word suggests
itself. His approach to patterns of syntax is very similar. Those of
Seneca's characteristic formulae which lend themselves to an oral,
paratactic style – the question-and-answer sequences, the exclama-
tions, the strings of parallel questionings and assertions – are
carried over intact. Other sentences show the translator (who had,
after all, some experience of handling Cicero) at ease with a
modest degree of complexity. But here, too, he is always willing
to modulate and transform, in the interest of an enhanced natural-
ness or clarity. Shifts of number, person, or mood, changes
from indirect to direct speech, or from synthetic to analytic presentation, overwhelmingly go with the grain of Cartagena's
target-language. The outcome is a plainer tale than the original:
more transparent but also slower-moving. Yet, as Olga Impey has
pointed out, a balance is still held between fidelity to the Castilian
reader and fidelity to the Latin author.[30] In this strongly persuasive
and communicative idiom, something of the pointed brilliance of

29 Mascagna, ed., *La Rethórica de M. Tullio Cicerón*, p. 31.
30 Olga Tudorica Impey, "Alfonso de Cartagena, traductor de Seneca y precursor del
   humanismo español," *Prohemio*, 3 (1972), 473–94, at 483–4. Further essential reading
   on the approach to translation represented by Cartagena is Peter Russell, *Traducciones y
   traductores en la península ibérica (1400–1550)*, Monografies de Quaderns de Traducció e
   Interpretació, no. 2 (Barcelona: Universidad Autónoma, 1985). I return to this topic, at
   greater length than space here allows, in Chapter 5 of my forthcoming book "*Libro
   llamado Fedrón*": Plato's "*Phaedo*" Translated by Pero Díaz de Toledo* (London: Tamesis, in
   press).

Seneca's style is clearly lost, yet much of its serious urgency survives.

These were the priorities of Cartagena's own response as a reader of Seneca. That response, articulated in the *De providentia* prologue and more fully in the polemic against Bruni, hardly allows us to separate Seneca's rhetoric from his message, much less to put the rhetoric first: "quam dulcibus suasionibus et acutissimis increpationibus nos ad virtutem provocet" ("with what gentle persuasion and very pointed reproof he incites us to virtue"). Cartagena's experience of Seneca was, above all else, the experience of an encounter with a great teacher: "praecordia incitantur ac uiscera contremiscunt, illum tamquam quendam magistrum timendo" ("the heart is stirred up and the inward parts tremble, fearing him as we might fear a schoolmaster").[31] And if, in his own version, something of the emotional force is sacrificed to the integrity of the lesson itself, we have to recall the sheer novelty of what Cartagena was doing. To translate Seneca was to address one's language to ethical problems as they were experienced within the self. Medieval Castilian literature could discuss such matters in terms of moral theology or clothe them in allegory. It could offer manuals of conduct and any number of ethical maxims to be pondered and absorbed. But there hardly existed, as yet, any model of continuous secular discourse that could convey the inner life of the moral subject. That was what Cartagena, as translator of Seneca, managed to create. In the longer term, it was what made these translations culturally important.

This pioneering quality ought not to be forgotten when we

---

31 These quotations in Birkenmaier, "Der Streit," p. 174; cf. prologue to *De providentia* in *Cinco libros de Séneca*, fols. 52v–53r: "Ca commo quier que muchos son los que bien ovieron hablado, pero tan cordiales amonestamientos, ni palabras que tanto hieran en el coraçon, ni así traigan en menosprecio las cosas mundanas, no las vi en otro de los oradores gentiles.... mas Séneca tan menudas e tan juntas puso las reglas de la virtud con estilo eloquente commo si bordara una ropa de argentería bien obrada de ciencia en el muy lindo paño de la eloquenciá" ("For though there are many who have spoken well, I never saw in any other Gentile orator such cordial admonitions, or words which so pierce the heart, or instill such a contempt for the things of this world.... but Seneca set down the rules of virtue in eloquent style, in such a dense and detailed way, as if embroidering a beautifully worked silver design of knowledge on the fair cloth of eloquence"). In responding to these qualities as a translator, Cartagena was sustained by his belief that the vernacular was capable of its own kind of eloquence; see *Duodenarium*, MS Burgo de Osma, Catedral 42, fol. 15v. I owe this reference to Mr. Gerard Breslin, who is currently editing this important text.

approach the contrast between Cartagena and another theologically oriented interpreter of Seneca, from the following century, the young John Calvin: for the first thing likely to strike us about that contrast is its sheer scale. It is as if, in the intervening one hundred years, classical scholarship had reinvented itself as a discipline. It had not quite done that. But by Calvin's time scholarly standards and practices which in Cartagena's day had been the exclusive concern of a very few Italians had spread throughout the literate minorities of many European nations. In another change of at least equal moment, printing had become established.[32] In a sense, the whole raison d'être of Calvin's Senecan commentary derives from these two developments. Its precise character was shaped by changes of a more specific kind.[33] The wider diffusion of knowledge about Tacitus had transformed the image of Seneca the man. And Erasmus, with his collected edition of Seneca's works (1529), had furnished the world with a reliable notion of what Seneca had actually written. Taken together, these factors go far to account for the differences between Cartagena's *De clementia* and that of Calvin. Yet those differences remain extraordinary – the more so, given other contrasts in their circumstances.

Cartagena, in 1432, was a mature scholar, at the leading edge of the intellectual development of his country; he was writing for

32 Together, these changes also led to a wide diffusion of Senecan texts among relatively undereducated laity – "homines infantes ac vix semi-grammaticales" ("immature men, scarcely half-formed in letters"), as Erasmus put it in his edition of 1515 (cited by Ford Lewis Battles and André Malan Hugo, eds., *Calvin's Commentary on Seneca's "De Clementia,"* Renaissance Society of America: Renaissance Texts Series, no. 3 [Leiden: Brill, 1969], p. 34).

33 For the immediate biographical setting, see Alexander Ganoczy, *The Young Calvin* (Edinburgh: Clark, 1988), pp. 71–6; also William J. Bouwsma, *John Calvin: A Sixteenth-Century Portrait* (New York: Oxford University Press, 1988), pp. 13–14. Bouwsma, "Changing Assumptions in Later Renaissance Culture," *Viator,* 7 (1976), 421–40, and Suzanne Selinger, *Calvin against Himself: An Inquiry in Intellectual History* (Hamden, Conn.: Archon Books, 1984), pp. 6, 150–1, show how Calvin himself belongs to a context of renewed interest in Stoicism in the early sixteenth century; cf. François Wendel, *Calvin: Sources et évolution de sa pensée religieuse,* Histoire et société, no. 9, 2nd ed., rev. (Geneva: Labor & Fides, 1985), pp. 13–14; Karl Reuter, *Von Scholaren bis zum jungen Reformator: Studien zum Werdegang Johannes Calvins* (Neukirchen-Vluyn: Neukirchner Verlag, 1981), pp. 89–104. See also Charles Partee, *Calvin and Classical Philosophy,* Studies in the History of Christian Thought, no. 14 (Leiden: Brill, 1977); Léontine Zanta, *La renaissance du Stoïcisme au XVIe siècle* (Paris: Champion, 1914).

the most privileged and sympathetic of lay patrons. Yet he felt obliged to write in the vernacular and to limit his presentation of Seneca to a basic underpinning of what was said in the text. In scholarly terms, even when allowance has been made for these self-imposed limitations, his work is less than reliable. It was, even so, an outstanding public success, compared with almost any scholarly production in the Castile of his day. What that meant, however, was merely some twenty-five surviving manuscripts and an unknown (but not necessarily large) number now lost.[34] In 1532, Calvin a recent graduate of a French-provincial law school, could find a commercial publisher and could anticipate a public for an exhaustive and detailed commentary that was wholly in Latin.[35] Cartagena appears to have worked from the single manuscript which chance and his patron had set before him. Calvin, though able to exploit Erasmus's edition as his base-text, could venture with confidence into textual emendation. His array of sources is of another order of magnitude altogether from Cartagena's, embracing dozens of names and hundreds of references. The most obvious result is that, even if we compare only those passages on which both writers have something to say, Calvin emerges as knowing far more.[36]

That impression may not, in every case, be wholly fair to Cartagena. On Julia's misconduct, for example, Calvin has more to say because he is less reticent as well as better informed (p. 191). He is also more willing to embark on theological argument, and several times cites Augustine in refutation of Stoic teaching, where Cartagena passes over the point (pp. 363–5, 367). As a rule, though, Calvin quite simply does know more. He can discuss Roman history and customs in far greater detail; he can pinpoint

34 See Blüher, "Der Streit," pp. 133–4n.
35 Battles and Hugo, *Calvin's Commentary*, p. 11. This edition of Seneca's text and Calvin's commentary is the source for all otherwise unattributed information cited here regarding the latter's handling of Seneca.
36 On the humanist scholarship of the *De clementia* commentary, see Wendel, *Calvin*, pp. 12–20, especially the reference on p. 20 to the self-confidence with which the work was offered to a humanist public; cf. also Bouwsma, *John Calvin*, p. 13: "a kind of sophomoric satisfaction," but a genuinely original aptitude for textual editing is also at work here (ibid, p. 117). On the range of Calvin's references, see also Ford Lewis Battles, "The Sources of Calvin's Seneca Commentary," in *Courtenay Studies in Reformation Theology*, I (Appleford: Sutton Courtenay Press, 1966), pp. 38–66. On Calvin's humanism generally, see Bouwsma, *John Calvin*, pp. 113–27, and the bibliography given on p. 240; also Ganoczy, *Young Calvin*, pp. 178–81.

Stoic arguments in terms of the tradition from which they come. He can take up topics of which Cartagena is evidently unaware: the dramatic origins and the poetic meters of maxims quoted by tyrants (pp. 225, 351); the philological background of *superstitio* (pp. 363–5); the medical implications of *lippitudo* (p. 373). If he offers no comment on *triumphus*, it is probably because he does not expect readers to need help with this term – and he does have plenty to say, as Cartagena does not, on the related matter of the Civic Crown (p. 333).

Yet the comparisons are not all to Cartagena's disadvantage. His explanation of what Roman soothsayers actually did (fol. 112v) seems more to the point than Calvin's diffuse attack on those who relied on them (pp. 133–4). Cartagena's metaphorical reading of the "prison" implied by "e ludo in carcerem" (fol. 157v) makes better sense than Calvin's rather forced literalism at this point (p. 327). Sometimes the two seem to be putting forward legitimate alternative emphases. Calvin's explanation of why Seneca's praise of Nero was well judged in psychological terms (p. 123) is not, in principle, better grounded than Cartagena's attempt to account for it historically (fol. 99v), though at least the Frenchman did not have to rely on Seneca's supposed correspondence with Saint Paul. When Calvin agrees with Erasmus that Book II is probably defective, he is being more scholarly than Cartagena; when he characterizes it, rather grudgingly, as full of "scholasticis argutiis" ("scholastic subtleties"), his dismissal of an element which Cartagena plainly valued merely reflects a different set of priorities (p. 336). The specific, even ad hominem, intentionality of Cartagena's translation can actually make his work appear more purposeful, more adequately addressed to its own situation than Calvin's academic exercise.[37] And there is, in the earlier writer's response to Seneca, an immediacy not matched in Calvin's rather frigid advocacy of his author's stylistic excellence and of his entitlement to a major place in Latin letters. The advantages with which Calvin must emerge from any comparison, at the present

37 This is perhaps to underestimate what Bouwsma calls his "emotional investment" in the work (*John Calvin*, p. 14). The recent Ph.D. dissertation of Michael J. Monheit, "Passion and Order in the Formation of Calvin's Sense of Religious Authority," Princeton University, 1988, relates the *De clementia* commentary to Calvin's search for the kind of authority which might underwrite religious beliefs and to his concern with the relationship between virtue and the passions.

time, are not those of literary competence or of intellect but rather those of relative modernity. And, even on that basis, the fact that a selection of Cartagena's Senecan versions (which did not, as it happened, include _De clementia_) went through five printed editions between 1491 and 1551[38] ought to make us curious to know more of this important, and still neglected, interpreter of Seneca.

38 _Cinco libros de Séneca_: editions of Seville (1491), Toledo (1510), Alcalá de Henares (1530), Antwerp (1548), Antwerp (1551). These editions include _Libro de la vida bienaventurada_, _Libro de las siete artes liberales_ (i.e., Letter 88 to Lucilius), _Libro de amonestamientos e doctrinas_, _Libro primero de la providencia de Dios_ and, under the title _Libro segundo de la providencia de Dios_, the translated selection from Manelli's _Compilatio_, with some minor related items. The list is not of the printers' devising; it reflects the content of one particular manuscript tradition, going back as far as Juan II's reign.

5

The Epicurean in Lorenzo Valla's
On Pleasure

MARISTELLA DE P. LORCH

A PREMISE

Words (*nomina*) – some key words in the articulation of
Lorenzo Valla's thesis – have a very special meaning in the
dialogue *On Pleasure*. The argument seems to develop on the basis
of certain words used with a very particular meaning – words like
honestas, honestum, utilitas, utile, fruitio – and to receive impetus
from certain verbs, often used in the first person singular: *aio atque
affirmo, sic statuo*, or *volo*. These words are very difficult to
translate. That is why, in trying to present my interpretation of
Valla's Epicureanism, I often feel forced to use the Latin terms,
together with their English translations. It is as if Valla had
invented them.

The key word in the whole argument is *voluptas*. I define
elsewhere (*A Defense of Life*, Chap. 2) Valla's process as "discovering a new meaning and function in an old word." In the present
essay I try to shed further light on this new meaning by focusing
on the character of the Epicurean as the main character of a drama
which opposes *voluptas* to *honestas*.[1]

---

1 I refer the reader, for all details on the content of the dialogue and on its textual history,
to the critical edition: L. Valla, "*De vero falsoque bono*," ed. Maristella de P. Lorch (Bari,
1970); to the English translation of it: *L. Valla's "On Pleasure*," by A. K. Heiatt and
Maristella de P. Lorch (introduction, notes, and appendix by Maristella Lorch) (New
York, 1976); and to my recent *A Defense of Life: L. Valla's Theory of Pleasure* (Munich,
1985), on which the present essay is based. For my bibliography on Valla in general and
on *On Pleasure* in particular, see my recent essay "Lorenzo Valla," in A. Rabil, ed.,
Renaissance Humanism: Its Foundations, Sources, and Legacy, 3 vols. (Princeton, 1988),

SETTING THE PROBLEM: WHAT IS VALLA'S *VOLUPTAS*?

Lorenzo Valla disseminated his dialogue *On Pleasure* in its first version, entitled *De voluptate*, from Piacenza in 1431. He must have conceived it, and probably wrote it down (at least in part) while in his hometown of Rome, where he lived close to the environment of the Curia. In 1430 he moved to Piacenza, called there by his friend Panormita, whom he introduced as the "Epicurean" in his dialogue. Valla was in his middle twenties.

The audacious thesis that "pleasure [*voluptas*] is the only good," which appears as part of the defense of the title, warns the reader that Valla's *voluptas*, and consequently his Epicureanism (and conversely his Stoicism), have only superficially something in common with the revival of the two doctrines, which took place around him among humanist colleagues, friends and enemies, partially but not totally in consequence of the discovery by Poggio Bracciolini of Lucretius' *De rerum natura* in 1417. The time was ripe for a defense of Epicurus as a philosopher, focusing not only on man but on the elusive concept of pleasure. This fact is witnessed by Cosma Raimondi and Filelfo, among others, around Valla, and, shortly after his death, by Marsilio Ficino: They all wrote on pleasure. Eugenio Garin, among other scholars, has documented the issue most convincingly.[2] It is to be noticed that

vol. 1. In the present chapter I quote the text from the Heiatt and Lorch edition, hereafter abbreviated *OP*. Citations to book, chapter, and paragraph number are given in parentheses in the text.

2 Valla's Epicureanism has been the object of varied interpretations, from Fiorentino and Mancini to Corsano, Gentile, Radetti, Garin, Trinkaus, and Siegel, just to mention a few; in recent years by di Napoli, Fois, Camporeale. A useful overview was offered by Cesare Vasoli in his "Nuove prospettive su Lorenzo Valla: Interpretazioni e commenti." Garin's "Ricerche sull'Epicureismo del Quattrocento," in *La cultura filosofica del Rinascimento* (Florence, 1979), pp. 72–92, analyzes different interpretations of Epicurus: Giovan Battista Buoninsegni (Florence, 1458); Maffeo Vegio, who in the *De educatione liberorum et eorum claris moribus* sees him as a master of human wisdom; Dominici, in the *Lucula noctis*. Epicurus attracted the attention of many writers in the early fifteenth century and shortly before then. Among them were Domenico di Bandino, and Benvenuto da Imola, in his commentary on the *Inferno*; a sixteenth-century Dominican writer, in *De divisione et laude philosophiae quae ad mortes pertinet* who considers Epicurus as the symbol of a radical form of naturalism. The Roman *Accademici* around Pomponio Leto were accused of following *voluptas Epicuri*. Among the most objective interpreters was Cosma Raimondi, a contemporary of Valla who lived in Pavia, whose *Defensio Epicuri contra Stoicos, Achademicos et Peripateticos* is reproduced in Garin's appendix (pp. 87–93). Garin deals with Ficino's *De voluptate* (1457), with Bruni's brief introduction to the corpus of Aristotle's writings (which prefaced the volume throughout the sixteenth

even before the humanists, Thomas Aquinas had articulated an interpretation of the concept of pleasure of which Valla might have been aware.[3]

In order to understand the role and function of Valla's Epicurean in the dialogue *On Pleasure* and the nature of the Epicureanism he professes, we must begin by paying due attention to the defense of *voluptas* as a word (*nomen*), which opens the first version of the dialogue, *De voluptate*, in 1431. The additions to the subsequent versions, primarily to Books II and III, stress the main point of the defense: Pleasure (*voluptas*) is the only good.

Si quis forte ex amicis gravibus ac severis . . . If some of my friends are shocked by the title *De voluptate* and should from the outset question my constancy and perhaps ask what strange desire [*cupido*] has prompted me to write on pleasure to which we have never been given nor wanted ever to seem to be inclined, to anyone who asks this question in sincerity and friendship I wish to give and must offer immediate satisfaction. He should know that I have never departed from my habitual behavior. In fact I have given proof of the fact that I have adhered to it. Nevertheless, I preferred to entitle these books *De voluptate* [On pleasure] because of the flexibility and pleasant quality of the word *voluptas* [molli quodam et

century), and with Bruni's letter on wealth to Tomasso Cambiatore, in which Bruni seems to resent Epicurus' traditionally praised austerity. Finally, in dealing with Valla, Garin declares that Valla goes beyond an ethical interpretation of Epicureanism by stressing a good, divine nature. Garin hints also at Filelfo's moral interpretation (letter to Andrea Alamanni, 1450). Garin concludes that around the middle of the fifteenth century one can distinguish three tendencies in the rebirth of Epicureanism, the first connected with a materialistic view of reality (the *dominae voluptates* of Panormita and the Roman group of Pomponio Leto); the second tendency considering Epicureanism as a most serious moral system (see Beroaldi's *De felicitate*); the third stressing the cosmic value of *voluptas*, its divinity. This aspect, according to Garin, is only slightly touched upon by Valla. It is present in Palingenio Stellato's *Zodiacus vitae*, which is full of admiration for Epicurus and Lucretius, and in the hymns of Marullo, whose Venus is Neoplatonic and Lucretian (some hymns are quoted at pp. 84–5). Of Epicurus, even the pious Giovanni Nesi and Cristoforo Landino speak favorably.

My own interpretation of Valla's Epicureanism does not use the comparative approach. It centers on Valla's work in itself as a drama. Valla's dialogue, however, is a creation whose originality becomes more evident if related to the often superficial judgments on Epicurus by his contemporaries.

Recently, a most interesting approach to *voluptas* in Valla was taken by Domenico Pietropaolo in a review of my book *A Defense* entitled "On the Dignity of *voluptas*," *Italian Quarterly*, 9 (Spring 1988), 66–75.

3 Thomas Aquinas, *Summa Theologica*, Ia.IIae.27.1. It is discussed at length by Umberto Eco, *The Aesthetics of Thomas Aquinas*, trans. Hugh Bredin (Cambridge, Mass., 1988), pp. 25–37, and, among others, by M. Mothersill, *Beauty Restored* (Oxford, 1986), pp. 332–66.

non invidioso nomine]. This title was chosen instead of *De vero bono* [On the true good], as I might have done, since through the whole book we expound the true good and declare it to be this very *voluptas*. This friend may ask: Do you say indeed that *voluptas* is the true good?

To the objection of these friends young Laurentius answers, "I say so and I declare it and declare it in such a way as to assert that there is no other good besides pleasure [*voluptas*]" (Ego vero aio atque affirmo et ita affirmo ut nihil aliud praeter hanc [voluptatem] bonum esse contendam). The style of the Latin text reveals Valla in his fighting mood, ready to defend the word which he fills with a new meaning.

Thus the defense of a word (*nomen*) becomes for the young humanist an existential cause which calls for the whole self at its service. What is at stake is not a concept professed by an ancient school of thought that deserves to be unearthed and explained with the objectivity of a newly established philological science of which Valla was a pioneer. It is the discovery of a reason for accepting life, with all its miseries, pain, and the continual threat of physical death, as a challenge, life as tension between opposing forces, culminating in a remuneration which guarantees the restoration of the losses. The defense of *voluptas* becomes thus a defense of life itself, the phenomenon "life" as we live it moment by moment, with faith, hope, and mostly love in and for ourselves as living beings as well as for the cause and guarantee of our life, Jesus Christ in the act of Redemption.

On Pleasure is an attempt to render through language a deeply felt conviction (*persuasio*) that life as we live it first and foremost with our senses is a good (*bonum*) – in fact, the only good (*unicum bonum*) – because of the continuous successful encounter (*connubium*, or marriage) of the senses with the object of their desire (*natura*). In the act of the union, the senses act as body and soul (*corpus et animus*) united, while the goodness of the external object is guaranteed by a divinely providential nature.

"Here is the cause that I have decided to prove," continues young Valla in his defense of the word *voluptas* at the opening of *De voluptate* in 1431: "If I succeed in this enterprise, as I ardently hope, it will not be absurd to have taken up the subject and to have given this title to our work" (*OP*, p. 151, my translation). We have indeed here a case of *persuasio*, or deeply felt conviction (in the sense in which Quintilian uses the word), a conviction which is an act of faith. Lorenzo Valla could only speak as a Christian. He therefore finds the most forceful example of faith, *fides*, as the quintessence of the Christian religion: "Christiana *persuasio* que proprie *fides* dicitur," the Christian *persuasion* which is appropriately called *faith* (emphasis added).[4]

The defense of the word *voluptas* in *De voluptate*, as well as in the subsequent versions of the dialogue, is followed by the identification of its defender. He is a David, a Jonathan, a biblical hero who receives his arms from Jesus Christ himself. Within this context the originality of the defense is to be identified with the arms young Laurentius receives from Jesus and with the object of his attack. The arms are the shield of faith (*scutum fidei*) and a sword which is the word of God (*gladio quod est verbum Dei*). The objects of the attack are the false virtues and the philosophers who support them. What are these false virtues if not the denial of the only true virtue (*vera virtus*) or faith, true in the sense that it nourishes pleasure/*voluptas* and feeds on it? Not faith in one's own reason and in a form of wisdom which derives from it but in the miracle of life as we are allowed to live it in harmony with the universe, with the plants and the animals, with our fellow human beings, with the simple husbandman of the Virgilian *Georgics*, who "cleaves the soil with a crooked plow" (Incurvo terram dimovit aratro) (*OP*, I.2.5.).

Young Valla's main art for carrying out his God-given task is the human word, the art that deals with the use of language – philology, which constantly tries to relate the word (*verbum, nomen*) to concrete reality (*res*). The oratorical art, according to Cicero and Quintilian, whom Valla reveres as masters of the true philosophy, philology and eloquence, is responsible for keeping a safe contact between man and the reality that surrounds him. Cicero and Quintilian tell us that at the very dawn of civilization eloquence, as the art of language, man's most powerful expression

---

4 For a detailed analysis of the issue, I refer the reader to my essay "Rome in Valla's *On Pleasure*," in P. Brezzi and M. Lorch, eds., *Umanesimo a Roma nel Quattrocento* (Rome, 1981), pp. 191–210. See also Pietropaolo, "On the Dignity of *voluptas*," in whose view the fulcrum of the argument in Book III is "the principle that the demonstration of a thesis presupposes an act of faith or belief in the validity of the logic of the proof. This being given, the problem for the orator is how to construct his discourse so that it can generate its own authority and persuade its listeners that it is a valid avenue to truth" (p. 70). Cf. Lorch, *A Defense*, Pt. I, Chap. 10.

of the self, and philosophy, as the love of wisdom, and those who practiced these human arts (the orators and philosophers) lived together in harmony. It was their harmonious collaboration which made it possible for men, who so far had been living as savages, to build the first social community.

This attractive ancient myth is the basis for our understanding of what motivates young Valla in his defense of *voluptas* and in his choice of the "Epicureans" as the carriers of the flag in the battle in which he is about to engage against *honestas* and the false virtues.

It was the divorce of eloquence from philosophy, the myth continues, that caused the degeneration of the social community. Young Valla's defense of *voluptas* is a fight against such degeneration. When the philosophers, the story goes, cut themselves off from society to pursue their search for the true and the good within a world of their own which had ceased to be useful to society, they began using the word, language, in the abstract. Then their art of reasoning did not work at the service of man's real life. The false virtues emerged – false in the sense that fortitude, justice, temperance, or the art of contemplation became names without substance, nourishing a principle of honesty or virtuousness (*honestum*) that shines by a light of its own and does not bring help or solace to man in his everyday fight for survival. Thus the philosophers became deceivers of humanity, misleading it into the belief that man could find in his own reason sufficient support for his existence.

The task of the young orator Valla in his defense of *voluptas* is to warn men – common men, but mainly the wise and the cultured ones who venerate as *auctoritates* or infallible minds the philosophers such as Aristotle or Plato – that the ethical system they propose, as well as the metaphysics and ontology behind it, are of no use in the existential search for what is indispensable to life as we live it. What is indispensable is faith, hope, and mainly, love. The essence of *voluptas* is actually love, or *caritas*, love for a God who, by becoming man, made us aware of what the true virtues and the true *honestum* are. True virtue transcends the human reason which the philosophers have deified.

This message is no novelty in the history of Western Christian thought. The orator, Valla, avows that the novelty and originality of his thesis is mainly in his approach to it or the method he follows. Although the sword he is given by Christ is the word of

God, young Laurentius, as a human being, has to make the best possible use of the human word, of a human language, of the art of eloquence which he has learned from Cicero and Quintilian, and of the science of philology. The originality of *On Pleasure* should be sought, within this context, first and foremost in the choice of its literary genre: that is, in a most original use of the dialogue.[5] This dialogue, although composed mainly of three long monologues, should be read as the blueprint for a drama, the dramatic confrontation between two antithetical conceptions of life: *voluptas*, queen of all virtues, *regina virtum*; and *honestas*, the servant of *voluptas*.

This approach, I am aware, involves some oversimplification. Reading the dialogue as a drama, however, has advantages which allow us to approach Valla's thesis and the cruces of its interpretation in a constructive way. First, it helps us to avoid the pitfall of trying to identify Valla's position with that of one of the three interlocutors – the Stoic, the Epicurean, or the Christian. If we read the work as the blueprint for a drama, we automatically consider Laurentius not as a character (as he actually describes himself) but as the director of the show, who leads *voluptas* through the joyful, optimistic defense of life of the Epicurean to the safe shores of a Pauline world of faith and a biblical *paradisus voluptatis*, a paradise in which pleasure is life as fertility and procreation. Second, it allows us to intuit the intimate connections among the three main characters who personify, in concrete terms, the drama of *voluptas* and *honestas*. The Epicurean comes onstage, in response to the Stoic's plea for a more generous nature. The Epicurean, in his turn, unveils a most generous nature which the Christian will, through the Bible, reveal to be the gift to man of a most generous God.

WHO IS THE EPICUREAN?

The Epicurean is not only the defender of nature and of life. He is the concretization of *voluptas* and of its intrinsic optimism. Like *voluptas*, the Epicurean reflects life in its mercurial essence, which longs only to affirm itself and finds, within itself,

5 See D. Marsh, *The Quattrocento Dialogue: Classical Tradition and Humanist Innovation*, review in *Renaissance Quarterly*, 24 (Winter 1981), 572–5.

its own control, guide, or moral principle. In contrast with pleasure, *honestas* satisfies man's self-centered intellectual ego by promoting and celebrating a good for its own sake. This process works perfectly in the abstract, but fails to produce results in the here and now of a specific human situation.

In the realm of *voluptas*, man lives making full use of his senses as well as of his *animus*, the two being indissolubly united. The realm of *honestas* can, at its best, produce self-centered intellectuals or *philosophi* who miss out on real life; at its worst, which is often the case, *honestas* is a flag that intellectuals display to cover up their secret human passions, such as their desire for recognition in society, or for power and riches.

Honestas is a mask that Valla's Epicurean tears off from the most respected heroes of Roman antiquity and from the *auctoritates* in the world of *philosophia* (Aristotle). In so doing the Epicurean acts as *orator*, "a man having a duty not to swear allegiance to the law of any school or belief" (*OP*, I.13.11), as a practical man who uses his tools for a practical concrete aim, searching for reality which can give meaning to man's life and help him to live. He is the transformer of dialectics from a *scientia sermocinalis* (science at the service of discourse) to a *scientia realis* (science at the service of reality). Hence the *philosophus* should stand at the service of the orator, "like a soldier or lower officer": "The orators spoke in the midst of the cities about the best and most important subjects before the philosophers started to chatter in their nooks and crannies. . . . It is the orators . . . who must be designated leaders and kings" (I.11.1). The Epicurean orator is not inspired in his action by a knowledge of right and wrong written down in a law but by a deep, inner, instinctive understanding of human nature, of human desires, needs, and passions. The art of the *orator* consists in finding the proper language to pin down some characteristics of human nature.

Specifically, a study of the Epicurean as a character in the drama in which *voluptas* contends with *honestas* presupposes an understanding not only of Valla's concept of *voluptas* but, as well, an awareness of Valla's inductive method, which is based on the unicity of the *exemplum* reflecting the unicity of an individual human experience. Among the examples, the most significant is the example set by Epicurus in the concluding section of the Epicurean *oratio* (II.39.13, 30.5). Epicurus, differing from Aristotle, is not regarded by the Epicurean orator as an *auctoritas*; in fact, he is not even an example to follow. He is simply a human being who lives life in its fullness, harmonizing *sensus* and *animus*. Hence he accepts death serenely, as the end of a life well spent. It is Epicurus's example which introduces us to the *exemplum* of Jesus Christ.

The Epicurean is rather a tool in the hands of the orator Valla, who intends to fill with a new meaning the ancient word *voluptas*. We are back at the defense of the *nomen* which opens *De voluptate*. That defense disappears in the subsequent versions (*De vero falsoque bono*, disseminated from Pavia in 1433, and *De vero bono* and *De vero falsoque bono*, disseminated later from Naples).[6] The elimination of the passage in no way weakens the original meaning Valla intended to give to *voluptas*. The additions to the original version of *De voluptate*, especially in Book III (or Act 3 of the drama) – stress the original meaning given in the defense, with further documentation and stronger connections between the point of view of the Epicurean and the Christian.

The Epicurean as an interlocutor in the dialogue – be he Panormita, of the *De voluptate*, first version in 1431, or Maffeo Vegio of the following version, *De vero falsoque bono*, of 1433, and *De vero bono* of 1444 – represents the Epicureans. This fact might be taken to imply that whatever his characteristics are, they should be considered an indication of Valla's Epicureanism. This is only so within the logic of the dialogue. I doubt that we can draw from our Epicurean's behavior a convincing conclusion which allows us to relate Valla's Epicureanism to that of Cosma Raimondi, Filelfo, Bruni, Buoninsegni, or Ficino on historical grounds.

The Epicureans are mentioned at the conclusion of the passionate Proem to the dialogue, where Valla speaks, in the first person, as a Christian soldier. This preface is a true witness to faith on the part of the young Laurentius, who proposes to defend the Christian republic like a biblical David or Jonathan: Jesus Christ himself, as stated earlier, offers him his sword with which to kill the *philosophi*, whom he defines, in the language of the Christian fathers, as *allophili*, foreigners: people of another stock or race. This is done "with the help of our faith and of the word of God" (*OP*, Proem, 6).

6 See the introduction to the critical edition, pp. xxx–lii.

What is at stake is the reality of *honestas*, which, we might intuit, is the principle of virtue. Yet, like *voluptas*, its meaning will come to light a bit at a time through the drama which Valla creates, himself an orator in search of a new philosophy. From the beginning we are made aware that *On Pleasure* is not a philosophical work in the traditional sense, because it does not present a theory through a logical sequence of arguments. As suggested before, it is to be read as a literary work of a special kind, a "dramatic composition" conceived by young Valla at a time in which humanists around him in Pavia (Ugolino Pisani and Panormita, among others) were laboring to create a theater before Plautus and Terence had become accepted models. A dialogue such as *On Pleasure* is an example of the need to dramatize ideas, a need felt by the most creative among the humanists.[7]

The concreteness of the *res*, the subject, *voluptas*, which is life itself in its mercurial essence, refuses to be channeled or repressed by outside forces, resisting all forms of theorizing. This recalcitrance induces the author to identify ideas with human beings, as well as to make use of all available rhetorical ploys: metaphors, allusions, visions, symbols, and most of all, irony. A study of the function of irony in *On Pleasure* is most pertinent to our understanding of the role that the Epicurean plays in the unfolding of the drama. Let us look first at the dialogue as a drama.

In Act 1, the Stoic sets the problem (*OP*, I.2–7). In Act 1, Scene 2, the Epicurean unveils the essence of *voluptas* as the union of the

---

7 University farces flourished at the University of Pavia during the period when Valla was working on *De voluptate* (1431) and on the second version, *De vero falsoque bono* (1433). I found in the Vatican Library a manuscript of the *Janus sacerdos* attributed to Panormita (cf. M. Lorch, "The Attribution of the Janus Sacerdos to Panormita: An Hypothesis" *Quaderni, urbinati di cultura classici*, 5 (1968), 115–35). Ugolino Pisani's *Philogenia* (1432) and *Repetitio Zanini Coqui* are witness to the fact that the Italian humanists felt compelled to resurrect the genre "theater" on the model of the classics. Well-constructed plays, the result of these attempts, appear in the second half of the century. In Pavia, at the time when Valla was working on the studium, Panormita, whom Valla introduced as the Epicurean in the dialogue, was lecturing on Plautus; in Ferrara, Guarino (also an "actor," although a secondary one, in the dialogue) yearned to possess a copy of the nine comedies by Plautus which had just been discovered. The thought of giving dramatic expression to an idea must have occurred almost instinctively to a creative, adventurous humanist like Valla. Another interpretation of the humanists' natural tendency to express themselves in dramatic form is the *Comediola Michaelida*, by Ziliolo Zilioli, the unfortunate courtier who was kept for fifteen years prisoner in the tower of St. Michael by Nicholas III of Este. (The comedy was edited by W. Ludwig and myself [Munich, 1976]).

senses with the soul (*sensus–animus*) with the external object, in
order to obtain the satisfaction of a desire (I.9 to end). In Act 2
the *reductio ad unum* takes place, or reduction of all expressions of
honestas / virtus (*fortitudo, gloria, iustitia, contemplatio, serenitas animi*) to *voluptas* (Book II) through examples.

Act 3 offers the solution of the drama by unveiling the essence
of the phenomenon *voluptas*, originally discovered to be union of
the senses with the external object, as *caritas*, Jesus Christ's love for
humanity, expressed through the Redemption (Book III). Act 3
could also be subdivided into two scenes. The first is strictly
connected with the Epicurean's criticism of *honestas* as a hypocritical attitude of the *philosophi* toward life / *voluptas*; the second is
concerned directly with the revelation of the essence of *voluptas*
through the exegesis of biblical texts, mainly from Genesis and
Psalms.[8] The second scene of Act 3 is crowned with a fiction or
painting, *fictio* or *pictura*, of paradise, as it can be envisioned here
on earth with the means available to the common man. Are irony
and laughter limited to the section of the drama in which the
Epicurean is the main actor or conceived as a presupposition of the
whole structure of the drama?

An answer to this question will come from the study of the
Epicurean as he develops his argument from Scene 1 to Scene 2,
but mostly by the use that the Christian "actor" (be he Niccolò
Niccoli, in the first, or Raudense in the second version of the
dialogue) makes of the Epicurean's argument and of his *ars
argumentandi*. In terms of the drama, the answer depends on how
the Epicurean relates to the Christian.

At the end of the Proem, after having articulated his thesis in the
bitter and rude style (*acre et abruptum*) that is typical of his way of
arguing, Valla singles out within the ocean of the enemy army,
as the lovers of *honestas*, the Stoics as the target of his attack.
Automatically he assumes the defense of the Epicureans. No
reason is given for the choice, which appears quite arbitrary. It is
an act of will, as was the defense of the *nomen voluptas*: "Satis nobis
videtur hosce [the Stoics] adversarios contra nos statuore, assumpto patrocinio epicureorum." (We decided to establish the Stoics as
our enemies, while we take up the defense of the Epicureans.)
(*OP*, Proem, 7). The reasons, Valla states clearly, will surface as

8 See Lorch, *A Defense*, Chap. 10, Pt. 2, pp. 259–60.

the drama unfolds, *postea reddam*. They will emerge in a natural way, unveiled by the nature of the *res, voluptas*.

We are here given an example of the inductive method at work. As an orator who took up the cause of *voluptas*, Valla's main concern is that the *res* should suggest the *verbum* or the verbum unveil the *res*. Since an Epicurean will be brought onstage to counteract a Stoic, the Epicurean's style will be the key to his message: "In fact, as for the exigencies of subject, what would be further from defending the cause of the pleasure than a sad, severe style and the behavior of a Stoic, when I am taking the part of the Epicurean?" (*OP*, Proem, 8).

The key to the creation of the character of the Epicurean is to be sought, as I have said before, in the nature of *voluptas* and in the "reason for the author's initiative" (*consilii mei rationem*). Valla's usual style (*acre vehemens incitatum*), which reflects his aggressiveness, will naturally give way to *remissiori et magis leto genere dicendi* (a more relaxed and agreeable way of speaking). The orator's task is to serve society by persuading – that is, leading – the audience toward the *res*. The style is therefore not an instrument but a natural way of giving in to the *res*, as Socrates would do. The *res* which is *fides* leads the author to the choice of the Epicureans in a conditional way, by contraposing their "way of living" to that of the Stoics. It is the sincerity and straightforwardness of the Epicureans that makes the *honestas* of the Stoics stand out like a thorn in their flesh: a falsity, an empty *nomen*, which does not reflect any *res*.

The character of the Vallian Epicurean grows out of this powerful non-rational assertion – most original, for one finds no analogy to it in any of the contemporary defenses of Epicurus and Epicureanism.

THE CHRISTIAN EXPLOITS THE FLEXIBILITY (*MOLLITIES*) OF THE EPICUREAN'S *VOLUPTAS*

The word *voluptas* had been under scrutiny by many humanists around Valla. He was aware of the use made of it by Seneca, Lactantius, and especially Cicero, whose definition his Epicurean quotes literally in *On Pleasure*. He quotes it in the *Apologia* and in the *Defensio*. Valla, however, discovered some-

thing else in the *nomen voluptas*.[9] Perhaps only by listening to
its echoes in Lucretius' poetry could he have intuited its deep,
pregnant meaning. Whether or not that is so, Valla used it in a
nontraditional way in order to find an explanation to problems
concerning *natura* as *rerum natura* and concerning human nature,
that is, problems concerning nature and man.

Hence Valla's "theory of pleasure" is a theory of being as well
as of human behavior, an ontological as well as an ethical theory.
The combination of the two qualities that the *nomen voluptas*
possesses, *mollities* and *non-invidiositas*, suggests a new content for
the old Ciceronian word, a kind of openness which will be basic
to Valla's theory of pleasure, matching the element of tension
both in its negative form of anxiety or labor, and in its positive
aspect of dynamism. It is within this openness that we discover
the connections between the Epicurean and the Christian.

The structure of the dialogue reflects this openness and a dynamic movement upward, from Act 1 to Act 3, which incorporates
and transforms Act 1. *Mollities*, which also characterizes the
Ovidian concept of metamorphosis, seems to suggest a priori
that the *voluptas* unveiled and upheld by the Epicurean is of
the same substance as the one represented by the Christian.
Although it will undergo a metamorphosis, this fact allows the
Christian to side with the Epicureans against the Stoics, when,
before leaving the stage (Act 3, Scene 2) and mixing with the rest
of the actors for the final banquet, he will declare, "Secundum
Epicureos iudico et contra Stoicos" (I decide in favor of the
Epicureans and against the Stoics) (*OP*, III.8.6.).

Before then, at the end of what we define as Scene 1 of Act 3
(Book III, Chap. 4), the Christian concludes his attack on the *falsa
honestas* for which he treasured the Epicurean's *ars argumentandi* (in
Act 2, or Book II). By focusing his attack on entire sections of
Aristotle's *Ethica Nichomachea*, as the Epicurean had done on the
subject of *contemplatio*, the Christian declares that every action
carries with it its "good" and "evil." This good and evil must be
judged on the basis of the *tempus* and the *tempora*, the time and the
circumstances pertinent to it (*OP*, III.4.22–3). The problem of
virtue and vice, which depend on the passions that each human

---

9 See Lorch, *A Defense*, Chap. 2, "*Voluptas*: Discovering a New Meaning and Function in
an Old Word," pp. 27–38.

action implies, can only be judged within the individual space created by the historical *tempus*. Hence the importance of *rhetorica* as a true articulation of human passions through language is reiterated. In poetry we witness the praise of excess, as Quintilian proves in his evaluation of Homer. The Christian shows why Aristotle has failed as an *auctor* in his theory of virtue, as the Epicurean did for the Stoic in Act 1. Speaking for the author, the Christian asserts that the discovery of the *divina res, voluptas*, as the *unicum bonum* is exclusively the task of rhetorical discourse. It can only be expressed by the use of *nomina*, designating individual passions and virtues, generated by the *res* itself. This language is the result of a thinking process that is solidly anchored in a historical reality, a concrete and specific human time and space, the *loci* and the *tempora*. How does this fact influence the Christian's evaluation of the Stoic and the Epicurean, as the two actors who preceded him in the unfolding of the plot? We should not forget that the subject of the plot is the unfolding of *voluptas*.

The Christian's argument develops from the Epicurean's. The Epicurean, in Act 2 (Book II), in discussing true and false glory, suicide for the love of one's country, false and true justice, and false and true serenity of mind, proved in a self-evident way (*evidentissima ratione*) that man's real motivations are self-preservation and self-assertion. In an analogous way the Christian proves that the laws that govern man's life must be discovered in life itself. The Christian stresses once more the fact that rhetorical language presupposes a *contemplatio oculis*,[10] a process of thinking connected with our physical vision. This mode of thinking calls for a metaphorical language, expression through images. Consequently, in the second scene of Act 3 the Christian considers *voluptas* as a *divina res* a way of thinking in images, expressed through metaphorical language.

Finally the Christian's *voluptas* emerges in a compact chapter (*OP*, III. 5) between a forceful and willful "Ita sic statuo" (and thus I declare) and an apologetic closing "Hec ut potui breviter" (I expressed myself as concisely as I could). It is part of a *demonstratio* based on self-evidence, *manifestis rationibus*, which consists of a sequence of examples chosen in a subtle way in order to make us

---

10 See Maristella de P. Lorch, "Virgil in L. Valla's Dialogue *On Pleasure*," *Acta of the Early Renaissance*, 9 (1982), 33–56.

discover human nature at its most instinctive and spontaneous stage: the person *I*, what happens to me, to children, to farmers, to lovers. The focus is not on the object that motivates human desire but rather on desire itself, *appetitum movere*, where *appetere* indicates the tension implicit in the longing for something. Laughter, a rejection of monotony and a search for the varied and the strange, even vainglory, are all found to be legitimate aspects of a pleasure, that in itself is most natural and hence good. *Iuvare sibi*, or seeking one's own advantage, triumphs, supported by a surprisingly long list of examples, the most eloquent being that of Socrates, condemned because he was considered obnoxious by the Athenians. This is a typical case of ignorance of a factual reality (Socrates was innocent), yet the motivation was real and thus it proves that *voluptas* is always expressed through *utilitas* or *quod juvat*.[11]

Methodologically speaking, in the unfolding of his argument the Christian builds, from beginning to end, on the basis of the Epicurean's unveiling of *voluptas* as the most natural and instinctive sensual phenomenon (Act 1) and of his destruction of the Stoic's *honestas* (Act 2). No sooner has the Stoic's *honestas* vanished, this time by force of the argument of the Christian, than *Christiana honestas* emerges as the *vera virtus*, filled with the highest potential dynamism, in a beautiful simile: "It ripens under difficulties as gold is refined in the fire and grain dried in the sun" (*OP*, III.6.1). The poets, with their natural visionary power, which has been forcefully employed throughout Acts 1 and 2 by the Epicurean (especially in the *laus voluptatis* which closes Act 1 with a hymn to the poets), made virtue shine in their heroes, as the quintessence of their heroism, "by burdening a character – Aeneas in Virgil, most characters in the *Iliad* and in the *Odyssey* – with hardships, dangers, and labors" (ibid.). The myth of Heracles illustrates this virtue at its best.

What the Christian calls *nova virtus*, a new virtue, is indeed the final answer to the problem raised by the Stoic at the onset of Act 1: "Does man stand a chance, surrounded by a *natura* in which evil prevails?" It is through his dynamic reaction to life's difficulties that man expresses his faith in life and his love for it. After having destroyed the *honestas* of the Stoics along the lines followed by the

11 Lorch, *A Defense*, p. 253.

Epicurean, the Christian justifies this virtue as a means to reach man's faith and love for life and hope in it. At this point Jesus Christ appears suddenly onstage as a carrier of the pivotal message *Christiana religio*, which in the *Elegantiae* was defined as *fides* ("Christiana persuasio que proprie fides dicitur").

It is clear now why the Christian, in passing judgment on the Stoics and the Epicureans – chosen by Valla as the most noble sects because mentioned in Acts (3.7.1) – although admitting that both were wrong, because they spoke according to the *dogmata* of certain *auctores* instead of listening to the word of God, should declare himself to be on the side of the Epicureans. The Christian's siding with the Epicurean closes the circle which was opened in the Proem with the willful statement "assumpto patrocinio Epicureorum" (having assumed the defense of the Epicureans). A verbal coincidence between two statements in *On Pleasure* – "hanc ipse contumeliam . . . fieri nomini christiano non ferens" (he, not bearing the outrage himself, bears the name Christian) (Proem I.4) and "ipse Deus . . . cui abs te contumelia fit per me ipse respondet" (God himself, whose outrage is made by you, responds himself through me) (III.7) – explains now, at the end of the drama, what in the Proem sounded like an arbitrary decision on the part of the author.

The myth with which one of the participants in the dialogue (Guarino, in the second and third versions) defines the difference between the Epicurean's and the Christian's performance confirms the connection of the Christian to the Epicurean. The difference between the two is in the type of eloquence they practiced. Like the swallow in Ovid's story of Procne and Philomena, the Epicurean speaks for the citydweller.[12] The Christian performs with the sylvan eloquence of the poets, "eloquentia nemorali et poetarum." Comparing an orator to an athlete, Guarino further observes that the best orator is the one who shows pliancy in addition to muscular strength. This pliancy recalls the flexibility (*mollities*) that was said to be typical of the word *voluptas* (*OP*, III.27.2).

Both the Epicurean and the Christian orator "have called upon all the powers, arms, and arts that help one to win." Both are excellent orators in the sense that they both approached the *res*, the

12 Ovid, *Metamorphoses*, VI.424.

subject, *voluptas*, versus the *honestas* of the *philosophi*, with the maximum of openness, so as to allow the *res* itself, *voluptas*, to suggest the most fitting kind of language for its expression. Poetry and poetical language appeal therefore to both, but in a different way. The Christian (in versions two and three in real life a Franciscan friar) is more "poet" than the Epicurean (in real life the poet Vegio), because the Christian receives his message from the poetry of the Bible instead of from Virgil, a poetry which reflected the deepest reality of human life, searched for the highest possible reward while accepting, in difficulties, misery and death as intrinsic parts of life.

The greatest sources of pleasure, and thus a concrete *locus voluptatis*, are indicated by the Christian to be the "libri qui canonici vocantur / in quibus ab initio mundi sanctissima historia texitur" (the canonic books, in which is woven the holiest history from the beginning of the world) (*OP*, III.17). Yet the poetic metaphors of the Bible, which are the Christian's instrument for unveiling the *divina res* in Act 3, Scene 2 (III.9–11), are conceived of in analogy to the Virgilian metaphors used by the Epicurean in Act 2. In both cases they are instruments for the unveiling of the *res*: It is the *res* that differs.

Voluptas is, for both the Epicurean and the Christian, the creator of human history. Moreover, there is no claim on the part of the Christian every time "he reads the books and repeats to himself what he has read" to have discovered who God is, but what life is and what is the right attitude for living it fully. Yet the Christian's relationship with God is analogous to that of the prophets: "Quecunque enim cum illis locutus est nobisqu singulis loquitur" (What he said to them he says to us). This personal intimate relationship between the individual man and God / life gives an indescribable sweetness to the reading of the Gospels, whereas in the Old Testament one feels the omnipresence of a God more generous than a father. Again, it was the Epicurean, in one of the most dramatic moments of Act 2 (17) who prepared the way for the Christian's interpretation of the relationship of man / God as son / father by destroying, with the powerful weapon of irony, a world dominated by the indifferently serene "Aristotelian gods" (*OP*, II.28).

The Bible is a *documentum* of human history which cannot be read, as "philosophical books" are read, with the use of reason alone. Expressed as it is *per allegorias et enigmata*, it does not allow us to understand human history rationally, but only to perceive its essence by intuition, as the Epicurean had done in interpreting Virgil's famous closing passage of Book II of the *Georgics*, "Beatus qui potest rerum cognoscere causas" (Blessed is he who knows the causes of things). As it happens in general with a poem whose deep meaning we intuit, each of us, in a different time and under different circumstances, understands it in a different way. Hence we can conclude that it was the Epicurean who had indicated that poetry is the best human document through which one may reach the sources of the essence of *voluptas*. The Christian exploits the issue.

Finally, the *fictio* with which the Christian climaxes the long *demonstratio* (Act 3, Scene 2) displays his ability in what Valla defines *poetice loqui*, or poetic eloquence. Most of the issues the Christian touches upon actually correspond to issues treated by the Epicurean, as his eye is kept constantly on the most concrete aspects of human life, on what is the deepest motivation of human action. Even the *fictio paradisi* seems to be inspired by the desire to externalize *voluptas* through an *oratio*, as the Epicurean had done in Act 1 with sensual *voluptas*. We move, so to speak, out of theory (Act 3, Scene 1) into its application. By force of a deeper intuition into the *res*, the Christian tries to find the significance behind the individual act of pleasure as faith in pleasure itself. We conclude that the Epicurean and the Christian fight under the same banner, because they labor under the same need of externalizing the phenomenon *voluptas* through the proper language.

EPICUREAN LAUGHTER AND IRONY IN THE DRAMA OF *VOLUPTAS*

The Epicurean is literally called onstage to bring solace to the despairing Stoic. He does so *arridens*, with a benevolent smile. Through him *voluptas* emerges first of all as independence, *libertas*, in the interpretation and use of all sources of *auctoritates*, then as expression of the joy of living. We discover further that laughter and irony are *voluptas'* most powerful means of creating the tension inherent in the movement of the argument and in the dialectic of the dialogue. It is the benevolent smile of the Epicurean, his humor and his wit, that set us going, by steps, in the celebration of the senses until we reach the heart of the *res*, the phenomenon *voluptas*, which is union of the senses with the object of their desire (I.31). Finally, as part of *voluptas* in action, the Epicurean's smile and irony introduce us to the celebration of lewdness and adultery, whose inherent tension finds a solution in the apostrophe of the Vestal Virgin, who affirms the natural right of procreation as the continuation of the race. Both episodes can only be understood in an ironic context. The dramatic tension finds its solution in the eulogy of *voluptas*, where the verb *permanere* suggests that we have reached a point of arrival, yet not an end. At the end of Book I, we know that this is but the first act of the drama.

The *arridere* of the Epicurean, and his constant irony, are an act of benevolence in taking up the problem in all of its magnitude – the unveiling of the human soul with its secret motivations ("involucra animi evolvere"). The *risus* indicates – and the Epicurean says so clearly – his openness and flexibility in contrast to the rigidity of the Stoic. For this the Epicurean personifies *voluptas*. Actually he makes us aware through his words, as well as his attitudes and actions, that Epicureanism is but a category open to all possibilities within the concept of *libertas*, independence in the use of sources. According to Valla, Cicero unfortunately exploited this use of sources only partially, while Valla himself exemplifies it admirably in his usage of Lucretius via Lactantius.[13] What does this fact imply in the development of the argument?

The unveiling of the essence of what is good (*bonum*), upon which rests a self-evident proof (*evidentissima ratio*) of the unicity of *voluptas* as a *bonum*, takes place within the defense of nature, which had been the object of a frontal attack on the part of the Stoic. Hence the Epicurean's motivation becomes more metaphysical than ethical. If we look at nature with the necessary openness (*mollities*) – that is in this case with a smile – *arridens*, we automatically dismiss the *atomorum temeritas*, or atomic theory, of Epicurus and Lucretius in favor of a providential principle of the cosmos. The Epicurean quotes here a line from *De rerum natura* which is quoted also by Lactantius. In fact, as I said before, our Epicurean's personality as a character in the drama emerges in an acrobatic *duel-à-trois* with Lactantius and Lucretius. He uses

13 See Lorch, *A Defense*, pp. 61–72.

Lactantius' text, thus establishing the proper distance from Lucretius on the *atomorum temeritas*, but at the same time sets Lactantius straight on the interpretation of the nature of lightning, which, he says, is not a punishment of the gods but a natural phenomenon (*OP*, I.10–13).

Thus *voluptas* and *honestas*, defined first on the theoretical level through a Ciceronian definition – in a passage of *De finibus* where Cicero attacks Epicurus – are immediately contrasted in the concreteness of the drama: On the one hand stands *voluptas–utilitas–fruitio*, or expediency, the instinctive sense of preservation; on the other, *honestas*, disconnected from all forms of expediency (*utilitas*), leaving man to stick it out alone, by himself, in the cold. The Epicurean and the Stoic take up arms, each for his general, Epicurus and Aristippus on one side, Zeno on the other. Yet, the Epicurean is firmly committed to *libertas* – that is, he must allow the *res* to reveal itself on its own. Before moving into battle, there is a pause in which young Laurentius is asked by "his" Epicurean to profess "an act of faith" in the success of the Epicurean's enterprise (I.16). Faith is the premise of all proofs. Laurentius, to be sure, professes faith in his own kind of Epicureanism.

The central part of Act I – ideologically its most relevant moment – begins with the discussion of what is good (*bonum*) according to the classical tripartite classification. The focus is, however, only on the *external goods*. Declared to be the *only* goods, they allow nature to emerge, within an ironical context, as an entity in itself. Man can either use it well, badly, or not at all. In any case, nature exists, even if man does not use it. Man reaches nature through the bodily senses, which are not *goods* per se but means to reach the goods, that is, *nature*.

The Epicurean affirms himself within this ironic, witty context, step by step, not as the winner of a cause but as the discoverer of a truth. He acts out his *oratio* by dismissing, as quickly as he can, the traditional definitions and diving into the flow of life. Irony and wit remain throughout the means to reach the scope. Speech and wine, declared to be the exclusive privilege of man, induce him to rival in speech with the marvelous effects of wine drinking when done in the proper way. We witness through subtle irony a love interaction between the Epicurean and *natura*. *Natura*, like an artist, *affingit* or creates the *goods*, while the bodily senses are visually rendered, stretching their neck like a crane, toward the good to be used (*uti, utilitas*) and enjoyed (*fruit, fruitio*). *Frui* intermingles with *uti*, until it seems to win it over. Man is superior to animals because of his *fruitio* of the natural goods. *Fruitio* and *utilitas* coexist in man's elementary forms of *voluptas* with the possibility of a greater *fruitio* only with the greater refinement of the senses, which comes through man's own ingenuity.

Finally *bonum* is revealed in its ur-form as an act of fruition in a "marriage" of what receives (the bodily senses) and what is received (the external goods). *Virtus* exists only so as to increase the *fruitio*, or capacity of enjoyment. *Voluptas* therefore reigns as a queen over the old Stoic/Ciceronian virtues.

The real novelty and originality of the idea expressed is in the way it comes to light, a drama acted out by one and only one actor onstage, the Epicurean. And the Epicurean's main characteristic is *risus*, or irony. We feel the effect of *risus* on the Epicurean's "acting" every step of the way. As we have seen, it takes first the form of a benevolent smile, an attitude of openness toward humanity, in order to allow man to trust him and open the secret *involucra*, or folds, of his heart. After all, is not *voluptas* hidden in man's soul? The Epicurean started by declaring himself *curiosissimus* concerning the Stoic *mysterii* or mysterious inventions, where *curious* means "attentive" but also "curious" (*OP*, I.9.1). In Book I, *risus* is first of all a spontaneous expression of joy as the "orator" sees his adversary's argument so easily acceptable (I.8).[14]

Secondly, as we have seen, *risus* is an excellent psychological instrument that easily open man's heart. Thirdly, and most important, *risus* is a wonderful instrument for looking for the "truth." It is in fact the basis for a new means or method for facing a problem. *Ridere* created *openness*, literally reflected in the Epicurean's words "I promised to throw open the most intimate secrets to the Stoics" (*OP*, I.9.1). At the same time it reveals and encourages an adventurous spirit, the joy of moving into the unknown with a sense of trust in "the unexpected" and in oneself.

From the very beginning the Epicurean's smile revealed itself as irony, that theoretical device which, around Valla, was being discovered mostly because of the appearance in Leonardo Bruni's Latin translation of Plato's dialogues. Socrates is mentioned re-

14 Ibid., pp. 58–9.

peatedly in *On Pleasure* – in the case of the Epicurean in connection
with his irony.[15] As irony, the benevolent attitude of the Epicurean reverses a given situation, thus creating the possibility for
new ideas to come to light. It is highly significant, for instance, for
our understanding of Valla's concept of *voluptas*, that, at the outset
of the Epicurean's "action" against the Stoic "in defense of nature
and the human race" (*OP*, I.9.1), the Epicurean should engage, as
we have mentioned earlier, in a debate with Lactantius[16] – without
mentioning him by name – alluding to a passage where Lactantius,
speaking, for once, favorably of Epicurus, mentions with great
concern that *homines pii et religiosi* are often hit by misfortunes,
while the impious often go free of them. What is the clue that
leads us to Lactantius? The expression *pie et religiose* (piously and
religiously), with which the Epicurean ironically defines his own
attitude toward the subject. The ambiguity of the expression
warns us, first, to expect from our Epicurean something different
from Lucretius. Yet, we soon find out he is not on Lactantius's
trail either. Where does he stand, then? The ironical tone allows
the author of the drama to connect our Epicurean in Act 1 with the
Christian in Act 3, who faces *pie et religiose*, in a literal sense of the
expression, the process of discovering *voluptas* as *divina res*. This
Epicurean / Christian connection is reinforced by the Epicurean's
mention of a divine nature, *Natura sive Deus* (with the support of
Ovid), a topic which will be fully exploited by the Christian.
Irony, in this case, leads us to a new concept of piety and religion,
pietas and *religio* (I.9).

We are on the wrong track if we expect Valla to spell out his
conclusions in rational terms. He will leave us, throughout, en-
meshed in the ambiguity of a drama whose tension is partially
created by the wonderful rhetorical tool of irony. Irony triumphs
at its best within a sociological framework, at the close of Act 1,
when *voluptas* is presented in "action," in the apostrophe of the
Vestal Virgin (to a Platonic senate made up of men and women),
which follows the Epicurean's attack on the *Lex Julia* that condemns adultery (*OP*, I.37–8). By paradoxically establishing the
"usefulness" of adultery, the Epicurean conveys the message that

15 A study of irony in the Middle Ages and the Renaissance was recently completed at
   Columbia University by Dwight Knox and is to be published by Cambridge University Press. Knox briefly discusses Valla's use of irony.
16 *Div. Inst.* VII.17.3.

sex is a powerful drive which cannot and should not be suppressed. Never divide those who love each other, nor, as Machiavelli
implies (also ironically) in the *Mandrake*, try to keep together those
who have nothing in common. The vestal, led onstage by the
Epicurean as the victim of a hypocritical society, portrays the
overflowing power of *voluptas* as the natural instinct of procreation, a torrent that sweeps away the bloodless philosophers.
Voluptas triumphs in the end as a self-evident truth, not drawn
from books but discovered in our deepest selves. It is present in
the myths of the poets, who have better insight than others into
the mystery of life.

Act 2 is indeed the *pièce de force* of the Epicurean acting as *eiron*,
as the Christian will plainly define him in Act 3. The kind of irony
used in Act 2 will be fully exploited by the Christian himself.
While the actors of the drama in Act 1 have been *Natura* and the
bodily senses, in Act 2 the appearance of Stoic *honestas*, which,
being a product of the human mind, produced an ample "bibliography," forces the Epicurean to deal with "ideas" and "ideals" –
that is, with the famous and infamous virtues. If in Act 1 his
language had been as simple as that of the Virgilian husbandmen
or *agricolae*, never marring the joyful atmosphere inspired by
voluptas, in Act 2 he has to deal with the language of the philosophers. And here is where irony must be applied to the
maximum.

Taking into account the fact that man strives for the preservation of his own life and for its fuller realization, the Epicurean is
at work in Act 2 to unveil the secrets of the human heart, the
motivation of human action. With the concreteness that *voluptas*
requires, he does not deal abstractly with man in general; he leads
onstage those specific individuals who have been chosen by the
Stoic / Ciceronian tradition as paradigmatic examples of those
virtues in which Stoic / Ciceronian *honestas*, the privilege of an
elite, finds its expression. Regulus, Curtius, Lucretia, Brutus,
Diogenes, mainly Aristotle, and also Plato with his mythical
Gyges will be forced to confess, one after the other, what was the
true personal, selfish motivation for their noble, virtuous action.
Only through their confessions will we discover what such ideals
as fortitude, glory, contemplation, and serenity of soul are in
reality: man's desire for self-preservation and self-realization.
The *res* will surface, from beginning to end, through the induc-

tive method, by means of examples. A thick forest of abstract concepts, such as *patria*, will be done away with simply by letting *voluptas* come to light. In the end, nature and the senses remain the protagonists of the drama, with the total exclusion of man-originated "intellectual" forces.

The only serious moment, indeed the only tragic moment of the act, is faced by the Epicurean at the conclusion of his part in the drama, when he must answer the question of whether *voluptas* can conquer death. Although death will win, the forces of life within man will be shown to be so powerful as to induce the Epicurean implicitly not to accept even the heroic death of Epicurus as an answer to the problem. Love and hope, and mostly faith in life, will triumph at the end of Book II with a eulogy of *voluptas* which creates human history and constitutes the basis of society. Faith, love, and hope are new forces discovered by the Epicurean to be existential necessities in a world dominated by the iron law of *utilitas*.

THE EPICUREAN AND EPICURUS

The personality of the Epicurean as actor in the drama of *voluptas* acquires a special light when he interacts with Epicurus. He does so through the manipulation of passages from Cicero, where Cicero, who otherwise condemns Epicurus, praises him for his moral fortitude in confronting death (*OP*, II.31). At the end of Book II we realize that what the Epicurean has stressed progressively from one episode to the next is, within the world of the senses, man's humanity, in its desires, aspirations, and ideals, from civic to intellectual life. In the end we are presented with an exceptional man, Epicurus, who serenely accepts human reason with all of its limitations. When the Epicurean says, "Epicurus meus vult" (My Epicurus declares), it is as if he said, "My human reason exacts."

The Epicurean incorporates Epicurus, like Aristotle, in the world of *voluptas* for what he could offer and for the way he offered it. Epicurus is absorbed by Valla's new *voluptas*, which, unlike the one in which he believed, knows no zones of darkness. "Let us serve pleasure as long as we can. Our possibilities are indeed great" is a motto that works for the senses as well as for human reason, the use of body and soul, *mens–animus*, in the satisfaction of our sensual desires. For a brief moment, our Epicurean has given in, and poetically so, to the temptation of *carpe diem*. "My Epicurus declares that after physical death nothing remains," he says, following this with a long paragraph that stresses, in its rhetorical structure, based on parallelisms, the indissoluble link of man–animal in death (*OP*, II.29).[17] Yet, unlike Cicero, the Epicurean stresses Epicurus's acceptance of what life is, in confronting death, without any form of resignation. Therefore he celebrates not Epicurus's fortitude in confronting death, as Cicero does, but his *mollities*, his pliancy, his flexibility, which Cicero had repeatedly attacked. In this *mollities* we discover the true relationship between our Epicurean and Epicurus. Having admitted this point of connection, the Epicurean moves on, on his own, along the path of his *voluptas*, which, with that assertion "possumus atque multum" (We are very powerful within the realm of the senses) announces man's superiority over the animals not because of reason but because of the senses. Valla's Epicureanism is revealed by this statement. Epicurus is nothing but a pawn, as Aristotle was, in the development of the drama *voluptas*.

Speaking of the poets, our Epicurean hints at the fact that from this *corporea voluptas* (bodily pleasure) "gradus quidam fit ad illas futuras" (a certain step is made toward those futures). Literally he speaks of the Elysian fields, creation of the poets, product of the hope of the poets. So, he seems to imply, why not believe the poets? Sensual *voluptas* is so powerful as to exact by its very force a continuation of the energy of life. The poets have a better intuition into life's mysteries than Epicurus.

It is this opening at the end of Act 2 that allows the Christian, in his final judgment of the Epicurean (*OP*, III.2)[18] to justify him as a *simulator*, one who dissimulates what he thinks. How can the discoverer of *voluptas* as life's original force be so "dissimilar from himself" as to follow the *humana ratio* of Epicurus up to the point of accepting with total death the end of that form of indestructible life that is *voluptas*? The Christian judges the case of the Epicurean within the laws of the rhetorical discourse that encourage *libertas* and *risus*, complete openness and ample use of irony, as the most efficient means to transmit an idea: "And therefore I suspect that

17 See my essay "'Mors omnia vincit improbe': L'uomo e la morte nel *De Voluptate* del Valla," in *Miscellanea di studi in onore di V. Branca* (Florence, 1983), 2:177–92.
18 Lorch, *A Defense*, pp. 257–8.

you were not serious but joking, as it is your habit, like Socrates, whom the Greeks called *eirona*" (*OP*, III.8.3).

What the Christian reproaches in the Epicurean is not the *iocum* or the *eirona*, nor the *simulate loqui*, a form of dissimulation through irony for which he should in fact be praised, but rather his having followed to an extreme degree the *philosophica ratio* beyond the point allowed by the *res* itself, the subject in question, *voluptas*. The *Epicuri ratio* represents the maximum of consistency of life and thought in the physical world of the senses, and thus it is the only form of commendable *philosophica ratio*. It does not, however, take into account the quintessence of the energy discovered to be *voluptas*, which cannot be made to vanish with the "death" of the senses. In the most natural way, called onstage by the *res*, Jesus Christ replaces Epicurus as the carrier of the final message: By offering humanity the paradigm of an all-encompassing form of love, based on *fides* as *persuasio*, Jesus Christ opens (see Proem) and closes the drama that is the drama of life to which Redemption guarantees a happy solution.

One final question remains. Was there a historical character in real life who inspired Valla in the creation of the character of the Epicurean? Perhaps it was the *bon-viveur* Panormita. However, the poetry of the *Hermaphroditus*, and what we know of Panormita from his letters and from the accounts of others, only partially explain the choice. Besides, Valla replaced him, in the second version, with the chaste poet Maffeo Vegio. The Epicurean of *On Pleasure* is an orator, a poet / philosopher, an inquisitive mind that searches for new ways to reach the truth, a clever user of the arms of *risus* and irony. As the author of the drama, Valla chose to distance himself from all the characters of the play. There is dissent among today's critics as to who among the two main interlocutors of the dialogue reflects Valla more directly. On the basis of Valla's appearance as a *miles Christianus* in the Proem, one would be tempted to identify him with the Christian "actor" who brings the drama to a happy solution. Yet the forceful power of assertion of the Epicurean, his love for the unexpected and the adventurous, and, most of all, the pleasure he takes in shocking the audience through his oratorical devices, lead us to the assumption that the Epicurean is the closest in nature to his creator.

6

Seneca's role in popularizing Epicurus in the sixteenth century

LOUISE FOTHERGILL-PAYNE

Current opinion tends to see Stoics and Epicureans as unlikely bedfellows, but in the sixteenth century this was not so. As a matter of fact, Seneca, the great *auctoritas* on Stoic ethics in the Renaissance, was also considered something of a spokesman for Epicurean thought.

In one of his *Essais* Montaigne gives us a telling insight into the way Seneca's moral philosophy was viewed at the time. For this sixteenth-century reader, the classical authors that best combined the useful and the pleasurable were Plutarch and Seneca. The reason why these two authors in particular suit his temperament best, Montaigne explains, is that the knowledge he seeks is there treated in bits and pieces ("à pièces décousues") that do not require long hours of reading for which he has neither time nor inclination ("qui ne demandent pas l'obligation d'un long travail, dequoy je suis incapable").

Concentrated short passages are for Montaigne also most useful, especially in the case of Seneca's correspondence with Lucilius, which forms "the most beautiful and profitable part of his writings" ("la plus belle partie de ses escrits, et al plus profitable"). Another advantage in reading Seneca and Plutarch, Montaigne argues, is that little effort is involved in getting started, and that these authors allow him to break off wherever he feels like it ("Il ne faut pas grande entreprinse pour m'y mettre; et les quitte où il me plaît"). Furthermore he finds their teaching simple but relevant ("d'une simple façon et pertinente"), as each represents the cream of philosophy ("cresme de la philosophie"). But while Plutarch expresses sweet Platonic thoughts, so well suited to polite society,

Seneca, both a Stoic and an Epicurean, distances himself from common practice ("Plutarque a les opinions Platoniques, douces et accommodables à la société civile; l'autre [Seneca] les a Stoiques et Epicurienes, plus éloignées de l'usage commun"). Apart from being off the beaten track, Montaigne concludes, Seneca's teaching is also more specifically practical and to the point ("plus commodes en particulier et plus fermes").[1]

The *essai* in which Montaigne evaluates his favorite nonfiction writers is titled "Des livres" (Of books). The heading is significant in that it reveals a delight in books and a pride in being well read. In this respect Montaigne is representative of the educated gentleman engaged in both the affairs of the world and the exercise of the mind but whose involvement in wordly matters does not allow for long hours of research. Hence his preference for authors who provide short answers to the "science" he seeks, in simple and relevant terms. Finally, while Plutarch's Platonic views may be more suitable for the contemplative life, for Montaigne, Seneca's Stoic / Epicurean thought, so novel, practical, and to the point, is more useful for the active life.

The fact that Montaigne, like so many other gentlemen-readers, chose Seneca as his preferred source for Stoic and Epicurean wisdom bears further examining. Why was Seneca, rather than Diogenes Laertius, Cicero, or Lucretius, so successful in transmitting Epicurus' teaching to that new brand of reader, the nonscholar?

Here I would argue that in order to popularize a set of ideas the transmitter must fulfill a series of conditions. In the first place, he must himself be popular, that is to say he should be an esteemed and "approved" author; his medium, in the form of language, style, and format, should suit the consumer; his message should be new, relevant, and accessible to "all"; and his formula should be subject to imitation, that surest sign of success. Last but not least, the recipient's mood and circumstances should be just right for the wholesale reception and dissemination of these ideas.

---

1 Pierre Villey, ed., *Les essais de Michel de Montaigne* (Paris, 1922), 2:112. My translation. By contrast, Montaigne finds Cicero's moral writings boring: "Mais, à confesser hardiment la vérité (car, puis qu'on a franchi les barrières de l'impudence, il n'y a plus de bride), sa façon d'escrire me semble ennuyeuse"; that is, "But, quite frankly (now that I have overstepped the limits of decency, there is no more holding back), I find his style downright boring" (p. 113).

This essay then, will bring together a number of apparently disparate elements in an attempt to throw some light on why Seneca was so attractive to the layman and what ordinary people may actually have extrapolated from Seneca's presentation of Epicureanism.

To that end I will first discuss Seneca's enormous popularity and the many translations that were made in response to the non-Latinist's demand to read Seneca in the vernacular. Then I will examine Seneca's style and the format of his message, as well as Seneca's actual words in introducing Epicurus in his prose work. Since the study and performance of Seneca's tragedies was an important factor in making Seneca's message accessible to "all," I shall finish with a brief consideration of a Neo-Latin school play written along Senecan lines, where Epicurus occupies center stage.

Seneca's fame as an approved author had been rising steadily throughout the Middle Ages,[2] but it was only in the fifteenth century, with the spread of literacy, that his message was to become available to the nonprofessional reader as well.[3] However, the layman's love of antiquity was not always matched by the ability to read the authors in the original. Hence the proliferation of translations that accompanied this renewed interest in the classical world.[4]

2 See my *Seneca and Celestina* (Cambridge, U.K., 1988). Chap. 1, "Towards a Senecan Tradition," pp. 1–25; Paul Faider, *Études sur Sénèque* (Ghent, 1921); R. R. Bolgar, *The Classical Heritage and Its Beneficiaries* (Cambridge, U.K., 1954); idem, ed., *Classical Influences on European Culture, A.D. 500–1500* (Cambridge, U.K., 1971); L. D. Reynolds, *The Medieval Tradition of Seneca's Letters* (Oxford, 1965); J. E. Sandys, *A History of Classical Scholarship* (New York, 1967); Karl Alfred Blüher, *Seneca in Spanien. Untersuchungen zur Geschichte der Seneca Rezeption in Spanien vom 13. bis 17. Jahrhundert* (Munich, 1969), translated into Spanish as *Séneca en España* (Madrid, 1983); T. Griffin, "Imago vitae suae," in C. D. N. Costa, ed., *Seneca* (London, 1974), pp. 1–38; Gilles D. Monsarrat, *Light from the Porch* (Paris, 1984).

3 See H. J. Chaytor, *From Script to Print: An Introduction to Medieval Vernacular Literature* (1945; reprint, London, 1966); Franz H. Bauml, "Varieties and Consequences of Medieval Literacy and Illiteracy," *Speculum*, 55 (1980), 237–65; M. B. Parkes, "The Literacy of the Laity," in D. Daiches and A. Thorlby, eds., *Literature and Western Civilization: vol. 2, The Medieval World* (London, 1973), pp. 555–77; Jeremy Lawrance, "The Spread of Lay Literacy in Late Medieval Castile," *Bulletin of Hispanic Studies*, 62 (1985), 79–94.

4 See Paul Chavy, *Traducteurs d'autrefois: Moyen Age et Renaissance*, 2 vols (Paris, 1988); Peter E. Russell, *Traducciones y traductores en la península ibérica (1400–1550)* (Barcelona, 1985), and my *Seneca and Celestina*, Chap. 1, pp. 11–16.

If the number of translations is an indication of success, then Seneca stands out as the most popular Latin author in the Middle Ages. The fact that an anthology of Seneca's letters was translated into French as early as the thirteenth century, that the French version was translated into Italian and Spanish, and that most vernacular versions had appeared in print by the end of the fifteenth century, speaks as much of Seneca's popular appeal to the non-Latinist as to the need to read him in the vernacular.[5]

Interestingly, the preface to the Spanish translation of one such anthology, itself based on an Italian version, reveals very similar reasons for Seneca's preeminence as a transmitter of Stoic and Epicurean thinking to those outlined by Montaigne a century later. In the manner of a blurb for a dust jacket, the *Prohemio* to the first printed edition of the anthology says,

Because Seneca belongs to the Stoic School, he holds that nobody can be happy without virtue and so . . . he mixes his own words with the *sententiae* of a philosopher named Epicurus, who said that pleasure was the highest good, but still in such a way that it turns to honesty.

[. . . mezcla entre sus dichos las sentencias de un filósofo llamado Epicuro, que dizía que el deleyte es soberano bien, pero toda vía en tal manera que él tornasse a honestad.][6]

Two things call our attention here: In the first place Epicurus strikes us as a relative newcomer, given that he had to be introduced by his name and his main tenet. In the second, we should

---

5 See M. Eusebi, "Le piu antiche traduzione francese delle *Lettere morale* di Seneca e i suoi derrivati," *Romania*, 91 (1970), 1–47. For a history of the Latin manuscripts, see L. D. Reynolds, *Medieval Tradition*. The manuscript history of Seneca's letters in translation has not been studied in detail yet. It is a fact, though, that anthologies of Seneca's *Epistulae morales* were among the first incunabula. The earliest Italian printed translation of Seneca's letters appeared in Venice in 1494 under the title *Epistule del moralissimo Seneca nuovamente fatte volgare*. (I am grateful to Letizia Panizza for this information.) In Spain, the earliest printed translation of the letters was based on an Italian anthology of seventy-five letters and was published in Zaragoza in 1496 under the title *Las epístolas de Séneca*, while a collection of nine treatises comprising *Les epistres morales* appeared in Paris in 1500 under the title *Les oeuvres de Sénèque translatez de latin en français* (Chavy, *Traducteurs*, pp. 842–1000). England did not participate in this first stage of vernacular editions. The earliest translation, by Thomas Lodge, *The Workes of L. A. Seneca both Morall and Naturall*, printed in London in 1614, belongs to a new wave of activity which produced more scholarly and unabridged translations of Seneca's works in the seventeenth century.

6 *Las epístulas de Séneca* (Zaragoza, 1496).

note the tolerant and almost conciliatory tone in which Epicurus' pleasure principle is set down: "pleasure turning to honesty."

In other respects too, the preface clarifies why Seneca rapidly became such a popular author in the Middle Ages. Apart from mentioning Seneca's use of Epicurean maxims, the translator also refers to Seneca's reputation as a "near-Christian," based on Jerome's ambiguous testimony in Book 12 of *De viris illustribus*. Paraphrasing Jerome's words, the preface concludes,

Deste sabio Séneca fizo San Héronimo muy especial mención enel libro que él compuso de los *Varones claros* por tales palabras: Anneo Lucio Séneca de Córdova fue hombre de gran continencia en su bevir: el qual yo no pusiera en el catálogo delos santos, si a ello no me provocaran aquellas epístolas que de muchos son leydas, de Paulo a Séneca y de Séneca a Paulo.

[Of this wise Seneca Saint Jerome made special mention in his book of *Illustrious Men* in the following words: Annaeus Lucius Seneca from Cordoba was a man of great moderation in his way of life, and I would not include him in the list of saints were it not for those letters from Paul to Seneca and from Seneca to Paul which are read by many.][7]

The legend of a friendship between Seneca and Paul of course gave the moral philosopher that extra aura of celebrity, aided by Jerome's testimony that Seneca's teaching, too, was nearest, and therefore most acceptable, to Christian dogma.[8] There is no doubt, then, that Seneca was an "approved" author and that his words carried weight. Added to that, his medium was eminently suitable for widespread acceptance in that the translations made his message readily available to the nonscholar, thus considerably enlarging the circle of his readership.

We should not underestimate the role of the translator in the transmission of classical wisdom. This was true especially at the

---

7 For a modern edition of the correspondence, see C. W. Barlow, ed., *Epistulae Senecae ad Paulum et Pauli ad Senecam (quae vocantur)* (Rome, 1938); see also E. Liénard, "Sur la correspondance apocryphe de Sénèque et de Saint Paul," *Revue belge de philosophie et histoire*, 11 (1932). 5–23; Kenneth M. Abbott, "Seneca and St. Paul," in Donald C. Riechel, ed., *Wege der Worte: Festschrift für Wolfgang Fleischauer* (Cologne, 1978), pp. 119–31; Reynolds, *Medieval Tradition*, pp. 82–9, 112–13. See also Letizia A. Panizza, "The St. Paul–Seneca Correspondence: Its Significance for Stoic Thought from Petrarch to Erasmus," Ph.D. diss., University of London, 1976.

8 *In Isaiam*, 4.11; *Patrologia latina*, 24:147.D., cited by Gordon Braden, *Renaissance Tragedy and the Senecan Tradition* (New Haven, 1985), p. 70.

end of the Middle Ages, when translators had to grapple with concepts for which there was then no equivalent in the vernacular. Added to that, the translator, as interpreter of these concepts, had the responsibility of making pagan wisdom acceptable to the demands of the Church. To this end, he provided a "custom-made" version of the original words by selecting and translating "safe" passages only and surrounding the text with exegetic glosses. Here, Seneca presented relatively few problems, as much of Stoicism had been incorporated into Christian thinking, with the exception of issues concerning fate, the *summum bonum*, suicide, and man's relation to God. The solution in such cases seemed to be either to explain in a marginal gloss the difference between pagan and Christian thinking or simply to leave out the offending passage. Hence the preponderance of carefully screened anthologies, a format which at the same time seemed to be more amenable to the "occasional" reader.

Seneca's style was eminently suitable for this sort of treatment. Since his writing could be cut up into what Montaigne so aptly called "pièces décousues," Seneca soon became a rich source for quotable maxims which, in turn, were brought together in miscellaneous sentence collections. Such was Seneca's prestige that many of these compilations carried the label "by Seneca" to lend them greater authority. Among these pseudo-Senecan compilations are the *Proverbia Senecae*, the *Tabulatio et expositio Senecae*, the *De moribus*, the *Formula vitae honestae* and its reworkings under the titles *Copia verborum* and *De quattuor virtutibus*.[9] These books, both in Latin and in the vernacular, kept their appeal for the average reader until well into the sixteenth century. As a result, Seneca became one of those authors who are more quoted than read and whose canon was pillaged, rearranged, and reinterpreted as best suited each consecutive generation of readers.

Both style and format then, as well as the familiarity of the vernacular language, were responsible for the wholesale dissemination of Seneca's ideas. In addition, his message, again in the

9 For an understanding of the medieval preference for anthologies and compilations of words of wisdom, see M. B. Parkes, "The Influence of the Concepts of *ordinatio* and *compilatio* on the Development of the Book," in J. J. G. Alexander and M. T. Gibson, eds., *Medieval Learning and Literature: Essays Presented to Richard William Hunt* (Oxford, 1976), pp. 115–41, and my *Seneca and Celestina*, Chap. 2, "Senecan Commentary as a Frame of Reference," pp. 26–44.

words of Montaigne, had something new to offer ("élognées de l'usage commun"). Indeed, in the fifteenth and sixteenth centuries, Seneca was considered to be a refreshing departure from caviling Scholasticism and dogmatic metaphysics. In an attempt to explain Seneca's enormous popularity in the Renaissance, G. M. Ross ventures,

The fifteenth and sixteenth centuries had seen not only the revival of virtually all the ancient systems of philosophy, but also a splitting-up of scholasticism into a number of opposed factions. This multiplication of unreconciled dogmatisms led to a wide-spread dissatisfaction with metaphysics in general encouraging both scepticism, which attacked dogmatic metaphysics on its own ground, and a preference for an untechnical, popular philosophy of life, precisely the area where Seneca had the most to offer.[10]

Seneca's influence on the humanists can hardly be overestimated. The growing neo-Stoic movement chose the Roman philosopher as its preferred teacher on ethics, while his eminently portable and quotable sentences continued to be a rich source of epigrams, mottoes, and other notable sayings. Erasmus' editions of Seneca's *Opera omnia* (1515 and 1529), now purged of apocryphal works, and his *Flores Lucii Annaei Senecae Cordubensis* (1528) do not replace but complement the medieval annotated anthologies and translations, all of which continue to circulate in print well into the sixteenth century. New commentaries also see the light: Guillaume Budé publishes his *De contemptu rerum fortuitarum* in 1520. In 1530 Zwingli publishes a commentary on Seneca's *De providentia*, and in 1532 Calvin's commentary on Seneca's *De clementia* appears. In the second half of the century Pierre Charron, Montaigne's friend and disciple, continues the tradition with his *De la sagesse* (1601), while in Holland Justus Lipsius champions Seneca's prose style, simultaneously spearheading a growing anti-Ciceronian movement. Among Justus Lipsius' most influential books are *De constantia* (1583), *Manuductio ad philosophiam Stoicam* (1604), and *Physiologia Stoicorum* (1604), all based on Seneca's philosophy. Concurrently, Caspar Schoppe is working in Germany on his *Elementa philosophiae Stoicae moralis* (1606). Meanwhile, in Spain Quevedo and Gracian voice Seneca's

10 G. M. Ross, "Seneca's Philosophical Influence," in C. D. N. Costa, ed., *Seneca* (London, 1974), pp. 116–65, quotation on p. 148.

Stoicism throughout their moral writings, and in England
Thomas Lodge produces the first English translation of Seneca's
complete prose works (1614).[11]

As a transmitter of Epicurean thought, Seneca's appeal to the
layman must have been precisely this practical approach to life
and, equally, the small doses in which he serves up his advice.
While, for instance, Diogenes Laertius systematically reconstructs
Epicurus' teaching in Book X of his *Lives of Eminent Philosophers*,
and Lucretius sings the praises of an Epicurean universe in a poem
of epic dimensions, or Cicero sustains a long argument with his
Epicurean friend in Books I and II of *De finibus*, Seneca's references to Epicurus' teaching are invariably short and most often
take the form of a maxim drawn from Epicurus. Seneca thus
caters fully to the busy humanists' thirst for knowledge by offering practical advice that proves useful, simple, relevant, new, and
to the point, or, in Montaigne's words, "profitables ... d'une
simple façon et pertinente ... éloignées de l'usage commun ...
plus commodes en particulier et plus fermes."

Of all Seneca's writings, his treatise on happiness and his
correspondence with Lucilius have always been favorites with the
general public. And it is precisely in these two books that Seneca
transmits Epicurus' message in the most positive way.[12]

Indeed, on reading *De vita beata*, one might even get the impression that Seneca approves of Epicurus' *voluptas*. In Chapters
12 and 13, for instance, he exonerates Epicurus of all blame for the
bad reputation his school enjoys. It is not Epicurus' fault, he says
in Chapter 13, but that of his followers, who find excuse for their
depravity in a philosophy which praises pleasure, and yet, he goes
on, Epicurus' "voluptas" is "sobria ac sicca" (sober and, literally,
"dry"). Then, at the risk of upsetting the Stoics, Seneca describes

11 For Seneca's influence in the sixteenth century in general, see the studies mentioned in
    note 2 (this chapter) and also Leontine Zanta, *La renaissance du Stoïcisme au XVI siècle*
    (1914; reprint Geneva, 1975); M. W. Croll, "Juste Lipse et le mouvement anticiceronien
    à la fin du XVI siècle," *Revue du seizième siècle*, 2 (1914), 220–42; G. Williamson, *The
    Senecan Amble: A Study in Prose from Bacon to Collier* (London, 1951); J. L. Saunders,
    Justus Lipsius: The Philosophy of Renaissance Stoicism (New York, 1955); Henry Etting-
    hausen, *Francisco de Quevedo and the Neostoic Movement* (Oxford, 1972).
12 All references to Seneca's work and pertinent translations are taken from the Loeb
    Classical Library editions. Hereafter citations are given in parentheses in text. For the
    Letters to Lucilius, citations are to letter number and section.

the sect as "sancta," "recta," and "tristia" (austere). Unlike his
fellow Stoics, he adds, he does not condemn Epicureanism as a
school of vice but simply states that it has a bad name, is of ill
repute and yet undeservedly so ("male audit, infamis est, et
immerito," *De vita beata*, 13.2).

In the fragment on leisure often appended to *De vita beata*,
Seneca openly takes issue with the Stoics on the question of early
retirement. Contrary to Zeno's teaching that man shall engage in
public affairs unless something (a corrupt state for instance) prevents him from doing so, Seneca sides with Epicurus, who says
that "the wise man will not engage in public affairs except in an
emergency." And again, anticipating the question "Why, in the
very headquarters of Zeno, do you preach the doctrines of Epicurus?" (*De otio*, I), Seneca maintains that "retirement in itself will
do us good; we shall be better off by ourselves" ("melioribus
erimus singuli," ibid., I.1).

It is in the *Epistulae morales*, however, that Epicurus is most
often quoted. In fact, each of the first thirty letters to Lucilius
closes with a "saying of the day," and each one of them is culled
from Epicurus' large storehouse of *sententiae*. In her sourcebook of
Seneca's canon, Anna Motto has listed 64 references to Epicurus,
as against 49 to Socrates, 45 to Cato, 33 to Plato, and so on in
descending order until we reach Aristotle, with a meager rating of
23.[13]

Seneca himself is aware of his preference for Epicurus when
quoting the sayings of famous philosophers. Quizzed by Lucilius
on why he quotes Epicurus in preference to his own school, he
counters that maxims do not belong to anybody in particular but
are common property ("Itaque nolo illas Epicuri existimes esse;
publicae sunt et maxime nostrae," 8.2). But here, precisely, lies
the danger of the quotation. As well as being a link in the endless
chain of passing on previous knowledge, a *sententia* can so easily be
quoted out of context and lead to pseudoknowledge. For Seneca,

---

13 See A. L. Motto, *Guide to the Thought of L. A. Seneca in the Extant Prose Work*
(Amsterdam, 1970). In her introduction she quotes Seneca's dictum "disputare cum
Socrate licet, dubitare cum Carneade, *cum Epicuro quiescere*, hominis naturam cum
Stoicis vincere, cum Cynicis excedere" (It is all right to argue with Socrates, to doubt
with Carneades, *to relax with Epicurus*, to overcome the human condition with the Stoics
and to overstep the bounds with the Cynics) (*De brev. vit.*, XIV.2), my italics and
translation.

however, the attraction of Epicurus' maxims lies in the impact of such a quotation, because "it is so surprising that brave words should come from a man who professes to be so effeminate. At least," he adds, "that is what most people think" ("quia mirum est fortiter aliquid dici ab homine mollitiam professo. Ita enim plerique iudicant," 33.2).

But again, there is a risk in quoting a rival at frequent intervals. As Seneca says, the element of surprise will make the Epicurean maxims stand out more boldly, but, for that very reason, the general public will get to know an *auctoritas* more by his name than by his canon and might thus indiscriminately approve of all his words, particularly where these seem to be sanctioned by a moralist of Seneca's stature.

As a matter of fact, Seneca agrees with Epicurus on many points, such as the minimal power of fortune over the affairs of men and the unconditional acceptance of nature as a model and guide for the happy life. Seneca also agrees with Epicurus on the virtue of the simple life and poverty, in diverse slogans such as "Cheerful poverty is an honorable estate" ("Honesta res est laeta paupertas," 2.6) and "If you live by nature, you will never be poor, if by public opinion, you will never be rich" ("Si ad naturam vives, numquam eris pauper; si ad opiniones, numquam eris dives," 16.7).

But all is not admiration and ungrudging agreement. On the topic of peace of mind, Seneca agrees with Epicurus that the body plays an important role ("Absolutum enim illud humanae naturae bonum corporis et animi pace contentum est," 66.46) but strongly disagrees that the pleasure resulting from virtue, and not virtue itself, makes one happy ("beatum efficiat voluptas, quae ex virtute est, non ipse virtus," 85.19). Unsurprisingly, it is on the topic of pleasure and pain that Seneca disagrees with Epicurus most. According to Seneca no pleasure can be derived from pain ("incredibilius est dicat Epicurus, dulce esse torreri," 66.18), and he denies that torture can be sweet ("audi Epicurum, dicet, 'et dulce est,'" 67.16). Letter 99 then goes into the issue more deeply, examining grief in its relation to pleasure ("quid habeat iucundum circa se et voluptarium," 99.29) to come to the conclusion that to admit any pleasure from grieving is sheer indulgence in self-pity bordering on exhibitionism.

Generally speaking, Seneca disagrees with Epicurus uncompromisingly on the precedence of pleasure over virtue (67.85 and 99) but he also discusses the Epicureans' lack of civic spirit (*De otio*, III.2) and indifference to God (90.35) as well as their twofold division of philosophy into natural and moral, regretting the absence of the rational (89.11).

In his last letter to Lucilius, Seneca discusses the role of the senses in determining what is good and what is bad, rejecting Epicurean thinking in this respect, because, as he says, "We Stoics maintain that it is a matter of understanding, and that we assign to the mind." But even here, Seneca is reluctant to condemn the Epicureans for their "sins," since "What wrong could such men be committing if they looked merely to the senses as arbiters of good and evil?" ("Quid autem peccant, si sensibus, id est iudicibus boni ac mali, parent?" 124.2).

On only one occasion does Seneca mention Epicurus' physics, and that is in Letter 72 but still as part of a discussion on attaining wisdom. "Those who lack experience," Seneca says, "will fall into that Epicurean void and infinite space" ("in Epicureum illud chaos decidunt, inane, sine termino," 72.9).

Nowhere in the letters is Epicurus' universe actually discussed. For that we have to turn to *De beneficiis*, which was not translated before the seventeenth century. In Book IV of that work, Seneca refutes the Epicurean view that the gods could not care less about man's happiness: "Itaque non dat deus beneficia, sed securus et neclegens nostri, aversus a mundo aliud agit aut, quae maxima Epicuro felicitas videtur, nihil agit" (*De Benef.*, IV.4.1). In a later passage in the same book, Seneca again mentions how the Epicureans have stripped God of all his weapons, of all his power, and thrust him beyond the range of fear (IV.19.1). As *De beneficiis* is about bestowing gifts and showing gratitude, Seneca wonders how Epicureans can ever be grateful to God, since, according to them, they owe him nothing, "being a mere conglomerate of atoms and mites that have met blindly and by chance. Why, indeed, should one worship God?" (IV.19.3).

For Seneca, as for Cicero, it is Epicurus' ethics which are at stake. But unlike Cicero, who uses logic to destroy his Epicurean friend's arguments in Book II of *De finibus*, Seneca is reluctant to enter into a full-scale philosophical debate.

As Seneca's sketchy approach to Epicureanism is particularly apparent in the *Epistulae morales*, it might be useful to examine the selection and composition of the medieval anthology which proved so popular until late in the sixteenth century.

The first thirty letters, which, with few exceptions, close with an Epicurean maxim, all appear in the same sequence as in the original. After Letter 30, however, the anthology picks and chooses, leaving out the more lengthy philosophical letters, among which figure precisely those which contain Seneca's refutation of Epicureanism. As a result, the average reader was introduced to Epicurus in a most positive manner, for the first half of the collection praises the philosopher as a wise and brave man, without any clear negative comment about his teaching.

Since the anthology was meant to be a guide to practical living, the reader would look in vain for an explanation of Epicurus' physics or a discussion of the differences between Epicureanism and Stoicism. For instance, absent are the more detailed letters on pleasure versus virtue (85), the parts of philosophy (89), and the attainment of knowledge through the senses (124). Consequently the *magister voluptatis* may well have seemed an attractive alternative, to a great segment of society, especially as Seneca qualifies this *voluptas* as "sobria ac sicca."

Of course, there were other reasons for a spreading neo-Epicureanism. Diogenes Laertius was translated into Latin by Ambrogio Traversari in the early years of the fifteenth century, corrected by Benedetto Brognoli, and printed in Venice in 1475. This was followed by Froben's 1524 Basel edition, and another one by Johannes Sambucus, published by Plantin in 1566, while yet another, especially commissioned by the pope from Thomas Aldobrandinus, appeared in Rome in 1592, followed by the anonymous *Physica et meteorologica*, published in Leiden in 1595.

In addition, the period's love for maxims was amply catered to by diverse Epicurean sentence collections, such as the *Epicuri loci a Cicerone interpretati* (Paris, 1557) and *Epicuri sententiae aliquot aculeatae ex Seneca* (Louvain, 1609), culled from both Cicero and Seneca's works. It should be noted, though, that all these editions were in Latin, thus preventing the man in the street from profiting from them. As usual, Seneca and Cicero were the main sources for Epicurean references.[14] Of these two however, Seneca proved to

14 Other books in which Cicero contrasts Epicurean and Stoic points of view are *Paradoxa Stoicorum, De fato, De natura deorum,* and *De divinatione.* See N. P. Packer, *Cicero's Presentation of Epicurean Ethics* (New York, 1938).

be the more powerful transmitter because of his tragedies, which had now been translated and imitated and were much admired throughout Europe.[15]

The sixteenth century was as fascinated with Seneca *tragicus* as with Seneca *philosophus*. As European drama became increasingly secular it looked for inspiration to the classical world, and here Terence and Plautus were the obvious models for the comic mood. For the tragic mood, however, playwrights did not seek out the Greek tragedians but instead turned to Seneca, whose reputation was by now well established as an *auctoritas* in moral matters.

Tragedy was then seen as a matter of moral edification, and there was no better teacher than Seneca to show the fatal outcome of uncontrolled and uncontrollable passions. Interestingly, it was not so much his dramatic technique which made Seneca a model for the tragic mood but the same *sententiae* which had made Seneca's prose so profitable to the average reader. In the sixteenth century, sententiousness was seen as a virtue and not a vice, and its appropriate use in tragedies was felt to enhance the impact of the tragic effect; in the words of Bartolomeo Ricci, "Tragoediae vero gravitatem, de qua nunc agitur, adiuvari imprimis gravitate sententiarum, nemo non intelligit" (Everybody knows that the weight of the tragedy . . . is increased by the weightiness of the *sententiae*).[16] And when it comes to *sententiae*, Ricci goes on, Seneca has no rival. Julius Caesar Scaliger, too, while conceding that Seneca had borrowed his plots from the Greek, nevertheless admires Seneca's style and especially the *efficacia* of his *sententiae*, which according to Scaliger are "quasi columnae, aut pilae quaedam universae fabricae illius" (They are as it were the pillars and columns of its whole construction).[17] Even Justus Lipsius, while not quite as much an admirer of Seneca's tragedies, approves

15 For a discussion of the influence of Seneca's tragedies on the development of European drama, see Jean Jacquot, ed., *Les tragédies de Sénèque et le théâtre de la Renaissance* (Paris, 1964); Donald Stone, Jr., *French Humanist Tragedy, a Reassessment* (Manchester, 1974); Eckard Lefevre, ed., *Der Einfluss Senecas auf das europaische Drama* (Darmstadt, 1978); Gordon Braden, *Renaissance Tragedy and the Senecan Tradition* (New Haven, 1985).

16 *De imitatione libri tres* (Venice, 1545), p. 22v. Cited by J. W. Binns, "Seneca and Neo-Latin Tragedy in England," in Costa, ed., *Seneca*, pp. 205–39, quotation on p. 231.

17 *Poetices libri septem* (Lyons, 1561), p. 223. Cited by Binns, "Seneca and Neo-Latin Tragedy," p. 230.

of Seneca's *sententiae*, often so excellent and miraculously acute ("probae, acutae, interdum ad miraculum"). Nevertheless he also regrets their tendency to become mere "sententiolae," by which he means obscure, meaningless, weak and feeble words ("fracta, minuta, quaedam dicta, obscura aut vana").[18]

Seneca's fame as a tragedian coincided with a rapid rise in public education throughout Europe. Because of the didactic value and "sententiousness" of the Senecan model, Seneca's tragedies as well as Neo-Latin plays written along Senecan lines were considered especially appropriate for instruction in rhetoric and moral philosophy. The performance of plays became a standard feature of education in the secular *scolae Latinae* and grammar schools, as well as in the numerous Jesuit colleges that were founded shortly after the establishment of that order in 1540.[19] Tragedies, in particular, were seen as beneficial for the young – in the words of Melanchthon, "both for preparing their minds for the numerous duties that life brings and the control of immoderate desires, and for giving them some training in eloquence" ("cum, ad commune faciendos animos de multis vitae officii, & de frenandis immoderatis cupiditatibus, tum vero ad eloquentiam").[20] Moreover, Jesuit school plays acted as a unifying force in European education, since colleges often exchanged scripts for performance in the annual *ludus literarius*, a one-day "open house" to which the whole town was invited. It is to one of these plays that I will now turn, as the subject matter is none other than a refutation of Epicurus' teaching.

Well before Epicureanism became an issue of serious academic debate in the seventeenth century, a Jesuit schoolmaster

---

18 "Animadversiones in Tragoedias quae L. Annaeo Senecae tribuuntur," in *Opera omnia*, 1:873. Cited by Binns, "Seneca and Neo-Latin Tragedy," p. 230.

19 See Foster Watson, *The English Grammar Schools to 1660, Their Curriculum and Practice* (1908: reprint, Cambridge, U.K., 1968). For France, see the select bibliography in I. D. McFarlane, *Buchanan* (London, 1981), pp. 548–50. For the Jesuit colleges the bibliography is immense and ranges from studies on individual colleges to research in curriculum and instruction. For the latter, see *Monumenta historica Societatis Jesu*, "Monumenta paedagogica," n.s., ed. Ladislaus Lukacs (Rome, 1965–74). See also F. Charmot, *La pédagogie des Jesuites, ses principes, son actualité* (Paris, 1943). For the Jesuit school drama, see Nigel Griffin, *Jesuit School Drama: A Checklist of Critical Literature*, Grant & Cutler Research Bibliographies and Checklists, no. 12 (London, 1976).

20 *Epistola . . . de legendis tragoediis & comoediis* (1545), cited by McFarlane, *Buchanan*, p. 200.

in Spain had taken things into his own hands.[21] The place was Jerez de la Frontera and the year 1586. The town was prosperous, thanks to a flourishing trade in the much-valued wine that we now know as sherry. Judging by the frequent references to European wines and coinage in the play, spectators were no strangers to current trends in Europe, and Epicureanism, it would seem, was rife in this wealthy community. But rather than denounce the new philosophy in a thunderous sermon from the pulpit, this Jesuit schoolmaster preferred to portray Epicureanism on the stage.

Playwrights, more than other authors, must be in tune with their audience's understanding and expectations, and drama, in the words of J. L. Styan, is a case of "feeling the pulse of an age or of a moment in time like no other art."[22] Clearly, the time, place, and sensibilities of the audience were ripe for a debate on Epicureanism. By presenting Epicurus in person, this schoolmaster must have expected the men, women, and children in the audience to recognize, identify, and, he clearly hoped, reject the new philosophy. Both Ciceronian and Senecan references in the play point to an audience that had at least a secondhand knowledge of relevant texts, while pokes at the citizens' aspirations to buy a title of nobility (a "hidalguía") hint at a possibly nouveau-riche audience. For the rich merchant, books and libraries were status symbols and being well read a sign of urbanity. Hence the urge on the part of the moralists to correct what they viewed as misconceptions or heresy and their need to reach the widest circle of society through the public entertainment of theater.

The play, called *Historia Filerini*, is written half in Latin and half in Spanish and is conceived as a long (four-act) morality play in the Senecan mold.[23] The structure of the play follows the same formulary rhetoric (i.e., rhetoric by example) which Richard Griffiths has observed in French Renaissance tragedy: "Formulary rhetoric, dealing as it did with separate entities in themselves, created a tendency towards making of drama a string of such 'set

21 See my "Jesuits as Masters of Rhetoric and Drama," *Revista Canadiense de Estudios Hispánicos*, 10:3 (1986), 375–87.

22 *Drama, Stage and Audience* (Cambridge, U.K., 1975), p. 11.

23 The piece occurs in two separate manuscripts, nos. 398–9 of the Cortes Collection in the Academia de la Historia in Madrid. All references are to MS. 399, which is in a much clearer hand and seems a fair copy of no. 398. The spelling has been modernized for this chapter.

pieces', often very loosely strung together."²⁴ In the *Historia Filerini*, these "set pieces" are references taken from both Senecan and Ciceronian sources and set out to act as *exampla* of a moral dilemma.

The plot is simple enough: Two young boys, in search of peace of mind, meet with diverse personifications and typifications of vices and virtues which lead them to penitence and admission to the order. Not a very exciting play, except that the first two acts are entirely taken up by a meeting with Epicurus, portrayed as an ugly old man who tries to indoctrinate the young people. The counter arguments are put forward by the allegorical figure Disillusion (Desengaño), who tries to steer the boys in the right direction with admonishments of gloom and doom. Of course, Disillusion is, first and foremost, a spokesman for the Church, but also, possibly, meant to be a corrective for much Senecan pro-Epicureanism.

Although Latin is used in Epicuro's opening exultation on the subject of pleasure and happiness, his introductory monologue is in Spanish. The choice of language is in itself significant, in that it shows the importance of translations in the transmission of ideas. The schoolmaster doubtless felt that the non-Latinists in the audience, whose knowledge of Epicureanism was mainly based on Seneca's translated words, needed to be put right as well.

The debate between Epicurus and his Stoic counterpart Desengaño starts with a few easily understood Latin verses:

Epicuro: O vitae rationem; quam sola
            praescribit voluptas, placidissimam.
Desengaño: O eggregiam haeresim; o intolerabilem
            dementiam.

But soon the dialogue changes into Spanish, this time on more specific points. Epicurus admonishes the boys to enjoy life before the world comes to an end and "heaven proudly reduces you to a heap of mud" ("el cielo de ti se alabe de haberte resuelto en lodo"), to which Desengaño offers the alternate, no less worrying perspective, namely that "for the immortal soul God has something different in store, with which He can, and knows how to punish"

24 *The Dramatic Technique of Antoine de Montchrestien: Rhetoric and Style in French Renaissance Tragedy* (Oxford, 1970), p. 35.

("al alma immortal le cabe / otra suerte con que Dios / castigarla puede, y sabe").

Epicuro's monologue then forcefully describes how gravity steers the unfeeling stones toward their center in swift motion; wild beasts are instinctively drawn to pleasure and contentment, and man will never tire of following nature to that same end. In another scene, asked by one of the boys to tell him more about his "secta y doctrina," Epicuro again starts his explanation with a reference to Epicurean physics. Showing how "Heaven in hasty flight this life destroys" ("con apresurado vuelo / va devanando la vida") he admonishes his young listener to enjoy life, avoid "pain and trouble," and find the roses among the thorns. His brief exposition is then followed by a hymn to nature. In vain does Desengaño remind the boy of God's wrath and divine justice; Epicuro denies the existence of any law except his own nature, any other God but his belly, and any greater glory than that of his throat. Only this God gives rewards ("beneficios"), and so only this one does he worship. On the topic of peace of mind, which, according to Desengaño, is not of this world, Epicuro contends that some earthly creatures, albeit few, enjoy peace ("que tambien los terrenales / aunque pocos, tienen paz"). When the other boy expresses his fear of God, Epicuro, predictably, dismisses this with the cynical observation that his Lord, up there in heaven, could not care less about what happens to us here below.

The main exposition, however, is not in words but in action, as the plot moves toward that *summum bonum*, an Epicurean banquet. The only hint of doctrine here is the role of the senses in beholding a gorgeous pastry (seen, no doubt, as the schoolboy's *summum bonum*). With surprising subtlety the schoolmaster transposes the Epicurean theory of knowledge to the mysteries of the sacrament: "May faithful souls now contemplate / delights atop a pastry plate. / O mysteries ne'er heard before / that all five senses should adore / and here find all their pleasure." ("Contemple el alma fiel / lo que encierra en si un pastel / o misterios nunca oidos / que todos cinco sentidos / hallan su deleite en él").

So far, the references to Epicurus all reflect the schoolmaster's own reading. Epicuro's hymn to nature reflects Lucretius' *De rerum natura*, references to animal instinct remind us of Cicero's *De finibus* (I.9.30), the denial of God's generosity echoes Seneca's

De beneficiis (IV.19), and the senses as a means of knowing the mysteries of a pastry does credit to the author's satirical talent.

As master of rhetoric, and thus schooled in Ciceronian style, it is not surprising that he turned to Cicero's *De finibus* for inspiration in his attack on Epicureanism. In staging his banquet, he clearly followed Cicero's maxim that those who dine well do not necessarily dine virtuously ("qui bene cenent omnes libenter cenare, qui libenter, non continuo bene," *De Fin.*, II.8.24). Cicero, in fact, gives some telling examples of profligate dinner parties in Rome to prove the great difference between pleasure and virtue (*De Fin.*, II.8.23–6). By looking around Jerez, the schoolmaster may well have found a confirmation of this dictum in the wine merchants' unreserved indulgence in food, wine, and luxury. Thus he highlights his banquet with long enumerations of succulent dishes, famous wines, and valuable currencies in Europe, as well as singing the praises of Spanish coins and worldly goods in general. To show that this pleasure is nevertheless not virtuous, he accompanies the banquet with drunken song and dance, crowned by Epicuro's blessing of his glass of sherry, after which the drunken old man rolls under the table.

The mood chosen to attack Epicureanism is ridicule, and the target is food and wine. By reacting in this manner to what the schoolmaster perceived as the accepted view, he not only attacked Epicurus but also Seneca, who specifically, in *De vita beata*, qualifies Epicurus' "voluptas" as "sobria ac sicca." Countless other moralists and teachers in Europe were no doubt reacting in the same way, in an attempt to stem Seneca's rising fame as a "fons sapientiae," as Justus Lipsius calls him affectionately (Letter 89 of *Ad Italos et Hispanos*).

But no schoolmaster nor preacher could curb Seneca's impact, as his words were now circulating in Justus Lipsius' prestigious edition of 1605 and a new wave of translations saw the light in the seventeenth century.[25]

Seneca's teaching on Epicureanism and Stoicism, offered up as "flores" in sentence collections, as "pièces dècousues" throughout

25 The seventeenth century saw many more *Senecae opera* complete with commentaries, such as the three volumes of *L. Annaei Senecae Opera, quae exstant, Integris Justi Lipsii, J.Fred. Gronovii, et Selectis variorum Commentariis illustrata* (Amsterdam: Elsevier, 1672). The second volume contains an "Index Rerum et Verborum" in which Epicurus figures prominently.

his prose, and as "set pieces" in his tragedies, continued to be quoted, used, and reinserted in diverse anthologies, thus creating a popular image of Epicureans (and for that matter of Stoics) which bears little resemblance to the original. This process of reduction, exclusion, transformation, and ultimately misinterpretation may well be the reason why, in the unschooled mind, Epicurus' "voluptas" would henceforth be associated with the delights of the table. The fact that these pleasures, if used in moderation and good taste, continued to be viewed as a virtue rather than a vice may be due to the perennial attraction of a well-prepared dinner, as well as to Seneca's benevolent attitude toward Epicurus' "voluptas."

7

Stoic contributions to early modern science

PETER BARKER

INTRODUCTION

Although there are clear indications of Stoic influence in a
variety of scientific contexts during the sixteenth and seventeenth
centuries, the Stoic contribution to early modern science has been
almost completely ignored.[1] The silence of modern historians
of science is a striking contrast to writers in other fields – for
example, politics and literature – where Stoic influences are widely acknowledged and accorded considerable importance.[2] I will
attempt here to do two things. First, I will make a case for Stoic

I would like to express my thanks to Mordechai Feingold for assistance with sources and
to David Lux, Roger Ariew, and Bernard Goldstein for critical comments on earlier drafts.
I would also like to thank the Special Collections Department of the University Libraries
at Virginia Tech, and its director, Glenn McMullen, for assistance in securing a copy of
Basso's *Adversus Aristotelem*.

1 An extreme example is J. L. E. Dreyer, *A History of Astronomy from Thales to Kepler*
(Dover: New York, 1953), pp. 159–60. After reviewing Stoic cosmology, Dreyer
concludes, "But the time was past for such hazy talk, and the Stoics remained outside the
development of science, Posidonius alone having the courage to prefer mathematical
methods to metaphysical arguments." Attempts to redress the disregard for Stoicism in
the history of science have been made by Peter Barker and Bernard R. Goldstein, "Is
Seventeenth Century Physics Indebted to the Stoics?", *Centaurus*, 27 (1984); 148–64;
Peter Barker, "Jean Pena and Stoic Physics in the Sixteenth Century," in Ronald H. Epp,
ed., *Spindel Conference 1984: Recovering the Stoics, Southern Journal of Philosophy*, 13 (suppl.)
(1985), 93–107; and perhaps most successfully by B. J. T. Dobbs, "Newton and
Stoicism," in ibid., pp. 109–23. Other relevant material appears in Peter Barker and
Bernard R. Goldstein, "The Role of Comets in the Copernican Revolution," *Studies in
History and Philosophy of Science*, 19 (1988), 299–319.

2 See, for example, Gerhard Oestreich, *Neostoicism and the Early Modern State* (Cambridge:
Cambridge University Press, 1982), esp. p. 38, and Gilles D. Monsarrat, *Light from The
Porch: Stoicism and English Renaissance Literature* (Paris: Didier-Erudition, 1984).

influence in early modern science; and second, I will discuss how this influence occurred. My conclusion will be that investigating the transmission of Stoic ideas obliges us to examine the wider cultural context of science. This is not an argument for externalism as a substitute for internal history of science; I take the position that Stoic influence can only be understood by combining both perspectives.

The recognition of Stoic influence has been hampered by the lack of a detailed historical argument for the transmission of specifically Stoic scientific ideas. I will therefore examine the transmission and later use of the Stoic ideas found in Cicero's book *On the Nature of the Gods* and the subsequent employment of these ideas in sixteenth- and early seventeenth-century discussions of the substance of the heavens by Jean Pena, Robert Bellarmine, Tycho Brahe, and Johann Kepler. Last, I will briefly describe the use made of these ideas by Sebastian Basso, to underline the eclectic way in which Stoic natural philosophy was combined with other positions – for example, atomism – during the late sixteenth and early seventeenth centuries.

It is obviously anachronistic to refer to Pena, Bellarmine, Brahe, Kepler, and Basso as "scientists." Nor would it be much improvement to call them "natural philosophers," although that label might have appealed to their seventeenth-century successors. The scope and limits of natural philosophy are controversial issues throughout the sixteenth century. A central question is the extent to which natural philosophy is to be subordinated to other fields of knowledge, and to what extent it may be pursued as an autonomous study. Perhaps grouping these people together may be sufficiently defended by pointing out that (with the possible exception of Bellarmine) later writers regarded them as contributing to what came to be called "natural philosophy" and, later, "natural science." Calling them scientists is no more than a shorthand way of recognizing their contributions to the history of the subjects we today call "scientific," notably astronomy and cosmology.

Describing the group in the categories of their own times, we find not one label but five. Pena was a university professor of mathematics. Bellarmine was an ecclesiastic. Tycho was an aristocrat. Kepler was a court mathematician. Basso was a physician. For Stoic themes to appear in the works of such a diverse group suggests a wide dissemination of these ideas – a climate of opinion, rather than a tightly articulated theory passed on by a small group of its exponents. I will be concerned to show that such a climate of opinion existed by the middle of the sixteenth century, and that it embodied specific cosmological ideas used by the figures to be examined.

THE AVAILABILITY OF STOIC COSMOLOGY

The combination of Stoic ideas with non-Stoic elements in early modern science is one factor impeding their recognition. But the problem of identifying Stoic ideas occurs even in ancient sources. The problem takes two forms. First, some of the best sources for Stoic scientific ideas in antiquity were themselves neither scientific nor Stoic. Second, some of the most influential ancient scientists employed Stoic ideas, but in combination with other views that were non-Stoic, making claims of influence ambiguous.

If a Stoic is someone who adhered exclusively to the doctrines of the Greek Stoics, then some of the best Latin sources, Cicero and Seneca, cannot be simply classified as members of that school. Both draw eclectically on ideas originating with Peripatetics, Skeptics, and Epicureans, as well as Stoics, but their comparisons of these schools make them useful sources. Stoic doctrine taught that right action presupposes an understanding of nature. The consideration of Stoic views on ethical questions therefore required the prior presentation of at least some Stoic physics, particularly cosmology and cosmogony. Thus, despite their eclecticism, Cicero and Seneca provide a good background in Stoic scientific ideas.[3]

After reviewing some of the things we can learn from these sources, I will present evidence for Renaissance interest in a key text: Cicero's *On the Nature of the Gods*. We shall see that Stoic

3 My main focus will be Marcus Tullius Cicero, *De natura deorum* (On the nature of the gods), (Loeb Classical Library, 1933). Improved translations of several passages appear in Brad Inwood and L. P. Gerson, *Hellenistic Philosophy* (Indianapolis: Hackett, 1988). See also H. C. P. McGregor, trans., *The Nature of the Gods*, introduction by J. M. Ross (Harmondsworth: Penguin Books, 1972). For Seneca, preferably see J. Clarke, *Physical Science in the Time of Nero: Being a Translation of the "Quaestiones naturales" of Seneca* (London: Macmillan, 1910), and Barker and Goldstein, "Is Seventeenth Century Physics Indebted to the Stoics?".

scientific ideas were preserved and transmitted outside strictly
scientific contexts but, by the mid-sixteenth century, were widely
available to anyone with what we would now call "scientific"
interests.

The central feature of Stoic physics is the *pneuma*, an all-
pervading medium which intelligently directs the cosmic cycle.
"Nature," according to Zeno of Citium, the founder of the Stoic
school, "is a craftsmanlike fire, proceeding methodically to the
work of creation".[4] The *pneuma* is a combination of fire and air. At
the beginning of each cosmic cycle it differentiates itself into the
other elements, creating a geocentric cosmos with earth at its
center, water above the earth, air above the water, and increasingly pure *pneuma* extending to the sphere of the fixed stars. Like
Aristotle's cosmos, the Stoic cosmos is bounded by this sphere.
But the Stoics offer a single, integrated physics of the heavens and
the earth, in contrast to the sharp separation of the substance and
physics of the heavens from the region below the moon in Aristotle
and later Scholastic cosmology.

Another important difference is the nature of the heavenly
bodies themselves. For Aristotle, the element fire is confined to
the region below the moon, where in common with other terrestrial elements its natural motions are finite and rectilinear.
Aristotle's heavens are composed of a fifth element, the natural
motion of which is circular and nonterminating. The planets,
including the sun, are somehow constructed from this element.
Their apparent motions are compounded from the eternal circular
motions of geocentric spherical shells of the fifth element, which
completely fills the heavens. A host of fundamental commitments
therefore obliges Aristotle to deny that the stars, or the planets,
including the sun, are in any way fiery, despite the light and heat
associated with them.[5]

The Stoics suffer no equivalent prohibition. Because fire forms
part of their fundamental substance, which also fills the heavens, it
is quite natural to account for the light and heat of the heavenly
bodies as celestial fires. This account also solves the problem of
explaining planetary motion when Aristotle's celestial spheres
are replaced by the Stoics' fluid *pneuma* as the substance of the

---

4 Cicero, *De natura deorum*, II.22.57. (Hereafter *Cicero*, with book, chapter and section.)
5 Aristotle, *De caelo*, II.7.

heavens. The *pneuma* is intelligent, and so is anything made from nearly pure *pneuma*, including the celestial fires. As intelligent agents, the sun and planets are capable of directing their own motions, and they are compared to living things in the air and waters, as creatures appropriate to their element.[6]

But fires need fuel, and this provides the final link in the chain of events directed by the *pneuma*. The universal substance circulates from the periphery of the cosmos to its center. The inward half of the cycle conveys the influence of the heavens to the inhabitants of the earth, and provides the physical underpinning for astrology and divination.[7] The outward portion of the cycle carries with it that nourishment – in the form of terrestrial materials – required to keep the celestial fires burning. These terrestrial materials are not replenished, and as time passes increasingly more of the substance of the world is locked up in celestial fires. Finally, the elements of the terrestrial region are used up, the sun absorbs the other celestial fires and itself reverts to undifferentiated *pneuma*. The entire cycle of differentiation, conflagration, and dissolution then repeats, as it has since eternity.[8]

All of this is well described in Cicero's *De natura deorum* (On the nature of the gods), probably written about 45 B.C., in the form of a dialogue between an Epicurean and a Stoic, with a rebuttal by an Academic Skeptic. Cicero employs his Stoic spokesman Balbus to present a series of arguments for the existence of the gods. The main argument now goes by the name of the "argument from design" (briefly, that the construction of the cosmos is evidence of a beneficent creator). In presenting these arguments Balbus is obliged to describe the design of the universe in some detail, and in doing so he lays out a good deal of Stoic cosmology. Stoic doctrines are also well represented in the criticisms of the Epicurean arguments given at the beginning of the dialogue. And although the interlocutor who rebuts both Epicurean and Stoic cases is presented as an Academic Skeptic, this is one place where Cicero, whatever the general difficulties of classifying his

---

6 On the intelligence of the universe, see Cicero, II.21.54–5; on the comparison of the planets with creatures inhabiting air and water, ibid., II.42.

7 On the influence of the sun and moon, see Cicero, II.19.49–50.

8 On the pneumal cycle linking the heavens and the earth, see Cicero, II.46.118–19 and II.33.83–4. An important recent study is A. A. Long, "The Stoics on World Conflagration and Everlasting Recurrence," in Epp, ed., *Recovering the Stoics*, pp. 13–37.

intellectual position, clearly sides with the Stoics. At the very end of the dialogue he undercuts his own interlocutor, saying "to me Balbus' arguments seem to be closer to a semblance of the truth."[9]

Cicero and Seneca are helpful reference points, because they label certain doctrines as Stoic, but the very popularity of Stoic doctrines makes them less visible in late antiquity. Scientific authors in this period often combine elements of Stoic ideas with other positions. Ptolemy, for example, accepts the Stoic theory of vision, as Mark Smith has recently shown,[10] although his cosmology is largely Aristotelian. Galen gives the *pneuma* a prominent role in his physiology[11] but in other respects is Hippocratic or Aristotelian. Stoic ideas suffer a similar fate in philosophy proper.[12] As an example of the assimilation of Stoic ideas in Neoplatonism, Verbeke points to the use of the doctrine of seminal reasons by Plotinus.[13] In all these cases, Stoic doctrines are integrated into a wider position that is not simply Stoic, and their Stoic antecedents cease to be noted. As Michael Frede has recently put it, speaking of Galen, "The only views one took to be distinctively Stoic were those one took exception to, while those views one found congenial were regarded as part of the common platonic heritage."[14]

Ptolemy, Galen, and the Neoplatonists evidently were vehicles for transmitting Stoic doctrines from antiquity to the Latin Middle Ages and beyond, but I will sidestep the difficulties of disentangling the various strands in these complex positions by examining cosmological doctrines that were both labeled as Stoic and transmitted without modification, through Cicero's *On the Nature of the Gods*, as a means of establishing a Stoic contribution to early

9 "[M]ihi Balbi, ad veritatis similitudinem videretur esse propensior." Cicero, III.40.95.
10 A. Mark Smith, "The Psychology of Visual Perception in Ptolemy's *Optics*," *Isis*, 79 (1988), 189–207.
11 Leonard G. Wilson, "Erasistratus, Galen and the *pneuma*," *Bulletin of the History of Medicine*, 33 (1959), 293–314.
12 Marcia M. Colish, *The Stoic Tradition from Antiquity to the Middle Ages* (Leiden: Brill, 1985).
13 Gerard Verbeke, *The Presence of Stoicism in Medieval Thought* (Washington, D.C.: Catholic University of America Press, 1983), pp. 4–5. Verbeke's exposition of the historiographical problems of separating Stoicism from its Neoplatonic matrix played a central role in determining my strategy in the present study.
14 Galen, *Three Treatises on the Nature of Science*, trans. R. Walzer and M. Frede (Indianapolis: Hackett, 1985), Introduction, p. xix.

modern science. I will pass over late antiquity and the Middle Ages, deferring to the studies of Colish and Verbeke,[15] and turn to the fourteenth century, when Cicero's work gained new importance as the basis for literary Humanism.

At the very beginning of the Humanist movement Stoic doctrines transmitted through Cicero reappear in the work of Petrarch. In particular, *On the Nature of the Gods* is extensively quoted in Petrarch's response to his Averroist detractors, probably completed in early 1368.[16] Here we are told, "Of all the writings of Cicero, those from which I often received the most powerful inspiration are the three books ... *On the Nature of the Gods.*"[17] This ringing endorsement is only slightly muted by the title of the work in which it appears: "The book of Francesco Petrarca the Laureate, on his own ignorance and that of many others." Extensive excerpts from Balbus' speeches are used to present the Stoic arguments from design,[18] although it should be noted that Petrarch sharply criticizes Cicero and his Stoic spokesman on other matters.[19]

In Petrarch's work we see that early in the Renaissance a book containing important Stoic scientific ideas was singled out for special attention. Although Petrarch was not interested in the cosmology described in *On the Nature of the Gods*, given his contemporary celebrity and influence his work must have placed these ideas before a wide audience. By the early sixteenth century Stoic ideas had become widespread. Just how widespread may be illustrated by the case of Rabelais, whose major works began to appear in the 1530s.[20]

15 Colish, *Stoic Tradition*; Verbeke, *Presence of Stoicism*.

16 H. Nachod, trans., "Francesco Petrarca: On his own ignorance and that of many others," in E. Cassirer, P. O. Kristeller, and J. H. Randall, eds., *The Renaissance Philosophy of Man* (Chicago: University of Chicago Press, 1948), pp. 47–133. The dedication to the grammarian Donato is dated 13 January 1368; ibid., pp. 47–8. On the circumstances of composition, see pp. 29–31.

17 Ibid. p. 79.   18 Ibid. pp. 81–6.

19 Petrarch criticizes Cicero for denying monotheism (ibid. pp. 87–91, esp. p. 91) and Balbus for impiety (p. 95).

20 *Pantagruel* appeared in 1532, and *Gargantua* in 1534. Two additional works in the series appeared before Rabelais's death in 1553, and a fifth posthumously. The attribution of the fifth work to Rabelais is controversial, but at least portions of it may be his. Compare N. H. Clement, "The Eclecticism of Rabelais," *PMLA* 42 (1927), 339–84, esp. p. 351, who regards the fifth book as the clearest exposition of Rabelais's ultimate beliefs, with M. A. Screech, *Rabelais* (London: Duckworth, 1979), who discounts it in his own approach to Rabelais.

Rabelais has been portrayed as an exponent of Stoic cosmology, most clearly in his monism and his employment of all-pervading fundamental substances, resembling the Stoic *pneuma*.[21] Recently, this reading has been qualified by drawing attention to Rabelais' commitment to evangelical Christianity.[22] It is also true that Rabelais alludes to everything ⊢ from Academic Skepticism, through Epicureanism, to orthodox Aristotelianism, Averroism, and, of course, Neoplatonism. Even reading Rabelais as a Christian eclectic, however, there are unambiguously Stoic elements in his work, giving some measure of the currency of these ideas in the second quarter of the sixteenth century. A variety of modern commentators agrees that Rabelais' central characters exemplify the Stoic sage's attributes of indifference (*apatheia*) and self-sufficiency (*autarcheia*).[23] From *Le tiers livre de Pantagruel* (1546) onward, Rabelais progressively introduces a Stoic worldview, identifying God and nature as an all-pervasive creative principle.[24] Also in the *Tiers livre*, Rabelais adopts the Stoic conception of the universality of natural law. Again Cicero's *On the Nature of the Gods* appears as a source.[25]

Petrarch and Rabelais are important not as examples of Stoic influence but rather as an indication of the extent to which Stoic themes pervaded general literary culture by the mid-sixteenth century. Whatever the status of Rabelais' own views on Stoicism, for example, it is clear that his readers were expected to recognize and respond to Stoic ideas. Petrarch's contemporary celebrity and Rabelais' wide readership establish that a large audience was either familiar with, or could reasonably be expected to have access to, works of Cicero like *On the Nature of the Gods*.

Although the examples of Petrarch and Rabelais establish the availability and widespread currency of Stoic cosmological ideas, nothing has yet been said to establish the entry point of these ideas

---

21 Clement, "Eclecticism of Rabelais."

22 J. C. Nash, "Rabelais and Stoic Portrayal," *Studies in the Renaissance*, 21 (1974), 63–82; Screech, *Rabelais*.

23 Clement, "Eclecticism of Rabelais," p. 351, n. 25, especially the portion concerning the *Tiers livre*; Nash, "Rabelais," p. 82.

24 See the passages from the *Tiers, Quart* and *Cinquiesme livres*, discussed by Clement, "Eclecticism of Rabelais," pp. 378–80.

25 Clement, "Eclecticism of Rabelais," p. 371. *On the Nature of the Gods* is cited by name at *Quart livre*, Chap. 28; cf. Cicero, III.56. For an allusion, see *Tiers livre*, Chap. 3, in François Rabelais, *Oeuvres complètes* (Paris: Pléiade, 1955), p. 341; cf. Cicero, III.14. Note that these particular references are not concerned with cosmology.

into scientific debates. Leaving aside the matter of astrology – which deserves separate treatment at much greater length – I believe the point of entry for Stoic ideas can be shown in one major scientific controversy: the sixteenth-century dispute on the substance of the heavens. The ideas and sources we have examined so far reappear in this debate, to which I now turn.

THE SUBSTANCE OF THE HEAVENS

During the sixteenth and early seventeenth centuries the Aristotelian account of the substance of the heavens was abandoned.[26] Although the use of Stoic ideas in these scientific contexts begins long after they had become commonplaces in educated general culture, such ideas became an important source of alternatives to the Aristotelian position, and their earliest appearances in the works of Pena and Bellarmine can be linked directly to Cicero's *On the Nature of the Gods*.

In a prefatory essay to a new edition of Euclid's *Optics and Catoptrics*, published in 1557 at Paris, Jean Pena (1528–58) argued, on the basis of Gemma Frisius's failure to detect atmospheric refraction, that a single medium extended from the surface of the earth all the way to the fixed stars. He rejected both the Aristotelian doctrine of the sphere of fire and the existence of celestial spheres composed of a fifth element, and he concluded that air filled the whole space they were thought to have occupied. A variety of evidence suggests that Pena's air derives from the Stoic *pneuma*, as described by Cicero in *On the Nature of the Gods*.[27]

As Pena describes it, the substance of the heavens functions and moves in the manner of the *pneuma*. He also calls this substance an "*animabilem spiritum*," "a spirit consisting of air," using the rare adjective *animabilis*, a variant of the more common *animalis*. Balbus uses the same unusual version of the adjective during his exposition of Stoic cosmology in Book II of Cicero's *On the Nature of the Gods*. According to the lexicons, this is the sole

26 William H. Donahue, "The Solid Planetary Spheres in Post-Copernican Natural Philosophy," in Robert S. Westman, ed., *The Copernican Achievement* (Berkeley and Los Angeles: University of California Press, 1975), pp. 244–75; William H. Donahue, *The Dissolution of the Celestial Spheres, 1595–1650* (New York: Arno, 1981).

27 Jean Pena, "De usu optices," printed as a preface to *Euclidis optica et catoptrica* (Paris: Andreas Wechulus, 1557). For details of Pena's life, opinions, and influence, see Barker, "Jean Pena."

appearance of the term in surviving classical Latin literature.[28] Later writers confirm the identification of Pena's air with the *pneuma*. Longomontanus, for example, associated Pena's proposal for the substance of the heavens with Stoicism.[29] We also know that Pena's patron Ramus, and his circle, were ardent Ciceronians. Pena's context and the testimony of near contemporaries both support the identification of Pena's air with *pneuma*, and the textual evidence points to *On the Nature of the Gods* as its source.

Within two decades the same doctrines reappear in another tradition where cosmological ideas were a matter for dispute – the tradition of commentators on Genesis. Between autumn 1570 and spring 1572 Robert Francesco, later Cardinal, Bellarmine gave a series of lectures, including a commentary on Genesis, at the University of Louvain. In these lectures Bellarmine supports his points by two kinds of arguments, the first based on the Bible and the church fathers, the second based on ancient pagan authorities. In the sections of his Genesis commentary discussing cosmology the ancient authorities are preeminently Stoic.

Bellarmine takes a quite uncompromising position on the substance of the heavens. The Scholastics are wrong; the heavens are corruptible and the substance of the heavens is fire. Bellarmine supports this position with the opinions of ecclesiastics from Basil to Bede. He then passes to the ancient authorities. Only Plato is mentioned before we are told, "All the Stoics held that the sky was fire according to the testimony of Marcus Tullius [Cicero] in Book II of his *On the Nature of the Gods*."[30]

---

28 Pena, "De usu optices," p. 2, 11.33–6: "Totum illud spatium, per quod . . . meant errantia sydera, esse hunc animabilem spiritum per rerum naturum sursum, quem spiramus, nec quicquam ab aere distingui." (This whole space, through which . . . the planets move, [is] an airy spirit, rising through all of nature, which we breathe, and which cannot be distinguished from air.) Compare Cicero, *De natura deorum*, II.36.91: "[T]erra sita in media parte mundi circumfusa undique est hac animabile spirabileque natura cui nomen est aer." (The earth is situated in the center of the world, and surrounded on all sides by this airy and breathable substance the name of which is 'air'.) On the variant form *animabilis* see *Thesaurus linguae Latinae* (Leipzig: Teubner, 1900–4), vol. 2, pt. 1, p. 73; C. T. Lewis and C. Short, *Harper's Latin Dictionary* (New York: Harper, 1907), p. 122; Austin Stickney, ed., *De natura deorum* (Boston: Ginn, 1901), p. 265, n. 107.5.

29 Donahue, "Solid Planetary Spheres," p. 270.

30 The passage ends "and of Macrobius in Book II, Ch. 10 of his comment on the *Somnium*." U. Baldini and G. V. Coyne, eds. and trans., *The Louvain Lectures of Bellarmine and the Autograph Copy of his 1616 Declaration to Galileo*, Studi Galileiani, vol. 1, no. 2 (Vatican City: Vatican Observatory Publications, 1984), p. 8.

Bellarmine again has recourse to Book II of *On the Nature of the Gods* to support the view that the stars are made of fire.[31] He goes on to recognize the necessity for adopting the Stoic account of the nature of planetary motion:

If we wish to hold that the heaven of the stars is one only, and formed of an igneous or airy substance, an hypothesis which we have declared more than once to be in accord with the Scriptures, we must then of necessity say that the stars are not transported with the movements of the sky, but they move of themselves like the birds of the air or the fish of the waters.[32]

The portions of Bellarmine's Louvain lectures on cosmological themes for which we have records ended before the appearance of the nova of 1572, which was to have such important consequences for cosmology. There is no discernible connection between Bellarmine and the most important writer on the new star, Tycho Brahe, but there is great similarity in their accounts of the substance of the heavens, and in Brahe's case there is a prominent connection with Pena.

Brahe's views on the substance of the heavens are scattered throughout his works. In an early letter he says that the substance of the heavens is fire and attributes the view to Paracelsus.[33] But following the treatment of refraction in his astronomical works Tycho refers to the substance of the heavens as "Aerem Elementarem et animalem,"[34] a phrase very like Pena's, and perhaps borrowed from Christoph Rothmann, who accepted Pena's view and described it to Tycho. Later, Brahe seems to have been at pains to separate his own views on the substance of the heavens from those of Pena. In a letter to Caspar Peucer, written in 1590, Brahe denied that aether and the element air mingle, a view he blamed on Pena. Like Bellarmine, Brahe then invoked biblical authority in support of fluid heavens, and the authority of Plato and Paracelsus that the element involved is fire.[35]

Perhaps the clearest statement of Tycho's views on the substance of the heavens appears in a letter to Rothmann from the year 1589. Here Tycho reviews a variety of opinions and concludes

31 Ibid., p. 18.   32 Ibid., p. 20.
33 Tycho Brahe, *Opera omnia*, ed. J. L. E. Dreyer, 15 vols. (Copenhagen: Libraria Glendaliana, 1913–29), 4:382–3.
34 Ibid., 2:77.   35 Ibid., 7:230–1.

that the heavens are a "most pure and most liquid aethereal substance," which should be distinguished from all the terrestrial elements. However, if a comparison with the terrestrial elements must be made, the substance of the heavens is closest to fire, although it is a fire that burns without being consumed. Paracelsus is again mentioned as an authority.[36] The stars and planets are formed from this substance, and just as in the terrestrial regions, living things that are governed by the element earth dwell in the earth; flying things dwell in air; fish occupy the waters; so too the sun and stars, which are fiery in nature, make their revolutions in the region of aetherial fire.[37] Indeed, the heavens hold the same relation to the earth as the soul to the body.[38] This is, of course, the same Stoic view of the planets referred to by Bellarmine when he writes of "birds and fishes," while the remark about the soul and the body concisely describes the Stoic *pneuma*.

Additional information on Tycho's cosmology appears in the anonymous preface to a 1591 work on meteorology. Tycho's authorship is established by the testimony of Longomontanus in the second Danish edition of 1644. Here, the region of the three terrestrial elements (earth, water, and air) is distinguished from the aether of the heavens along conventionally Stoic lines, and, perhaps most significant for the purpose of classification, air is invoked as the mechanism of astrological influence.[39]

36 Brahe to Rothmann, 21 February 1589, in Brahe, *Epistolarum astronomicarum* (Uraniburg, 1596), pp. 137–51: "Caeli videlicet, substantiam esse Aetheream & liquidissimam, purissimamque; quandam materiam, supra omnem elementorum naturam exaltam.... Sin autem alicuius Elementi naturam Caelo affingere non admodum absurdum videretur, ego potius illud Igneum, quam Aereum esse concederem, prout ab Paracelso traditum est, qui illud Quartum & Igneum elementum noncupat.... Aetheris nomen apud Graecos illi attributum quasi Ardens seu Igneum, idque perpetuo cum sit inconsummabilis" (pp. 137–8).

37 "Ex quo enim Stellae & lumina caelestia Ignea flammantiaque appareant; ... Animantia, quae de Mumia Terrae constant in Terris degere: Volucres de Mumia Aeria in Aere: Pisces de Mumia Aquea in Aquis versari, & sic de caeteris, ut ob id verisilius appareat, Solem et Stellas, quae Ignis cuiusdam incombustibilis speciem repraesentant, in Aethere Igneo et incombustibili, (unde etiam Aetheris nomen apud Graecos illi attributum quasi Ardens seu Igneum, idque perpetuo cum sit inconsummabilis) suas Revolutiones exercere" (ibid. pp. 137–8). Tycho's repeated phrase "de Mumia Terrae... Aeria... Aquea ... " is obscure. In the text I offer a paraphrase which I believe captures the sense, but the parallel with Bellarmine's talk of "birds and fishes" is clear in any case.

38 "Existimo enim Caelum eam habere ad Elementarem Mundum rationem, quae est animae ad corpus" (ibid., p. 138).

39 John Christianson, "Tycho Brahe's Cosmology from the *Astrologia* of 1591," *Isis*, 58 (1968), 316–17.

In an earlier work I treated Brahe's separation of his position from Pena's, and his parallel insistence that the substance of the heavens is fire rather than air, as evidence that Brahe avoided the Stoic theory.[40] But the elementary structure of the *pneuma* is ambiguous, as Bellarmine recognizes when he describes the substance of the heavens as "formed of an igneous or airy substance." Brahe's final position is the same as Bellarmine's, which we have already seen is connected with the Stoic sections of Cicero's *On the Nature of the Gods*. Brahe's division between the substance of the heavens, considered as fire or aether, and the lower elements, including air, also appears in Stoic sources.[41] Hence his labeling of the substance of the heavens as fire may be a more refined version of the Stoic doctrine than that found in Pena. Although I have found no direct references to a source like *On the Nature of the Gods* in Tycho, the parallels with Bellarmine are striking, and Tycho clearly invokes the Stoic view of planetary motion, and a celestial substance that plays a similar role to the *pneuma*. Taken together with the cosmography in the book on meteorology, and particularly the account of astrological influence by contact action, this suggests, at least, that Tycho was drawing on Stoic ideas readily available at the time.

The works of Bellarmine and Brahe show that by the end of the sixteenth century Stoic ideas were increasingly adopted as an alternative to Aristotelian cosmology, although their application here comes late in comparison to their appearance in literary contexts. The case of Brahe suggests that by the time they appear in scientific contexts Stoic ideas had again become so well known that they did not require a label. These works also show the typical range of these ideas: unification of the physics of the heavens and earth; fire, or some version of the *pneuma*, as the substance of the heavens; the nature of astrological influence; and an alternative account of planetary motion. Another measure of the currency of Stoic ideas may be gained by noting that Kepler contrasts his *virtus motrix* not with the Aristotelian spheres but with the Stoic view that the planets are intelligences that move themselves.

40 Barker, "Jean Pena," p. 101.
41 On the distinction between the substance of the heavens and the terrestrial elements, see Cicero, II.36.91 and II.17–18, 26, 39. Cf. the portion of Diogenes Laertius, *Vitae Philosophorum*, 7.132–60, trans. Inwood and Gerson, in *Hellenistic Philosophy*, p. 97.

Kepler also accepts a version of the doctrine that air is the substance of the heavens, and his view can be linked to Pena and the Stoics.[42] In the *Mysterium cosmographicum* of 1596, Kepler had solved the outstanding problem of the spacing of the planetary orbits by his well-known construction inscribing the Platonic regular solids between the spheres carrying the planets. At the same time he had abandoned the solid spheres of the earlier cosmologies. The spheres of the planets were now to be understood geometrically, rather than physically, although the heavens still contain a continuous fluid substance responsible for physically constraining planets to their proper positions and also for carrying astrological influences. Kepler's celestial substance is another variation on the life-sustaining Stoic *pneuma*, and Kepler alludes to the Stoics in a typical witticism. In the 1596 *Mysterium* Kepler says,

> By what bars, what chains, what heavenly adamant, has this Earth, which in complete accord with Copernicus we designate to be in motion, been put into its orb? Truly, with that air which (fermented and mixed with vapors) all men drink in all around the surface of the Earth, which we penetrate with our hand or our body but do not separate or pull apart, though it conveys the heavenly influences right into our bodies. For this [air] is the heaven in which we and all the bodies of this world live, and move, and have our being.[43]

The identification of the substance of the heavens as air, and of its function as a support of life and a medium for conveying celestial influences, are all familiar Stoic themes. In the last line of the passage we find a playful reference to Saint Paul, who had used the phrase "in which we live, and move, and have our being" to refer to the spiritual source of Christianity. Kepler could reasonably expect his audience to recognize the allusion. The passage in the Christian Bible where Paul uses these words also contains that

---

42 Barker, "Jean Pena," pp. 101–3. See also Johannes Kepler, *Dioptrice* (Cambridge, U.K.: Heffer, 1962), Praefatio, pp. 1–11.

43 *Johannis Kepleri astronomi Opera omnia*, ed. C. Frisch, 8 vols. (Frankfurt: Heyder & Zimmer, 1858), 1:159, text to note h: "Vel haec Tellus, quam omnino cum Copernico vehi statuimus, quibus vectibus, quibus catenis, quo adamante coeleste in orbem suum inserta est? Eo nempe quem omnes circumcirca in superficie Telluris homines haurimus (fermentatum et commixtum vaporibus) aerem: quem manu, quem corpore penetramus, neque tamen discludimus aut semovemus, cum sit influxum coelestium in media corpora vehiculum. Hoc enim coelum est in quo vivimus, movemur, et sumus nos et omnia mundana corpora."

book's sole overt reference to the Stoics. But Paul is attacking the Stoics and Epicureans, a point that would not have been lost on Kepler's audience. Kepler ironically employs the phrase "in which we live, and move, and have our being" to describe the very position attacked by its biblical author.[44]

Kepler diverged from the Stoic account of the substance of the heavens chiefly in his rejection of the "birds and fishes" doctrine that the planets intelligently direct their own motions. As the passage quoted earlier shows, planets are constrained to particular orbits, and the physical influences responsible are passed through the celestial substance. In notes added to the 1621 edition of the *Mysterium*, he explicitly rejects the view that souls direct the motions of the planets[45] and clarifies his own view:

> What is an orb or a heaven? What except air? And what is air? What except a variety of immaterial body which imparts motion to the planets as it rotates? ... let us concede that our [terrestrial] air is a material body, permeable by powers producing magnetism, motion, heat, light and so forth, hence it may be a vapor not totally different in kind from [celestial] air, but rather distinguished by degrees of thickness from the surrounding expanses of [celestial] air.[46]

The need for a substance to transmit Kepler's long list of powers may be understood as a simple corollary of the idea that physical action requires a physical agent, which also motivates the Stoic physics of contact action. The differentiation of the underlying

---

44 For Paul's words, see Acts 17.28. Kepler wittily continues the allusion in the 1621 note on this passage, saying "Ludere placuit in voce aeris *paulo* audacius" (emphasis added): "It pleased [me] to play a little audaciously with the word 'air.'" The adverb "little" (*paulo*) is a pun on the name Paul, so the passage might also be read, "It pleased me to make an audacious Pauline play with the word 'air.'"

45 For the rejection of souls directing the planets' motions, see particularly the second and third notes following Chapter 20 in the 1621 *Mysterium*. The latter reads in part, "Earlier ... I believed absolutely that the cause that moved the planets was a soul, because I was imbued with the opinions of J. C. Scaliger on the subject of movement-causing intelligences. But after I reflected that this cause of motion becomes more feeble with distance, and that the light of the Sun also diminishes with distance from the Sun, I came to the conclusion that this force was something corporeal."

46 Kepler, *Opera*, 1:161 (Chap. 16, note h): "Quid orbis vel coelum? Quid nisi aer? Et quid aer? Quid nisi species immateriata corporis, quod motum planetis infert, in gyratione versantis? Atqui seposito lusu concedemus aerem nostrum esse corpus materiatum, permeabile a facultatibus magneticis, motoriis, calefactoriis, illuminatoriis et similibus: ut sit vapor non toto genere diversum ab aere, sed saltem gradubus crassitiei distinctus a circumfusis aeris campis."

medium through condensation and rarefaction is also a Stoic theme. The power capable of producing motion is the *virtus motrix*, Kepler's provisional solution to the physical cause of planetary motion and an antecedent of Newton's account.[47]

As a final example of the use of Stoic cosmological ideas, let us consider Sebastian Basso, a doctor of medicine, educated at the Jesuit Academy at Pont-à-Mousson, founded in 1572 near Nancy.[48] He is known as the author of *Twelve books of natural philosophy against Aristotle, in which the hidden ancient physiology is restored and Aristotle's errors are refuted with sound reasons*, published at Geneva in 1621.[49] The two dates 1572 and 1621 practically exhaust current knowledge of Basso's biography. He is described in almost all the secondary literature as an early seventeenth-century atomist.[50] Although atoms appear throughout his book, in matters of cosmology Basso is a Stoic. All the main themes we have so far identified in sixteenth-century references to Stoic cosmology reappear in Basso.

In his *ad lectorem*, Basso lists the second book of Cicero's *On the Nature of the Gods* prominently among his sources for non-Aristotelian cosmology.[51] Most important, Basso accepts the Stoic account of the substance of the heavens, recognizing that this

---

47 See especially the fifth and seventh notes following Chapter 16 and the fifth note following Chapter 22, added in the 1621 edition of the *Mysterium*.

48 Basso identifies himself as a doctor of medicine on the title page of his book (see n. 49, this chapter). On his education, see J. R. Partington, *A History of Chemistry*, 3 vols. (London: Macmillan, 1961), 2:387.

49 Sebastian Basso, *Philosophiae naturalis adversus Aristotelem, Libri XII*. I have consulted the Elzevir edition, published in Amsterdam in 1649. All subsequent page references are to this edition. Partington reports, and P. M. Rattansi rejects Jocher's claim that the first edition of Basso's book appeared at Rome in 1574: Partington, *History of Chemistry*, 2:387, n.5; P. M. Rattansi, "Sebastian Basso," in C. C. Gillespie, ed., *Dictionary of Scientific Biography*, 12 vols. (New York: Scribner, 1970), 1:425. In addition to Rattansi's failure to trace this edition, we may discount its existence for two other reasons. First, the date 1574 would require that Basso published a book only two years after beginning his higher education at Pont-à-Mousson, which as noted, was founded in 1572. Second, it would require that his criticisms of Toletus' *Physica* (Basso, *Philosophiae naturalis*, pp. 203–4) were published in the same year as that book appeared, i.e., 1574.

50 A recent example of the majority view – that Basso was an atomist – is C. Meinel, "Early Seventeenth Century Atomism: Theory, Epistemology and Insufficiency of Experiment," *Isis*, 79 (1988), 68–103. An exception to this view is Peter Barker and Bernard R. Goldstein, "Is Seventeenth Century Physics Indebted to the Stoics?", in which Basso is offered as an example of a seventeenth-century scientist using Stoic ideas.

51 Basso, *Philosophiae naturalis*, "ad lectorem," p. **3.

abolishes the celestial – terrestrial distinction central to Aristotelian physics. Rejecting the vacuum, Basso insists that however finely divided the parts of a material substance may be, there must be another substance, or universal spirit, filling the spaces between its smallest parts. This universal spirit fills the heavens and pervades the earth, and is explicitly identified with the Stoic *pneuma*.[52] Like Bellarmine, Basso links the Stoics with Plato as ancient authorities supporting his opinion.[53] Like Tycho, he gives fire as the substance of the heavens[54] and accounts for astrological influences by direct action.[55] But, unlike any of the earlier figures we have considered, he accepts the entire Stoic pneumal cycle in which the substance of the heavens flows both inward to the earth and outward, nourishing the stars.[56] Also, like Kepler, he replaces the Stoic view of planetary motion with something like a mechanical account. The planets are swept along by the fire in the heavens, which they both absorb and emit.

Basso is clearest on the elementary constitution of the universal spirit when he discusses the nature of planetary motion. The firmament is a solid sphere. Beneath it, the whole of the space through which the planets move is filled with the universal spirit, the nature of which is fire.[57] The sun and stars emit *igniculi* – little fires, or sparks – which descend to the earth and are replenished by a corresponding upward flow.[58] It seems, therefore, that the heavens contain a fiery spirit, carrying fiery corpuscles.[59] The planets are neither lighter nor heavier than the pneumal fire in the region where they move. But rather than directing their own motions, they are swept along by the fire in the heavens, which they absorb "through innumerable windows."[60] Ducts or conduits within the planet direct the fire that enters it. Mechanically

---

52 The entry "Spiritu Stoicorum quid" appears in the index to Basso's *Philosophiae naturalis* and refers to the discussion of the universal spirit that unifies the elements, which begins on p. 300. Basso also refers directly to the Stoics in the text.

53 Ibid., pp. 301–2.   54 Ibid., p. 465.

55 Ibid., pp. 503–7.   56 Ibid., pp. 464, 473.

57 Ibid., p. 465: "Si meminerimus quae superius probavimus esse duntaxat sphaeram unam solidam, scilicet firmamentum, sub quo clarum est, ex Veterum sententia, ignem implere vastissimum illud spatium quo Planetae suos faciunt circuitus."

58 Ibid., p. 473.

59 Note that the element fire consists of the universal spirit plus *aculei subtillissimi* (ibid., p. 304), literally, "most subtle pricks (or stings)." Partington, *History of Chemistry*, 2:388, renders the phrase as "fine and sharp corpuscles."

60 Basso, *Philosophiae naturalis*, p. 466.

actuated doors control access to these channels. With some doors open and others closed, the planet moves in the same direction as the predominant influx of celestial fire.[61] Hence the planets move as a side effect of the pneumal cycle: The "birds and fishes" account is explicitly rejected.[62]

Basso's use of Stoic ideas is eclectic, but his rejection of the view that planets move like birds through the air or fishes through the water suggests an important point about the status of the Stoic contribution to natural philosophy at the beginning of the seventeenth century. The nature of planetary motion became a problem as soon as the Aristotelian account of the substance of the heavens was seriously questioned. The birds and fishes account found in Bellarmine and Bŕahe is an initial response. The rejection of this position, in the works of Basso and Kepler, may be seen as a response to a response, and shows the development of a self-contained dialectic. By the beginning of the seventeenth century, then, Stoic natural philosophy was so well assimilated that disputes had become possible about questions internal to the position.

The responses of Kepler and Basso to the problem of planetary motion were to have interesting sequels. Kepler's *virtus motrix* foreshadows Newton's universal gravitation. The *virtus motrix* originates in the sun, its influence is universal and diminishes with distance, and the planets on which it acts "possess the ability to resist a motion applied externally, in proportion to the bulk of the body and the density of its matter."[63] Basso's cosmology places him among the founders of the mechanical philosophy. All material change is a result of the motion and rearrangement of atoms, directed by the *pneuma*.[64] The details of his account of planetary motion draw on ideas from contemporary mechanics. Atomism and mechanical explanations are prominent features of later mechanical philosophies, including the work of Descartes. Indeed, the early Descartes might be caricatured as Basso without the

---

61 Donahue, *Dissolution of the Celestial Spheres*, pp. 160–1, translates the relevant passages from Basso.

62 Basso, *Philosophiae naturalis*, p. 434.

63 Kepler, *Mysterium cosmographicum*, trans. A. M. Duncan (New York: Abaris Books, 1974), p. 171, n. 5. See also p. 169, n. 4; p. 171, n. 7; p. 219, n. 7.

64 Basso, *Philosophiae naturalis*, p. 387.

pneuma. Although Descartes mentions Basso, and is linked to him by several modern commentators, it is perhaps more accurate to represent Basso as an early contributor to the general position of which Descartes came to be the leading exponent.[65]

CONCLUSION

I set out to do two things. The first was to present a case for the influence of Stoic ideas in early modern science. In this essay I have tried to show the explicit connection between Cicero's *On the Nature of the Gods* and sixteenth-century claims that the substance of the heavens is air or fire, the constituents of the Stoic *pneuma*.

My second goal was to examine the way Stoic influence appears. Stoic ideas enter science from the broader intellectual culture. This culture received Stoic ideas transmitted through continuous traditions of theology, law, and medicine, and the study of their influence quickly takes us outside the history of science.[66] I have concentrated on cosmological ideas that can be found in a single book, Cicero's *On the Nature of the Gods*. But even here Stoic scientific ideas were preserved and transmitted in nonscientific contexts, where the book was used by writers like Petrarch and Rabelais, and theologians like Bellarmine. Only after Stoic ideas had become broadly available outside science did they begin to appear in the work of natural philosophers such as Pena, Brahe, Kepler, and Basso and contribute to modern science. Here too, understanding the Stoic contribution to early modern science requires that we consider factors outside the history of science itself. This conclusion assumes even greater significance if we turn from Stoic contributions to the content of science – for example

---

65 C. Adam and P. Tannery, eds., *Oeuvres de Descartes*, 12 vols. (Paris: Vrin, 1974), 1:156–67. In a letter to Isaac Beeckman, dated 17 October 1630, Descartes refers to Basso as just one among a long list of "renovators," none of whom he endorses (p. 158). Presumably Descartes himself wishes to be understood as an innovator, suggesting a metaphilosophical discontinuity, whatever continuity exists at the level of natural philosophy.

66 During the sixteenth century, interest in astronomy was usually subordinated to interest in astrology, and interest in astrology was usually subordinated to interest in medicine. See Barker and Goldstein, "Role of Comets."

ideas about the elements constituting the heavens – to metascience and the use of Stoic concepts such as natural law and providence.[67]

A final point concerns the manner in which Stoic ideas reappear. Stoic ideas are eclectically combined with other ideas. While the orthodox Stoic view takes the earth to be the center of planetary motions, and Basso accepts this view, Brahe combines one Stoic-derived celestial substance with a geo-heliocentric world system, and Kepler combines another with heliocentrism. Basso combines Stoic cosmological ideas with atomism, despite the well-known opposition of the ancient Stoics to atomism. Stoic ideas therefore re-entered scientific debates not as a complete system, distinguishable from an Aristotelian or an Epicurean system, but in eclectic combinations with ideas that in antiquity had been regarded as antithetic.

The eclectic use of Stoic ideas in no way diminishes the importance of their contribution. Kepler's *virtus motrix* stands as an antecedent to Newton's gravitational theory, while Basso's planets swept along by the fluid heavens foreshadow the Cartesian account. These views, then, provide the starting point for the seventeenth-century dispute between action-at-a-distance and action-by-contact, and that, I hope, is a contribution of sufficient importance to interest even a skeptic about the Stoics.

67 The scientific relevance of "providence" in this period has yet to be adequately appraised. A valuable starting point can be found in the work of Margaret J. Osler. See her contribution to this volume (Chapter 8, "Fortune, Fate, and Divination"), and "Providence and Divine Will in Gassendi's Views on Scientific Knowledge," *Journal of the History of Ideas*, 44 (1983), 549–60. See also Barbara J. Shapiro, *Probability and Certainty in Seventeenth Century England* (Princeton: Princeton University Press, 1983).

8

Fortune, fate, and divination: Gassendi's voluntarist theology and the baptism of Epicureanism

MARGARET J. OSLER

Pierre Gassendi (1592–1655) is most frequently remembered for introducing the philosophy of the ancient atomist Epicurus into the mainstream of European thought.[1] Gassendi's version of Epicurean atomism and his adaptation of Epicurean hedonism exerted major influences on seventeenth-century developments in science and political philosophy.[2] Before European intellectuals could embrace the philosophy of Epicurus, however, his views had to be purged of the accusations of atheism and materialism that had followed them since antiquity.[3] Gassendi, a Catholic priest, assumed the task of baptizing Epicurus by identifying the theolo-

This essay was written while I was an Annual Fellow at the Calgary Institute for the Humanities, to which I am grateful for providing release time for research and writing.

1 See, for example, Bernard Rochot, *Les travaux de Gassendi sur Épicure et sur l'atomisme* (Paris: Vrin, 1944); René Olivier Bloch, *La philosophie de Gassendi: Nominalisme, matérialisme, et métaphysique* (The Hague: Nijhoff, 1971); and Richard H. Popkin, "Gassendi, Pierre," in Paul Edwards, ed., *The Encyclopedia of Philosophy*, 8 vols. (New York: Macmillan and Free Press, 1967), 3:269–73.
2 For Gassendi's ethical and political views, see especially Lisa T. Sarasohn, "The Ethical and Political Philosophy of Pierre Gassendi," *Journal of the History of Philosophy*, 20 (1982), 239–60, and Lisa T. Sarasohn, "Motion and Morality: Pierre Gassendi, Thomas Hobbes, and the Mechanical World-view," *Journal of the History of Ideas*, 46 (1985), 363–80. See also Barry Brundell, *Pierre Gassendi: From Aristotelianism to a New Philosophy* (Dordrecht: Reidel, 1987); Marco Messeri, *Causa e spiegazione: La fisica di Pierre Gassendi* (Milan: Angeli, 1985); Lynn Sumida Joy, *Gassendi the Atomist: Advocate of History in an Age of Science* (Cambridge: Cambridge University Press, 1987); Howard Jones, *Pierre Gassendi (1592–1655): An Intellectual Biography* (Nieukoop: De Graaf, 1981); and Richard W. F. Kroll, "The Question of Locke's Relation to Gassendi," *Journal of the History of Ideas*, 45 (1984), 369–60.
3 For Epicurus' views on the gods and religion, see J. M. Rist, *Epicurus: An Introduction* (Cambridge: Cambridge University Press, 1972), pp. 140–63, 172–5.

gically objectionable elements in his philosophy and modifying
them accordingly.[4] For example, he denied the Epicurean doctrines of the eternity of the world, the infinitude of atoms, the
mortality of the human soul,[5] and the existence of the *clinamen*, or
swerve, which Epicurus had introduced in order to account for the
impact of atoms in an infinite universe and for the freedom of the
human will.[6]

Gassendi's *Syntagma philosophicum* assumes the form of a traditional "complete" philosophy in three parts, respectively entitled
"Logic," "Physics," and "Ethics." It contains a thorough exposition of philosophy and the history of philosophy.[7] Gassendi's
virulent anti-Aristotelianism, evident in his first published work,
the *Exercitationes paradoxicae adversus Aristoteleos* (1624), makes
it clear that one of his aims in the *Syntagma philosophicum* was
to produce a complete alternative to Aristotelianism in order to
provide foundations for the new science. He did so by endorsing
a Christianized version of Epicureanism, in which he intended to
discuss all of the traditional issues in philosophy – logical, natural,
and moral.[8]

4 Bloch provides the following inventory of the objectionable components of Epicureanism: polytheism; the corporeal nature of the gods; the absence of any providence; the
   denial of creation; the infinitude of atoms and their eternal existence; the mortality of all
   that is born; the negation of all incorporeal reality except the void; the infinity of the
   universe; the plurality of worlds; the claim that chance is the cause of the world; the
   natural character of the first cause; the negation of all finality in biological systems; and
   the materiality and mortality of the human soul. See Bloch, *Philosophie*, p. 300.
5 See Margaret J. Osler, "Baptizing Epicurean Atomism: Pierre Gassendi on the Immortality of the Soul," in Margaret J. Osler and Paul Lawrence Farber, eds., *Religion, Science,
   and Worldview: Essays in Honor of Richard S. Westfall* (Cambridge: Cambridge University
   Press, 1985), pp. 163–83.
6 Although there is evidence that Gassendi advocated the philosophy of Epicurus as early
   as 1628, he did not defend Epicureanism in print until the 1640s. During 1641 and 1642,
   he wrote an extensive series of letters to the new governor of Provence, Louis-
   Emmanuel de Valois. These letters contain a sketch of his later works on Epicureanism.
   Gassendi published two works on Epicureanism during his lifetime, *De vita et moribus
   Epicuri* (1647) and *Animadversiones in decimum librum Diogenes Laertii* (1649). Two other
   works appeared only in his posthumously published *Omnia opera*, 6 vols. (Lyon, 1658),
   the brief *Philosophiae Epicuri syntagma* and the massive *Syntagma philosophicum* (hereafter
   cited in text as *SP*, with volume and page number of the *Opera*). These works comprise
   the major sources available for analysis of Gassendi's restoration of Epicureanism. For
   the publication history of Gassendi's works, see Rochot, *Les travaux*, Chaps. 7–8.
7 For one interpretation of the historical aspect of Gassendi's work, see Joy, *Gassendi the
   Atomist*.
8 In a recent study, Barry Brundell demonstrates in detail the ways in which Gassendi
   replaced Aristotle's philosophy with that of Epicurus. See *Pierre Gassendi: From Aristotelianism to a New Natural Philosophy* (Dordrecht: Reidel, 1987).

If Gassendi's primary goal was to baptize Epicureanism, the theological framework within which he did so was voluntarist. Gassendi's voluntarism informed his philosophy of nature, his theory of knowledge,[9] and his ethics. The voluntarist strain underlying Gassendi's thought was the thread that bound his natural philosophy, theory of knowledge, and ethics into a coherent whole. It also provided him with grounds for choosing among the various ancient philosophies in his search for foundations for the new science. I shall argue in this chapter that his account of fate, fortune, and divination, in Book III of the "Ethics," can best be understood within the context of the voluntarist theology with which he wished to rehabilitate Epicurus.

Book III of the "Ethics," the last part of the *Syntagma philosophicum*, is entitled "On Liberty, Fortune, Fate, and Divination."[10] In this concluding section of his magnum opus, Gassendi discussed the major classical philosophies of nature and man in the context of freedom – human and divine. Gassendi was one of the last Renaissance humanists, at least in the domain of natural philosophy. Gassendi's humanism is evident in two principal ways: (1) his using ancient models, on the assumption that they are the great fundamental philosophical texts with which to work in order to formulate his own position; and (2) his style of writing, which is marked by frequent allusions to and quotations from classical authors.[11] It was the humanist streak in his intellectual approach that led him to seek a classical model for his philosophy. The

9 See Margaret J. Osler, "Providence and Divine Will in Gassendi's Views on Scientific Knowledge," *Journal of the History of Ideas*, 44 (1983), 549–60.

10 Significantly, Gassendi originally intended this section of the *Syntagma* to be placed in the "Physics," as the conclusion to his discussion about God's role in the universe. For the dating of the "Ethics" and for the history of this section, see Louise Tunick Sarasohn, "The Influence of Epicurean Philosophy on Seventeenth Century Ethical and Political Philosophy: The Moral Philosophy of Pierre Gassendi," Ph.D. diss., UCLA, 1979, Chap. 5.

11 I follow Kristeller, here, in defining Renaissance humanism, not in terms of a particular philosophical view, but in terms of the humanist regard for the authors of classical Greece and Rome. "It was the novel contribution of the humanists to add the firm belief that in order to write and to speak well it was necessary to study and to imitate the ancients." Paul Oscar Kristeller, *Renaissance Thought and Its Sources*, ed. Michael Mooney (New York: Columbia University Press, 1979), pp. 24–5. For a concise discussion of Renaissance humanism, see Paul Oskar Kristeller, "Humanism," in Charles B. Schmitt et al., eds., *The Cambridge History of Renaissance Philosophy* (Cambridge: Cambridge University Press, 1988), pp. 113–37. I am grateful to Letizia Panizza, who helped me to clarify my understanding of Renaissance humanism.

classical options available to him included Aristotle, Plato, the Stoics, and the Epicureans. His passionate anti-Aristotelianism ruled out using the Stagirite as his classical model. His choice among the remaining three great schools of antiquity was grounded on his theological assessment of their philosophies. In his discussion of fortune, fate, and divination, Gassendi employed explicitly theological criteria to choose among these classical options. He selected Epicurus because the ancient atomist's views were the ones that Gassendi believed could most easily be modified along voluntarist lines.

While questions about fate, fortune, and divination may, at first glance, appear rather remote from the primary concerns of seventeenth-century natural philosophy, in fact they involve metaphysical issues central to the articulation of the mechanical philosophy: the extent of contingency and necessity in the world, the nature of causality, and the role of providence and the extent of human freedom in a mechanical universe. Since antiquity, natural philosophers had dealt extensively with questions about fate, fortune, and divination. These issues had acquired a particular urgency with the rise of Christianity and the perceived need to reconcile natural philosophy with Christian theology.[12]

Gassendi's argument in Book III of the "Ethics" rests on two important principles: that a proper understanding of the world must include divine freedom, creation, and providence and that the possibility of ethics – moral choice and judgment – requires human freedom. The first of these principles embodies Gassendi's voluntarist account of God's relationship to the creation. His main expositions of this view appear in the "Physics," as part of his general discussion of "The Nature of Things Universally," and in the "Ethics," in the context of discussions of freedom, both human and divine (SP, 1:283–337 and 2:821–860).[13]

Gassendi unequivocally believed that God is the creator of the world and that he continues to rule it with "general providence

12 For the medieval and Renaissance background to Gassendi's discussion of these issues, see Antonino Poppi, "Fate, Fortune, Providence and Human Freedom," in Schmitt, *Cambridge History of Renaissance Philosophy*, pp. 641–67.

13 The "Ethics" was translated into English in the 17th century. See François Bernier, *Three Discourses of Happiness, Virtue, and, Liberty, collected from the works of the Learn'd Gassendus, translated out of French* (London: Awnsham & John Churchil, 1699).

and also special providence for humanity" (SP, 1:311).[14] God is radically distinct from his creation, and we cannot know him, since, lacking any imperfections, he is not like anything denoted by the words we might use to describe him (1:296–7). God's reasons are unknown to us (1:317–18). We do know one thing about God, however: He is free from any necessity or limits. "[T]here is nothing in the universe which God cannot destroy, nothing which he cannot produce; nothing which he cannot change, even into its opposite qualities" (1:308). Indeed, God can do anything, short of violating the law of noncontradiction (1:309).

Within these bounds, namely the stipulation that nothing God creates can impede his freedom, God makes use of second causes to carry out the ordinary course of nature.

It is his general providence which establishes the course of nature and permits it to be served continuously. From which it follows, as when either lightning or other wonderful effects are observed, God is not on that account suddenly summoned, as if he alone were its cause and nothing natural had intervened.... [H]owever, aside from him, particular [causes] are required which are ... not believed to be uncreated; but are believed to be hidden from our skill and understanding. (1:326)

Although God makes use of second causes, he can do without them if he pleases. Totally agreeing with one of the fundamental tenets of voluntarist theology, Gassendi maintained that God can dispense with second causes altogether if he chooses (1:317).[15] Moreover, God could have created an entirely different natural order if it had pleased him to do so (2:851).

The world, having been created by God, continues to depend upon him.

It depends no less certainly on its author than a light depends on its source; wherefore, as light cannot be observed without the sun from which it was created, the world cannot be preserved ... without God by

---

14 Lisa T. Sarasohn generously shared her translation of significant portions of Book III of Gassendi's "Ethics" with me. Translations are mine, unless otherwise indicated.

15 On this issue, Gassendi sounds much like Ockham, the paradigm exponent of voluntarist theology. "Whatever God can produce by means of secondary causes, He can directly produce and preserve without them." William of Ockham, *Ordinatio*, Q. i, N *sqq.*, in Ockham, *Philosophical Writings*, ed. and trans. Philotheus Boehner (Edinburgh: Nelson, 1957), p. 25.

whom it was produced.... And the world, which would be nothing without God, has nothing from itself whereby it could subsist on its own and stand without God.... [T]he world would be reduced to nothing if God were to cease supporting it. (1:323)

In other words, there are no immanent principles or reified natures or laws according to which the world might run autonomously. There is only God, on whom the individuals comprising the world absolutely depend. Consequently, there is no necessity in the creation which might limit God's freedom of will and action.

Truly he is free, since he neither is confined by anything nor imposes any laws on himself which he cannot violate if he pleases.... Therefore, God ... is the most free; and he is not bound as he can do whatever ... he wishes. (1:309)

If divine freedom provided the theological foundations for Gassendi's philosophy of nature, human freedom was the cornerstone of his ethics.[16] Moral choice and moral judgment depend, Gassendi argued, on the possibility of those choices and actions being made freely and deliberately. Actions done by accident or by necessity do not merit praise or blame (2:821).[17] Human freedom is an essential consequent of voluntarist theology, for if any human action were necessarily determined, there would be some element of necessity in the universe, a necessity which would restrict God's freedom.

Having laid the groundwork for his discussion – namely, a voluntarist conception of God's relationship to the natural world, and a concept of human nature incorporating free will – Gassendi proceeded with his analysis of the notions of fortune, fate, and divination. These concepts, central to the ancient philosophies which he played off against each other, appeared to challenge his Christian, voluntarist, providential view of God, nature, and man

16 Sarasohn, "Ethical and Political Philosophy of Gassendi," pp. 258–60.
17 Gassendi's concept of freedom, or *libertas*, is central to what follows, and he began by stating, "Freedom consists in indifference" (2:823.) That is, the will and intellect are said to be free if they are equally able to choose one or another of possible options and are not in any way determined to one or the other. Real freedom, understood as indifference, belongs only to beings that possess a rational intellect and stands in contrast to what Gassendi called "willingness," or *libentia*. Willingness characterizes the actions of boys, brutes, and stones, agents lacking the capacity for rational choice (2:822). Sarasohn discusses Gassendi's concepts of *libertas* and *libentia* in her "Ethical and Political Thought of Gassendi," p. 259.

by calling for the elimination of either creation, providence, and/or free will from their accounts of the world. By insisting on the primacy of his ethical and most especially his theological principles, Gassendi articulated a criterion for choosing among the alternative ancient philosophies.

The concept of fortune or chance had played a central role in the philosophy of Epicurus, both in his physics and in his ethics. According to Gassendi, Epicurus and his followers had sometimes written as if fortune or chance were a kind of cause.[18] Gassendi denied their claim, because there could be no autonomous causes in the world other than God.

Gassendi began his discussion of fortune or chance by adopting a working definition: Fortune is an unexpected consequence, a cause by accident. To illustrate his meaning, he cited the classic example of a man who discovers a treasure while digging in the ground to plant a tree. Finding a treasure is a totally unexpected consequence of his act. While the digging preceded the discovery, it was not the cause of the discovery, except accidentally. The discovery of treasures is not the usual or natural outcome of digging in the ground. Such unexpected consequences are called "fortuitous" in connection with agents that act freely; they are called "chance" in connection with inanimate objects (2:828).

According to Gassendi, fortune and chance both are expressions of contingency in the world. "Chance" is the name given to the kind of contingency which describes an event that may happen or may not happen in the future. An event which is said to be caused by chance or fortune is one which results from the unexpected concourse of several apparently unrelated causes. In the case of the unexpected discovery of the treasure, there is the concourse of the original burial of the treasure with the present digging in the ground (2:828). Each event is the perfectly natural outcome of a series of causes. The two series are unrelated, however, and so their concourse is unexpected: Therein lies the element of chance or fortune. "Fortune is truly nothing in itself . . . , only the negation of foreknowledge and of the intention of the events" (2:829). Thus there is nothing intrinsically mysterious about fortune. Certain misunderstandings make the concept of fortune problematic. Those who call fortune divine – a position which even

18 For the views of the Epicureans, see Rist, *Epicurus*, p. 51.

Epicurus had rejected – are ignorant of the causes of the events in question (2:829). Epicurus had, however, equated fortune with chance and denied that "there is divine wisdom in the world" (2:830). He had thus compared life to a game of dice. In this regard he had erred, according to Gassendi, because he had failed to appreciate that divine providence touches every aspect of nature and of human life (2:830–1). Fortune and chance, understood as unanticipated outcomes of unexpected concourses of causes, could easily be incorporated into an orthodox philosophy of nature by including divine providence and divine will among the class of "causes." Opposing Epicurean materialism and emphasizing the limits of human knowledge, Gassendi believed he could still reinterpret one of the important components of the atomic philosophy in a theologically suitable fashion.

If fortune and chance were concepts Gassendi encountered in the philosophy of Epicurus, fate was a concept which had played a central role in Stoicism. Where fortune and chance raised questions about the nature of contingency and causality in the universe, fate pointed to the complex problem of free will and determinism. "Fate," for the Stoic philosophers, "is a word for describing (quite neutrally) the state of affairs that was, is, and will be."[19] It describes the causal nexus of a deterministic universe.[20] Some understandings of fate, as Gassendi interpreted or misinterpreted them, seemed to incorporate an element of necessity which would obstruct both divine and human freedom. It is this necessitarian interpretation of the notion of fate to which Gassendi was primarily opposed.[21]

19 J. M. Rist, *The Stoic Philosophy* (Cambridge: Cambridge University Press, 1969), p. 122. Cicero discussed the concept of fate extensively, and Gassendi frequently cited his discussion of this and related topics. See Cicero, *De fato* (Loeb Classical Library, 1960), and *De natura deorum* (Loeb Classical Library, 1956). Both of these works center on debates between Stoics and Epicureans.
20 "Since the entire universe is governed by the divine *logos*, since, indeed, the universe is identical with the divine *logos*, then the universe, by definition, must be reasonable. The *logos* organizes all things according to the rational laws of nature, in which all events are bound by strict rules of cause and effect. Chance and accident have no place in the Stoic system. The causal nexus in the universe is identified with both fate and providence; fate, in turn, is rationalized and identified with the good will of the deity." Marcia L. Colish, *The Stoic Tradition from Antiquity to the Early Middle Ages*, 2 vols. (Leiden: Brill, 1985), 1:31–2.
21 See Josiah B. Gould, "The Stoic Conception of Fate," *Journal of the History of Ideas*, 35, (1974), 17–32.

At the beginning of his discussion, Gassendi observed that there are two chief views about fate: that it is something divine and that it is merely natural. Among those who regarded fate as something divine, Gassendi counted the Platonists and the Stoics; among those who considered it to be nothing more than natural necessity, he cited the advocate of atomism and hard determinism, Democritus (SP, 2:830–2). For the former group, fate had been defined as "the eternal God or that reason, which disposes all things from eternal time, and thus binds causes to causes, so that they affect from their series whatever they affect" (2:830). Thus Plato had sometimes considered fate to be part of the soul of the world, sometimes "the eternal reason and law of the universe." Likewise, the Stoics Zeno and Chrysippus had defined fate as "the motive force of matter, and the spiritual force and governing reason of the order of the universe." Seneca had gone so far as to identify fate with the god Jove (2:830). Despite the apparently theological and providential components of this Stoic interpretation of fate, Gassendi found their necessitarianism objectionable

because they are especially seen to defend what Seneca said, "The necessity of all things and actions which no force breaks." For, he says, "they act according to their law nor are they moved by any entreaty, they are not bent by any misery or grace; they follow an irrevocable course, they flow from destiny" . . . , indeed this necessity seems to be of such a kind that it completely removes the liberty of all human action and leaves nothing in our judgment (since what would be left would not happen by necessity, and something could happen outside of Fate or Decree). (2:831)

Fate, in this necessitarian sense, would undermine the basic principles on which Gassendi's universe rested. Since it denies free will, there would be no meaning in life. There would be no place for plans or prudence or wisdom, since everything would happen according to fate, "and all legislators would be either fools or tyrants, since they would command things which always were to be done, or what we absolutely cannot do, by which name all commands and exhortations are ridiculous and superfluous" (2:831–2). No action would be subject to moral praise or censure, since no action could result from a free act. All contingency in the universe would be eliminated, and all divination, prayer, and sacrifice would be rendered useless (2:832). The Stoic view, which identified fate with divinity, introduced into nature an inexorable necessity which destroys the possibility of human freedom.

Among those who considered fate to be something merely natural, Gassendi distinguished two groups of thinkers: those, like Democritus, who had interpreted fate in terms of natural necessity; and those, like Aristotle and Epicurus, who had taken a more moderate view. Democritus's conception of fate was similar to that of the Stoics, shorn, however, of any remnant of theology. For Gassendi, Democritus's version of atomism was the paradigm example of materialistic, reductionist, necessitarian determinism. "Democritus taught . . . that Necessity is nothing other than . . . the motion, impact, and rebounding of matter, that is, of atoms, which are the matter of all things. Whence it can be understood that 'Material Necessity' is the cause of all things which happen" (2:834). Since for Democritus everything, including the human soul, is composed of atoms, there is no room for real freedom in the universe he described (2:835). Not only freedom but also error would be impossible in such a world (2:834). There would be no room for divine providence if everything were necessarily determined by the motions and collisions of atoms (2:840).

Gassendi found a more moderate account of fate in the writings of Aristotle and Epicurus, who had understood this concept in terms of natural, but not inevitable, necessity. In contrast to Democritus, who had maintained absolute necessity, inasmuch as nothing could impede a cause from producing its effect, Epicurus had believed that the necessity in nature is not absolute (2:837). Rather, Epicurus had maintained, there are three ways in which things happen: by necessity, by plan, and by fortune (2:837). Like Aristotle, Epicurus had identified the necessary with the natural, rather than with the unchangeable, the inevitable, and the absolute (2:836–7). Second, Epicurus had made the logical point that it is impossible simultaneously to hold that all statements are either true or false and that there is absolute necessity in nature, because to do so would entail giving truth-value to statements about future contingents, something he regarded as impossible. Following Aristotle, in his famous discussion of tomorrow's naval battle (*De interpretatione* IX. 19a30–33), Epicurus "admitted this complex as truth, 'Either Hermachus will be alive tomorrow or he will not be alive.'" But Epicurus could not accept the possibility that either one of the disjuncts – "it is necessary that Hermachus be alive tomorrow" or "it is necessary that Hermachus not be alive tomorrow" – could be true. For "there is no such necessity in nature" (2:837).[22] Evidently Epicurus believed that necessity could only apply to statements about the past and present: The events described have already occurred or not occurred, and so statements describing them have a determined truth-value. On the other hand, statements about the future have an undetermined truth-value, so we cannot reason about them with necessity.[23] Therefore, since we cannot have knowledge of future contingents, it follows that there is no such necessity in nature.[24]

As for events that occur by plan or by fortune, these involve human free will, which Epicurus had tried to preserve by adding the *clinamen*, or random swerve, to the Democritean atoms. Indeed, according to Gassendi, Epicurus had introduced the swerve with the explicit intention "that it would shatter the necessity of fate and thus ensure the liberty of souls" (SP, 2:837).[25] Even though Epicurus had believed that all events in the natural world can be explained by the motions and collisions of atoms and even though he had maintained a materialistic conception of the soul, the addition of the swerve ostensibly imparted a certain unpredictability to natural events and freedom to human actions.

Despite the fact that Gassendi praised Epicurus for attempting to secure the freedom of the will, nevertheless the ancient atomist was not entirely free from criticism. In Gassendi's eyes, he had

22 Cicero discussed Epicurus' ideas about future contingents at some length in *De fato* XVI. 36–8.

23 The concept of necessity employed by Epicurus is not the same as that of twentieth-century modal logic. It was, however, common to the ancient and medieval discussions of the sticky issue of future contingents. See Calvin Normore, "Future Contingents," in Norman Kretzmann, Anthony Kenny, and Jan Pinborg, eds., *The Cambridge History of Later Medieval Philosophy* (Cambridge: Cambridge University Press, 1982), pp. 258–381. See also the introduction to William Ockham, *Predestination, God's Foreknowledge, and Future Contingents*, ed. and trans. Marilyn McCord Adams and Norman Kretzmann (New York: Appleton-Century-Crofts, 1969), pp. 1–33.

24 Similar ideas can be found among some of the Stoics. See Rist, *Stoic Philosophy*, p. 122.

25 Gassendi's analysis of the relationship between the swerve of atoms and free will is borne out by modern scholarship, although the main source for this doctrine appears to have been Lucretius, whom Gassendi cited extensively, rather than Epicurus. See A. A. Long, *Hellenistic Philosophy: Stoics, Epicureans, Sceptics*, 2nd ed. (Berkeley and Los Angeles: University of California Press, 1986), pp. 56–61; Rist, *Epicurus*, pp. 90–9.

failed to allow room for divine will, divine providence, and an immortal human soul. Moreover, the swerve of atoms does not really suffice to overcome the necessity of fate. According to Epicurus, the swerve accounts for voluntary motions, since they are ones "to which no region is determined, no time prefixed" (2:838). It is the unpredictability of the swerve which Epicurus had thought preserved free will. But, Gassendi objected, the swerve does not really overcome the necessity of fate, since events would still always happen by the same chain of necessary consequences. "What always happens by the same necessity would happen by a variety of motions, impacts, rebounds, swerves in a certain external series, like a chain of consequences" (2:838). Indeed, since Epicurus had regarded the soul as consisting of atoms, its choices would simply become part of the long causal sequence determining events in the material world.

Gassendi found all the traditional accounts of fate to be wanting, largely for theological reasons. The deterministic, reductionist atomism of Democritus left no room for either divine providence or free will:

Therefore, the opinion of Democritus must be exploded inasmuch as it can by no means stand with the principles of the Sacred Faith (indeed because of having removed from God the care and administration of things) and is thus manifestly repugnant to the light of nature by which we experience ourselves to be free. Certainly it is against him whom the most powerful objections ought to be made. (2:840)

Epicurus must be criticized as well, despite his valiant attempt to preserve free will, for by denying the possibility of knowing future contingents, he had denied God such knowledge. He "thus supposes that there is no creation of things and no divine providence" (2:840). Although the Stoics and Plato could be defended insofar as they affirmed divine creation, providence, and human free will, nevertheless they had reduced liberty to *libentia*, or willingness, and thus lacked an adequate concept of human freedom (2:846).

In order to be able to embrace the evident facts both of causal order and contingency within the bounds of his mechanical philosophy, Gassendi translated the classical notions into concepts which adequately reflected his commitment to divine and human freedom. He undertook to articulate a Christian reinterpretation of fate and fortune, thus providing a providential understanding of these concepts, just as Augustine had done centuries earlier.[26]

> Indeed, it can be said in one word, that to the extent that Fate can be defended, so can Fortune, if we will agree that Fate is the decree of the divine will, without which nothing at all is done; truly Fortune is the concourse of events which, although unforeseen by men, nevertheless were foreseen by God; and they are the connected series of causes or Fate. (2:840)

His reinterpretation of fortune and fate left plenty of room for the exercise of divine freedom and providence.

The fact that certain events appear to be fortuitous in no way impairs divine omniscience.

> From which you see, when the word "Fortune" was mentioned earlier, it indicated two things, the concourse of causes and the previous ignorance of events; Fortune can thus be admitted afterward with respect to man but not God; and on account of this first, nothing stands in the way of our saying that Fortune is a part not only of Fate, but also of divine providence, which foresees for man what he cannot foresee [for himself].[27] (2:840)

Given that all events are included in God's decree, how is it that some appear to be determined by a known sequence of causes, while others appear to be fortuitous? In an extended but questionable analogy between fate and the civil law, Gassendi explained that, just as the law refers to acts, such as theft or murder, which are not legal, so there are events which are contained in fate but are not fatally determined (2:841). For example, although the rising of a particular star results from a sequence of causes having its origin in divine decree, and an evening thunderstorm equally results from a similar series of causes, the simultaneous occurrence of the star's rising and the storm's breaking has not necessarily been decreed from the beginning. It is a fortuitous happening, occurring within the general framework of divine decree.

Reconciling fate with divine providence was not so difficult for

26 Vincenzo Cioffari, "Fate, Fortune, and Chance," in Philip P. Wiener, ed., *Dictionary of the History of Ideas*, 4 vols. (New York: Scribner, 1973), 2:230; Antonino Poppi, "Fate, Fortune, Providence, and Human Freedom," in Schmitt, *Cambridge History of Renaissance Philosophy*, p. 642.

27 See note 17, this chapter, for Gassendi's concepts of *libertas* and *libentia*.

Gassendi. A prima facie stickier problem was to reconcile fate with free will. Here again he turned to theology for his solution.

We call Fate, with respect to men, nothing other than that part of Divine Providence which is called Predestination by theologians . . . in order that predestination and thus Fate can be reconciled with liberty. . . . Theology teaches that God produced necessary and free causes, and thus both are subject to divine providence. (2:841)

Bringing in the vexatious doctrine of predestination seems an odd way to clarify anything. Indeed, Gassendi acknowledged that the problem of reconciling predestination and divine foreknowledge with human freedom had troubled both philosophers and theologians since antiquity. He stated the theological problem as follows:

Either God knew definitely and certainly that Peter would deny Christ, or he did not know. It cannot be said that he did not know, because he predicted it, and he is not a liar: and unless he knew he would be neither omniscient nor God. Therefore, he knew it definitely and certainly. Therefore it could not be that Peter would not deny. If God knew and Peter did not deny . . . , it would be argued of God that his foreknowledge was false and that he was a liar. If Peter cannot deny, then he is not free to deny or not deny. Therefore he is without freedom. (2:841)

In order to resolve this difficulty, Gassendi invoked the traditional Scholastic distinction between absolute necessity and necessity by supposition.

For example, that double two is four or that yesterday comes before today is absolutely necessary, although that you lay the foundations of your house or leave the city is not necessary: Nevertheless if you suppose that you will build your house or that you will be in the country, then for you to lay the foundation or leave the city is, I say, necessary from supposition. Since truly it is manifest from this distinction, that absolute necessity hinders that by which a certain action is elicited, however that which is from supposition does not hinder (for he who will lay down a foundation absolutely can not lay it, and he who will leave the city can not leave), thus they are clearly understood. Peter's future denial was seen by God necessarily, but nevertheless by a necessity from supposition, because of which nothing of liberty is taken away. . . . [T]hus although it was determined from the beginning that he would deny him, he does it freely in whatever manner he did it; afterward, since he did it, it was necessary.[28] (2:841)

Gassendi's talk about the "absolute necessity" of mathematical truths here should not lead us to conclude that he abandoned his voluntarism. In his debate with Descartes, he defended an empiricist, probabilist account of the epistemological status of mathematics.[29] Gassendi's concept of necessity contains a temporal component. The act of denial does not become necessary, despite being foreseen by God, until Peter commits it. Once it has occurred, it is part of the past which cannot be undone.

If indeed when it is said that Peter denied necessarily, this necessity is understood not as something that was truly in Peter antecedently which forced him to act, but only now that it is in this time which is as in the past and cannot not be past, thus the thing which is done by him is done in whatever way and cannot be not done by him. (2:841)

Necessity of this kind in no way impinges on God's freedom and omniscience.

Moreover, since God is omniscient, he foresaw that Peter would deny; yet the foreknowledge of the denial followed the foreknowledge of free determination, and thus he foresaw only that Peter would deny, since he foresaw that Peter would freely determine himself to denying. Whence it is, what can be said, that Peter did not deny because God foresaw it, but thus God foresaw since Peter would deny. (2:841–2)

God's foreknowledge of events does not cause those events to happen, but he has foreknowledge because they will happen. Since some of those events are the acts of free agents, there is no contradiction between God's foreknowledge and human freedom.[30] Gassendi continued, "even if everything is contained in Fate, nevertheless it is not all done by Fate, ... [such as those things] which happen contingently or freely and fortuitously" (2:844). That is to say, even if everything is included within the

28 See also Pierre Gassendi, *Disquisitio metaphysica seu dubitationes et instantiae adversus Renati Cartesii metaphysicam et responsa,* edited and translated into French by Bernard Rochot, *Recherches métaphysiques, ou doutes et instances contre la métaphysique de R. Descartes et ses réponses* (Paris: Vrin, 1962), pp. 470–2. English translation mine.

29 Gassendi, *Disquisitio,* Rochot edition, pp. 468–73.

30 Gassendi addressed at some length the question of how to interpret the doctrine of predestination. He rejected the view that the members of the elect and the reprobate had

domain of divine decree, that inclusion does not eliminate human freedom. God created free agents, as well as determined ones.

If discussions about fate and fortune really concerned the roles of contingency and necessity in the universe, divination raised questions about the nature of causality in the world. Divination had played a central role in Stoic thought, where it had been invoked to provide ostensibly empirical evidence for a deterministic account of the universe.[31] While Gassendi rejected Stoic fatalism and the more recent naturalism associated with it, he defended certain forms of divination on theological grounds.[32] In the final chapter of Book III of the "Ethics," entitled "The Meaning of Divination, or the Foreknowledge of Future and Merely Fortuitous Things," Gassendi explicitly opposed Epicurus, whose blanket denial of the possibility that knowledge of future contingents might be compatible with human freedom had led him to reject the possibility of any kind of divination. That this view is false, Gassendi argued, can be demonstrated by the fact that the biblical prophecies proved true (2:847). Moreover, some ancient views about divination had invoked demons as causal agents to explain how divination works. Despite the fact that other ancient

been chosen from eternity. Instead, he opted for the more liberal position, namely, "God knows it from eternity, but consequently for his decree, and he does not make his decree without foreseeing what you will do. For which reason, your act of will precedes, in God's foresight and decree, your predestination or condemnation, divine foreknowledge of your perpetual happiness or unhappiness, not because these antecedent causes and consequences are temporary but because we understand them in a human way.... Meanwhile it can further be inferred that there is here nothing preceding the will because of which the will is not free" (SP, 2:844).

31 S. Sambursky, *Physics of the Stoics* (New York: Macmillan, 1959), p. 66.
32 The immediate context of Gassendi's concern with divination, as well as that of his contemporaries Mersenne, Naudé, la Mothe le Vayer et al., was its notoriously naturalistic treatment in Pietro Pomponazzi's (1462–1525) *De naturalium effectum admirandorum causis sive de incantationibus* (1556) and *De fato, de libero arbitrio et de praedestinatione* (1520). Pomponazzi sought to undermine Christian doctrine philosophically in favor of a mechanistic, Stoic metaphysics and ethics. The reaction to him was exacerbated by his role in debates about the immortality of the soul. See Antonino Poppi, "Fate, Fortune, Providence, and Human Freedom," in Schmitt, *Cambridge History of Renaissance Philosophy*, pp. 653–60, and Eckhard Kessler, "The Intellective Soul," in ibid., pp. 500–7. See also Paul Oskar Kristeller, *Eight Philosophers of the Renaissance* (Stanford: Stanford University Press, 1964), Chap. 5; Jean Céard, "Matérialisme et théorie de l'âme dans la pensée padouane: "Le *Traité de l'immortalité de l'âme* de Pompanazzi," *Revue Philosophique de France et L'étranger*, 171 (1981), 25–48; Bloch, *La philosophie de Gassendi*, pp. 310–11; and Edward John Kearns, *Ideas in Seventeenth-Century France* (New York: St. Martin's, 1979), pp. 14–15.

philosophers shared Epicurus' rejection of divination, "the opinion of Epicurus is the least reprehensible, because he mocked the credulity and superstition of the Gentiles as much about divination as about demons; . . . at least he did not assert that demons generally exist" (2:848).[33] Epicurus' views were thus less unacceptable than the alternatives.

Demons concerned Gassendi, because the question of their existence bore on the deeper question of the nature of causality in the world. Rehearsing and rejecting various ancient doctrines about demons – that they are particles of the *anima mundi*, that they have a corporeal nature, that they are halfway between men and gods, that they move the heavenly spheres, that they are of some particular number or another (2:849–51) – Gassendi concluded that "whether this or another interpretation of the philosophers would be true, this at least holds, that demons were admitted by the philosophers" (2:851). The problem with all of these views, from Gassendi's standpoint, is that they removed divine activity from the ordinary workings of the world. "They judged that it was alien to the divine majesty to care for all particulars himself and without any servant," (2:851), thereby impugning divine power by implying that God uses ministers to carry out his will because of some defect in his nature. Gassendi countered by saying that "God uses ministers, not because of disgrace, impotence, or need, but because he wished it for the state of things which is the world. He judged it congruous. If he had wished to institute another order, he would not have done a disgraceful thing, nor would it testify to any impotence or need" (2:851). Unlike the highest prince in his realm, to whom the philosophers had compared him, God is present everywhere in the world, not just to the ministers he has designated. The philosophers had mistakenly substituted the activity of these demons for God's general providence. They had even invoked demons to supplant "special providence, or that which is with respect to man" (2:851–2).

In fact, Gassendi believed that there do exist various orders of angels and demons in the universe, an opinion he considered to be supported by "sacred Scripture and . . . explained by theologians."

33 In many ways Gassendi's chapter on divination can be understood as a commentary on Cicero's *De divinatione*, which Gassendi cites extensively. This book of Cicero's along with *De fato* and *De natura deorum*, provides a thorough and interesting account of the debate between Epicureans and Stoics in classical times.

(2:851). But there are also many false superstitions about the activity of these creatures, exploits which are exaggerated by the poets. There are

many little stories with which your ears are thickly filled; from which frequently, if you take away the fraud of impostors, the tricks of the crafty, the nonsense of old women, the easy credulity of the common people, something difficult is true which you will discover. It seems also that it must be said about this kind of filthy magic, by which the unhappy kind thinks himself carried away by he-goats . . . and afterward put to sleep by narcotic salves, these unfortunates would dream with a most vivid imagination that they were present in a most evil assemblage. (2:852–3)

Although such temptations and possessions do exist – Scripture, the lives of the saints, and the successful practice of exorcism attest to that fact – the point is to attend to our own spiritual and moral state, our relationship to God by virtue of his special providence, rather than to excuse our sins by pointing to evil demons (2:852). A proper understanding of demons had not been available to the ancient philosophers, who did not possess either the true faith or sacred Scripture.

Gassendi had embarked on this long discussion of demons because some ancient advocates of divination had appealed to demons to explain their practices.[34] Since it is sometimes possible to predict the future, as Scripture attests, one must consider "whether the prediction was made by the intervention of demons or the craftiness of the soothsayers or the credulity of those who asked for it" (2:853). Although there are genuine cases of prophecy, many predictions are made of things which have natural causes "incapable of impediments, such as eclipses, risings of the stars, and other things of this kind, which depend on the determined disposition and constancy of the motions of the heavenly bodies" (2:853). Considering this issue within the traditional Stoic doctrine that there are two kinds of divination, Gassendi maintained that genuine divination either depends on art – astrology or the ancient interpretation of signs such as the flight, song, and feeding of birds, or the casting of lots, or the interpretation of

34 It should be noted that the most important philosophical account of divination came from the Stoics, whose account was thoroughly materialistic, owing nothing to the personal agency of demons. See Sambursky, *Physics of the Stoics*, pp. 66–71.

dreams – or it does not.[35] In the closest approximation to a joke in the ponderous *Syntagma philosophicum*, Gassendi railed against "geomancers, hydromancers, aeromancers, pyromancers, and others, . . . and last those astromancers or astrologers who . . . seek foreknowledge from the stars" – all of them practitioners of "artificial divination" (2:854).[36] If astrology, which holds the principal place among the arts of divination, is "inane and futile, the others ought to be no less inane and futile" (2:854).[37]

Gassendi thus denied that divination by art, the kind the Stoics valued most, is divination at all, because it is nothing but the observation of regular sequences of events, whether or not we understand the causes of those sequences. In fact, Gassendi argued, any genuine divination would involve there being events which do not have causes (2:853). Otherwise, nothing more would be involved in divination than the kind of conjectures used in any of the sciences which make predictions about future events. In all conjectural knowledge, we attend to the known causes of events and predict what will probably happen. Such predictions are conjectural, based on reasoning about our observed knowledge of the world. Divining is no different from this kind of conjectural science, except that it is frequently deficient "of ratiocination and consultation" (2:855). The Stoics had agreed with this reasoning but had identified inductive methods with divination by art, which is based on an empirical understanding of the deterministic nexus of the world.[38]

---

35 Ibid. For the Ciceronian roots of this distinction, see *De divinatione*, I.12, 24, 72–92, and II.26.

36 Apparently this list of various sorts of diviners had a long history. Isidore of Seville wrote as follows in his *Etymologies*: "Varro dicit divinationis quattor esse genera, terram, aquam, aerem et ignem. Hinc geomantiam, hydromantiam, aeromantiam, pyromantiam dictam" (Varro said that there are four kinds of divination: earth, water, air, and fire. Hence they are called geomancy, hydromancy, aeromancy, and pyromancy) (Isidore of Seville, *Etymologies*, ed. W. M. Lindsay [Oxford: Clarendon Press, 1911], VIII.9, line 13.) I am grateful to Haijo Westra for bringing this point to my attention.

37 For a full account of Gassendi's rejection of astrology, see his "Physics," sec. 2, BK. 6, "De effectibus siderium," in *Opera omnia*, 1:713–52. This part of the *Syntagma* was translated into English in the seventeenth century. See Petrus Gassendus, *The Vanity of Judiciary Astrology*, or *Divination by the Stars* (London: Humphrey Moseley, 1659). For the context of this polemic, see Jacques E. Halbronn, "The Revealing Process of Translation and Criticism," in *Astrology, Science, and Society: Historical Essays* (Wolfeboro, N.H.: Boydell Press, 1987), pp. 197–217.

38 Sambursky, *Physics of the Stoics*, p. 67.

Gassendi's discussion of fortune, fate, and divination, in Book III of the "Ethics," was part of his articulation of the ultimate terms of explanation of the mechanical philosophy of nature which he developed as a conceptual framework to replace the Aristotelian foundations for science. In this part of his work, he dealt extensively with the concepts fundamental to describing the causal order of the world: necessity, contingency, and causality. His commitment to voluntarist theology and to human freedom provided criteria for his choice of Epicurus, from among the various classical philosophers, to serve as his model. Although Epicureanism suffered from a number of primarily theological flaws in Gassendi's eyes, his philosophy could be reworked along Christian, and more specifically voluntarist, lines, as exemplified in Gassendi's analysis of fortune, fate, and divination. This final section of Gassendi's massive *Syntagma philosophicum* illuminates the goals and methods embodied throughout his major philosophical endeavor.

9

Epicureanism and the creation of a privatist ethic in early seventeenth-century France

LISA TUNICK SARASOHN

By the seventeenth century in France, as Nannerl Keohane eloquently argues, the ideal of civic responsibility and commitment characteristic of the later Renaissance was being replaced by an advocacy of the private life. This trend was particularly strong among the libertines, the French freethinkers who challenged traditional intellectual authorities, who "took seriously the Epicurean counsel to avoid the business of the world. They conformed to its demands as far as necessary, and found their true pleasure in the company of friends."[1]

The most prominent Epicurean in France, Pierre Gassendi, joined his libertine friends in believing that philosophy could flourish only if it were free from the restraints of both the world and authority. To Gassendi, living a retired life freed the wise man from the time-consuming demands of the public life and at the same time allowed him to philosophize freely, obligated to no one but himself. In 1637, Gassendi wrote a remarkable letter to Galileo, telling the Italian to regard his house arrest as a "most desired and fortunate retirement" and that "wise men desire nothing more than to be removed from the turbulence of the court and the tumults of the city, and ... from the profane crowd, which like a many-headed beast knows nothing human, and hopes for nothing but lying, invidiousness, perfidy, and other such things."[2]

---

1 Nannerl O. Keohane, *Philosophy and the State in France: The Renaissance to the Enlightenment* (Princeton, 1980), p. 146.
2 Gassendi to Galileo, October 1637 in Gassendi, *Opera omnia*, 6 vols. (Lyon, 1658), 6:95. All translations are by the author unless otherwise indicated.

As early as his first published work, an anti-Aristotelian tract entitled the *Exercitationes paradoxicae adversus Aristoteleos* (Grenoble, 1624), Gassendi had indicated that the moral teachings of Epicurus held a special fascination for him, particularly the Epicurean principle that pleasure, or the tranquillity of the soul, is the highest good.[3] Gassendi was also interested in atomism, the cosmological underpinning of Epicurus' system, probably because of its congruence with the new scientific model of the universe that Gassendi endorsed.[4] By 1628, Gassendi was embarked upon a lifelong project of vindicating and rehabilitating Epicureanism, which, in the course of several works, he gradually transformed into a philosophy of his own.[5] This achievement made Gassendi a major intellectual figure in his own time, widely considered the peer of René Descartes.[6]

Epicureanism was the most challenging and difficult ancient philosophy that Gassendi, a Catholic priest, could have chosen to rehabilitate.[7] While the physical system was open to Christian rearrangement by introducing God as the creator and sustainer of

3 In the preface, describing his plan for the *Exercitationes* (never actually completed), Gassendi writes, "Finally, Book VII deals with moral philosophy.... In one word, it teaches Epicurus' doctrine of pleasure by showing in what way the greatest good consists of pleasure and how the reward of human deeds and virtues is based upon this principle." Quoted in Craig B. Brush, ed. and trans., *The Selected Works of Pierre Gassendi* (New York, 1972), p. 25.
4 There were many reasons – scientific, ethical, and personal – for Gassendi's attraction to Epicureanism. Several authors have suggested that Gassendi's rejection of Aristotelianism, which he was forced to teach while a professor of philosophy at Aix from 1616 to 1622, encouraged him to look for an alternative natural philosophy. See, for example, Barry Brundell, *Pierre Gassendi: From Aristotelianism to a New Natural Philosophy* (Dordrecht, 1987), pp. 48–54; Bernard Rochot, *Les travaux de Gassendi sur Épicure et sur l'atomisme, 1619–1658* (Paris, 1944), pp. 36–8; and my unpublished dissertation, "The influence of Epicurean Philosophy on Seventeenth Century Ethical and Political Thought: The Moral Philosophy of Pierre Gassendi," UCLA, 1979.
5 Gassendi's Epicurean works are the following: *De vita et moribus Epicuri libri octo* (Lyon, 1647); *Animadversiones in decimum librum Diogenes Laertii, qui est de Vita, moribus, placitisque Epicuri* (Lyon, 1649); *Syntagma philosophiae Epicuri* (Lyon, 1649); *Syntagma philosophicum*, in *Opera omnia*, 6 vols. (Lyon, 1658), vols. 1–2.
6 Gassendi wrote two polemics against Descartes's *Meditations*: the "Fifth Objections," appended to the *Meditations*, and the *Disquisitio metaphysica* (Amsterdam, 1644), a massive critique of Cartesian philosophy, written in response to Descartes's reply to the "Fifth Objections." Gassendi maintained the empiricist point of view against Descartes's rationalism.
7 Epicureanism had been considered anti-Christian since antiquity. The church fathers had vehemently attacked Epicurus' denial of divine providence and the immortality of the soul. During the Middle Ages, Epicureanism was regarded largely as a synonym for

the atoms, the ethical teachings were much more resistant to such amelioration.[8] The central doctrine that pleasure is the highest good had often been attacked in antiquity, and Epicurus' social philosophy, in particular, had been especially harshly attacked: Epicurus had taught that "we must release ourselves from the prison of affairs and politics"; that "the most unalloyed source of protection from men . . . is in fact the immunity which results from a quiet life and the retirement from the world"; and, finally that "the greatest fruit of self-sufficiency is freedom."[9]

Cicero was typical in his reaction, asserting that Epicurus denied both civic responsibility and human fellowship: "Justice totters or rather I should say, lies already prostrate, so also with all those virtues which are discernible in social life and the fellowship of human society."[10] The second-century A.D. Stoic philosopher Epictetus went even farther: "In the name of God, I ask you, can you imagine an Epicurean state? . . . Your doctrines are bad, subversive to the state, destructive to the family, not even fit for women. Drop these doctrines, man. You live in an imperial state; it is your duty to hold office."[11]

In the "Ethics" of his *Syntagma philosophicum*, Gassendi defended his ancient teacher against the charge of deserting public responsibility for the sake of individual pleasure: "Epicurus did not say this [i.e., that one should live the retired life] absolutely, but only on the condition, 'unless something intervenes.'"[12]

debauchery and atheism, until the Italian humanist Poggio rediscovered Lucretius in 1417. For the history of Epicureanism in the Middle Ages, see Kurt Lasswitz, *Geschichte der Atomistik von Mittalter bis Newton*, 2 vols. (Darmstadt, 1963), vol. 1. Pietro Redondi has recently argued, in his controversial work *Galileo Heretic*, trans. Raymond Rosenthal (Princeton, 1987), that the secret agenda behind Galileo's condemnation of Copernicanism was an effort to protect himself from the more serious charge of heresy, stemming from his atomistic ideas. Atomism, with its denial of Aristotelian substantial forms and accidents, was considered by some members of the Church a direct and pernicious threat to the orthodox theory of the Eucharist and transubstantiation. While Redondi makes a most interesting argument, it is difficult to understand why, if this threat were real, Gassendi, who was a much more recognized atomist, experienced no difficulty whatsoever with the Church during the course of his career. This question deserves further exploration.

8 Gassendi's rehabilitation of Epicurean atomism is found in the "Physics," in *Opera omnia*, vol. 1.

9 Epicurus, *The Extant Remains*, trans. Cyril Bailey (Oxford, 1926), pp. 98–9, 115, 119.

10 Cicero, *De officiis*, trans. W. Miller (London: 1938), pp. 399ff.

11 Epictetus, *The Discourses* (New York, 1928), II.55.

12 Gassendi, *Syntagma philosophicum*, in *Opera omnia*, 2:762.

Thus, if danger threatens one's country, a wise man will fight to defend it, but if no danger exists, it is foolish for a man who is not brave to become a soldier. Furthermore, he writes, "Epicurus ought not to be charged with shame, because he established a kind of life more innocent and more peaceful; and there was a society of men, who, having laid aside ambition, pursued tranquillity. This assuredly always may be permitted without shame, and especially it may be permitted in our own time."[13] The private life is "a calm and useful condition" which results in "the quiet of the soul and the freedom to study."[14] In short, Gassendi endorses the Epicurean maxim that it is better "to live apart," enjoying a contemplative and happy life.[15]

Gassendi's defense of the private, contemplative life was itself part of a long debate on this subject. In antiquity, Epicurus had responded to the active political ideals of Plato and Aristotle, just as Cicero and the Stoics reacted to Epicureanism and other Hellenistic philosophies that advocated individualist rather than corporate values. This debate over the relative value of the private and public life continued in the early Middle Ages in the discussion of the relative worth of the secular and regular clergy, or whether the cenobitic or eremetical life was closest to the Christian ideal.[16] In the later Middle Ages, Cicero, whose works were only partly known, was seen as the example par excellence of the sage who advocated the contemplative life, free from the business of the world – although some Scholastics in the thirteenth and fourteenth centuries embraced ideals of active involvement in the republic, just as new monastic orders, like the Franciscans and Dominicans, left the cloister for a life of Christian service in the world.[17]

Gassendi was familiar with both the ancient and medieval forms of this debate, and he was clearly influenced by the Renaissance humanists' emphatic discussion of these themes. As Hans Baron has demonstrated, Petrarch rediscovered, to his horror, the public Cicero, whom he castigated in his letters and treatises, crying out, "Why did you involve yourself in so many contentious and useless

---

13 Ibid., p. 764.   14 Gassendi to Valois, in *Opera omnia*, 6:105.
15 Gassendi, *Syntagma philosophicum*, in *Opera omnia*, 2:717–19.
16 Gerhart B. Ladner, *The Idea of Reform* (New York, 1967), pp. 341–5.
17 For Cicero's role in the Middle Ages and the Renaissance, see Hans Baron, *The Crisis of the Early Italian Renaissance* (Princeton: 1955, 1966), pp. 121–4, and Quentin Skinner, *The Foundations of Modern Political Thought*, 2 vols. (Cambridge, U.K., 1978), 1:54–7.

quarrels and forsake the calm so becoming to your age, your position and the vicissitudes of your life? What vain splendors of fame drove you ... into a death unworthy of a sage?"[18] In contrast to Cicero's avid republicanism, Petrarch admired the "monarchism" established by Caesar, which allowed the scholar to pursue a contemplative life under the protection of the ruler.[19] At the end of the fourteenth century, however, with the rise of civic humanism in Florence, humanists like Salutati and Bruni rejected Petrarch's endorsement of the *vita contemplativa* and instead advocated the *vita activa civile*, which they linked with the elevation of the active will or voluntarism over the passive intellect.[20] According to Skinner, these humanists were also reacting to the uselessness of Scholastic philosophy as a practical guide to life: This reaction explains "the growing belief that a life devoted to pure leisure and contemplation (*otium*) is far less likely to be of value – or even to foster wisdom – than a life in which the pursuit of useful activity (*negotium*) is most highly prized."[21] This is the terminology, if not the conclusion, that Gassendi was to use in pursuing the debate.

In yet another exchange in this centuries-long dialogue, later humanists, such as Pico and Ficino, who were attracted to Neoplatonic intellectualism, once again rejected civic participation for the joys of contemplative solitude, perhaps reacting to the absolute triumph of the *signori* in the mid-fifteenth century.[22] Gassendi cited Pico as one of the people he admired most among the moderns, along with the northern humanists Vives, Montaigne, Lipsius, and Charron, the last three of whom continued the debate between *otium* and *negotium*.[23] According to Keohane, after the

---

18 Quoted in Baron, *Crisis*, p. 122.

19 Ibid., p. 123, and Skinner, *Foundations*, pp. 54–7.

20 Eugene F. Rice, *The Renaissance Idea of Wisdom* (Cambridge, Mass., 1958), pp. 30–48, and Anthony Levi, *French Moralists* (Oxford, 1964), pp. 40–51. Gassendi discussed the relationship between the intellect and the will in the "Ethics," where this theme is expressly connected with his political theories. On these ideas, see L. T. Sarasohn, "The Ethical and Political Philosophy of Pierre Gassendi," *Journal of the History of Philosophy*, 20 (1982), 239–59, and "Motion and Morality: Pierre Gassendi, Thomas Hobbes and the Mechanical World-View," *Journal of the History of Ideas*, 46 (1985), 363–79.

21 Skinner, *Foundations*, p. 108.   22 Ibid., pp. 115–16.

23 Gassendi names his intellectual influences in the "Letter to du Faur de Pibrac," quoted in Brush, ed., *Selected Works*, p. 5, and in the preface to Gassendi's *Exercitationes paradoxicae adversus Aristoteleos*, in *Opera omnia*, 3:99. See Skinner, *Foundations*, pp. 217–18, and Keohane, *Philosophy and the State*, pp. 129–43, on the influence of Montaigne, Lipsius, and Charron.

"all-persuasive" influence of Montaigne, it was the similarly eclectic and skeptical Charron who had the most influence on the
thinkers of the early seventeenth century in France. Charron's
"vision of the schizophrenic sage who finds the appropriate composition of his character in the strict separation of his internal and
external life, who walks through the world with equanimity but
as an alien, . . . was taken up eagerly by the libertines."[24] After
reading Charron, Gassendi was led to reiterate in a letter to a
friend the Lucretian sentiment

How sweet it is when the winds are raging on the sea to behold the great
travails of others from the shore! How I contemplate and how I laugh
at the artificial histrionics the entire universe presents! How I seem to
foresee petty quibblings before me which you hint at, for I am not
so ignorant, especially of the history of France, that I do not have a
presentiment where things are headed.[25]

Keohane suggests that the historical reasons for the rejection of
the active life were the recent scarring experience of the French
religious wars, the evolution of the politiques' attitude that it was
necessary to separate private conviction and public policy, and
the amorality of the absolute monarchy, which contributed to the
"crisis of authority" of the seventeenth century and resulted in
the "disassociated," or alienated, savant of the early seventeenth
century.[26] But at this time, it was rarely possible for anyone,
regardless of his philosophic leanings, to disassociate himself
entirely from involvement in the public life of the state; there was
not yet the clear distinction between the state and society which is
so characteristic of our own culture. Patronage served as a mediat-
ing institution between state and society, intermingling the two in
a complex nexus of mutual obligations and responsibilities.
    What happens to the distinction between the private, contem-

24 Keohane, *Philosophy and the State*, p. 143. Keohane also emphasizes the importance of
    the neo-Stoic Lipsius, who urged the sage to avoid public calamities and seek the
    tranquillity and constancy found in the private life (pp. 131–3). Keohane and Levi both
    point out that it is difficult to distinguish Stoic, Epicurean, Skeptic, and Neoplatonist
    influences in the syncretic and eclectic thought of early seventeenth-century intellectuals
    (ibid., p. 29; Levi, *French Moralists*, p. 40).
25 Gassendi to du Faur de Pibrac, April 1621, quoted in Brush, ed., *Selected Works*, pp.
    5–6.
26 Keohane, *Philosophy and the State*, pp. 122–3.

plative life and the active, public life in these circumstances? Supported by patrons during most of his life, Gassendi responded by integrating divergent social imperatives and philosophic beliefs in both his personal life and his writings.

Gassendi's Parisian friends considered him to be "more than the historian of Epicurus, he was his disciple; even more he was his 'man.'"[27] To be someone's "man" meant something very specific in the social terminology of the seventeenth century: It was an indication that an individual had freely associated himself in a patronage relationship, adopting the role of client to the individual patron he served. Roland Mousnier calls this a relationship between protector and "creature," "which implied reciprocal affection and trust, total devotion and unlimited service on the part of the *créature*, protection and social advancement on the part of the protector."[28] This relationship was also described as friendship, and, as Richard S. Westfall has pointed out, emotional friendships were distinguished from instrumental friendships, where the friend or patron was expected to provide concrete benefits for the client.[29] Patronage did not always or necessarily preclude real ties of personal affection between client and patron, but "technically, clientage was a bond between men of unequal rank, usually a less affectionate relationship in which a client needed what he received from a patron."[30]

This sense of friendship was by no means alien to Epicurus himself, who, in the words of John Rist, believed "friendships arise in order that very tangible and specific benefits can be obtained."[31] For Epicurus, however, friendship was part of the private life and provided a safe haven from the vicissitudes of public involvement. Such a friendship in the seventeenth century had the opposite result: It often led to more, rather than less, public participation, for patronage was a quasi-public institution.

27 René Pintard, *Le libertinage érudit* (Paris, 1943), p. 152.
28 Roland Mousnier, *The Institutions of France under the Absolute Monarchy, 1598–1789*, 2 vols, trans. Brian Pearce (Chicago, 1979), 1:106.
29 Richard S. Westfall, "Science and Patronage: Galileo and the Telescope," *Isis*, 76 (1985), 13.
30 Sharon Kettering, *Patrons, Brokers, and Clients in Seventeenth Century France* (New York, 1986), p. 15.
31 John M. Rist, "Epicurus on Friendship," *Classical Philology*, 75 (1980), 123.

According to Sharon Kettering, "Patron–broker–client ties and networks were a way of organizing and regulating power relationships in a society where the distribution of power was not completely institutionalized. These ties and networks were both intra- and extra-institutional and operated inside, across, and outside formal power institutions."[32] Seventeenth-century patrons gave pensions to and found positions for their protégés, but in return they expected the client to wait upon them at need, whether within the household or at court. As Kettering describes this relationship,

> Clients were especially vulnerable as the inferior party in unequal relationships because they had to do most of the adjusting. There was a loss of self-respect in having to curry favor and the demonstrations of esteem and self-abasement that a client was required to perform were a public ritual of dominance and submission in an age that placed great emphasis on the outward symbols of rank and power.[33]

Gassendi was uncomfortable with the ambiguities inherent in friendship and patronage.[34] Nevertheless, the material and societal conditions of his age necessitated his participation in the patronage system. Gassendi came from a humble background: His parents were prosperous peasants from the village of Champtercier, near Digne, in southern France. His intellectual abilities were recognized at an early age and won him rapid social advancement.[35] By the age of nineteen, in 1613, he was appointed principal of the College of Digne. In 1616, he was elected theological canon of the cathedral church at Digne and ordained as a priest. The duties of canons included advising the bishop, performing mass at the cathedral, and governing the diocese if the bishop was absent. Theological canons were supposed to be experts in divinity, both teaching and preaching about Scripture regularly.[36] In 1635 Gassendi completed his ecclesiastical rise by becoming dean of the Digne cathedral.

Gassendi had done extremely well for himself. In the secular

---

32 Kettering, *Patrons*, pp. 72–3.   33 Ibid., p. 26.

34 This might be contrasted with the attitude of Galileo, who understood and readily conformed to the demands of the patronage system. See Westfall, "Science and Patronage," pp. 13–18.

35 A detailed biography of Gassendi can be found in Gaston Sortais, *La philosophie moderne depuis Bacon jusqu'á Leibniz*, 2 vols. (Paris, 1922), 2:1–10.

36 Mousnier, *Institutions of France*, pp. 287–8.

hierarchy of the Church, the canonical office was next in prominence to that of bishop and, by the eighteenth century, was almost
entirely staffed by members of the nobility, who often could look
forward to even higher clerical preferments. Canons received an
income commensurate with their dignity, and this, although not
large, was enough to allow them to live in some state. In short,
"canons enjoyed an income which, though modest, was secure, a
rank in society, eligibility for distinguished functions, and dignity
and rights of precedence."[37] However, canons were required to be
in residence at their cathedrals for most of the year, and Gassendi
indicated, in his history of the Digne cathedral, that canons who
were absent for more than a month from the cathedral were
deprived of their incomes.[38] For a time Gassendi was able to
supplement his clerical income with a professorship at the nearby
University of Aix (1617–23), but his university career ended
when the Jesuits took over the university in 1623.[39]

Although Gassendi enjoyed his clerical duties at Digne, he also
found them onerous, allowing little time for scholarship.[40] Thus,
if Gassendi wanted more time for his studies, if he wanted to
travel and meet other scholars, and if he wanted to utilize libraries
and materials not available in southern France, he would have to
have a patron.[41] Even though the obligations of clientage might
likewise circumscribe Gassendi's life of contemplative scholarship,
it was the only option open to him. In one of his earliest extant
letters, written in 1630, Gassendi clearly indicated his awareness
of the paradoxical nature of seventeenth-century friendship or
patronage:

Why is there this divine tie unless life is lived more delightfully, when
friends assist each other mutually and delight in giving services? And if

37 Ibid., p. 328.
38 Pierre Gassendi, "Notitia ecclisiae Diniensis," in *Opera omnia*, 5:693–4.
39 Lynn Sumida Joy, *Gassendi the Atomist* (Cambridge, U.K., 1987), p. 29, describes
   Gassendi's career after he lost his position at the University of Aix: "Without a
   university appointment, he joined the ranks of private scholars and assumed a social role
   which placed a quite different set of demands on him. Although he continued to fulfill
   his long-standing ecclesiastical duties as theological canon (1614–34) and dean (1634–
   55) of the Catholic cathedral at Digne, his intellectual career now depended on the
   support of individual wealthy benefactors and on a network of other private scholars
   capable of aiding his research and judging his work."
40 Pierre Gassendi, *Lettres familières à François Luillier*, ed. Bernard Rochot (Paris, 1944),
   pp. 53, 77–8, 94.
41 Joy, *Gassendi the Atomist*, p. 29, makes this same point.

someone exacts something from another by compulsion, then pleasure, which is required from friendship, is lost. Everything ought to be free, and if someone yields to a friend, so that he returns duty alone, then he will have nothing from him but pain: He is not then a true friend.[42]

In 1618, while a student at the University of Aix, Gassendi met his original patron, Nicolas-Claude Fabri de Peiresc, with whom Gassendi managed to recreate the Epicurean ideal of a pleasurable, free, yet expedient friendship. For a while, at least, Gassendi avoided the more onerous aspects of seventeenth-century patronage. In his biography of Peiresc, Gassendi described their relationship:

For his love truly, to me, was so great, that it is easier for me to conceive it in my mind, then to express the same in words, and it may suffice to say, that I account it a great happiness, that he prized me so dearly, and that it was his pleasure to have me so frequently with him, and to make me privy to all his private thoughts and inclinations and besides other matters, to utter his last words, and breathe out his very soul itself, into my bosom. . . . For, seeing that as oft I think, speak, or hear of that man, I feel my mind filled with a most intimate and sweet passion of joy and pleasure.[43]

Peiresc was a member of the Nobility of the Robe. He served as a councillor to the parlement of Aix, but his chief importance was as a patron of arts and letters. A noted humanist and naturalist himself, he "perhaps more than any other person of his period lays a firm foundation for the Republic of Letters, as it will be found to exist during the next hundred years."[44] All of the notable intellec-

---

42 Gassendi to Golius, September 1633, in Gassendi, *Opera omnia*, 6:38.

43 Pierre Gassendus, *The Mirrour of true Nobility and Gentility: Being the Life of the Renowned Nicolaus Claudius Fabricius Lord of Peiresc*, trans. W. Rand (London, 1657), pp. iii–iv. I have modernized the spelling when necessary for clarity.

44 Harcourt Brown, *Scientific Organizations in Seventeenth Century France* (Baltimore, 1934), p. 5. Robert S. Westman, "The Reception of Galileo's *Dialogue*, A Partial World Census of Extant Copies," in *Novita celesti e crisi del sapere*, ed. P. Galluzzi (Florence, 1984), p. 336, when discussing Peiresc's role in helping publish a Latin translation of the *Dialogue*, remarks, "Peiresc was but one of a new breed of men in late 16th and early 17th century France who had been trained either for the law or the clergy and who found in the promotion of natural knowledge a way of expressing their rise to higher social ranks as *conseillers* to the *parlements* or members of the diplomatic corps or abbots of wealthy religious houses. The prominence of lawyers and abbots in the 17th century French scientific scene is, to my knowledge, without precedent in Europe and England during this period." There does indeed seem to be a unique patronage role played by Nobility of the Robe at this time. See the example of François Luillier mentioned later in

tuals of the early seventeenth century either visited him in Aix or corresponded with him, and according to Gassendi, "Hence he proved a large sanctuary always open to learned men, for to him all had recourse that had business at court . . . and no man ever went away, whose patronage he did not cheerfully undertake."[45]

In terms of the history of science, Peiresc was the focal point of a nascent scientific academy; in terms of social history, Peiresc was a broker of patronage, who served as an intermediary between the great patrons of the court and the intellectuals who sought their favor, directing the flow of patronage from high to low. According to Gassendi, Peiresc had "the affection and esteem of the masters of the exchequer, because he never demanded anything for himself, but was only an intercessor for good and deserving men."[46]

Peiresc took an active interest in Gassendi's work on Epicurus. Gassendi informed him of the progress of his study, and Peiresc procured manuscripts and books to help Gassendi with his research.[47] In addition, through the years the two performed scientific experiments and observations together. Peiresc would have been the last to ask Gassendi to sacrifice any of the advantages of the private life to fulfill public obligations, for he himself delayed taking up his duties at Aix "because he feared . . . once obliged to [hold] office, he should soon be deprived both of the liberty to study and the opportunity for travelling."[48] Gassendi's

this chapter. For more on Peiresc's role in the Galileo affair, see my article "French Catholic Reaction to the Condemnation of Galileo, 1632–1642," *Catholic Historical Review*, 74 (1988), 34–54.

45 Gassendus, *Mirrour of true Nobility*, pp. 171–2,

46 Ibid., p. 207. See Orest Ranum, *Artisans of Glory: Writers and Historical Thought in Seventeenth Century France* (Chapel Hill, 1980), pp. 22–4, for an account of other men who played a similar role as "brokers" of patronage to men of letters in the sixteenth and seventeenth centuries. Robert Harding, "Corruption and the Moral Boundaries of Patronage in the Renaissance," in *Patronage in the Renaissance*, ed. G. F. Lytle and S. Orgel (Princeton, 1981), pp. 52–3, argues that patronage was coming into disrepute in the early seventeenth century, because patrons were promoting unworthy clients.

47 See Gassendi's letter of April 1631 to Peiresc, describing the plan for his work on Epicurus, in Nicolas-Claude Fabri de Peiresc, *Lettres de Peiresc*, ed. P. Tamizey de Larroque, 7 vols. (Paris, 1893), 4:249–51.

48 Gassendus, *Mirrour of true Nobility*, p. 79. Joy has described Gassendi's biography of Peiresc at length, because she considers it crucial to Gassendi's historical methodology. She notes that Gassendi himself defined Peiresc as a "private" person, whose intellectual deeds were as worthy of memorializing as the deeds of kings and princes. According to Joy, Gassendi distinguished Peiresc's duties to church and state from "Peiresc's activities

relationship with Peiresc shows that for a few Frenchmen, fortunate in their patron, patronage allowed for political withdrawal and philosophic freedom. This could only happen, however, if the tie between patron and client remained emotional as well as instrumental, and, I suspect, if they shared an equal commitment to the intellectual life.

Peiresc introduced Gassendi into the intellectual community of Paris, where Gassendi met François Luillier, another patron of letters, who was councillor to the parlement of Metz and a master of accounts. Gassendi traveled and lived with Luillier between 1629 and 1632, and then again from 1641 to 1648. Gassendi described Luillier as his "comrade, friend, and patron."[49] As in his relationship with Peiresc, this was another association of equals, regardless of a difference in social class. Unlike Peiresc, however, who was universally respected, Luillier had a much more dubious reputation: He was not only a freethinker but also a moral libertine. Gassendi's friendship with Luillier, among others, has led the modern historian of libertinage, René Pintard, to characterize Gassendi as a secret libertine.[50] I would suggest that their friendship had more to do with patronage than with moral outlook. Gassendi wrote to a friend, "You ask me how only with Luillier I can live most freely? Since he loves me sincerely, so he is not the opponent of that most honest condition [of freedom]."[51]

Gassendi contrasted this "freedom" with the lack of freedom he would experience if he accepted another man, of higher rank, as his patron: "There is another who solicits me exceedingly, so that I may vow to him brotherhood, and indivisible and perpetual society.... I will never escape him if I go with him to Paris, for he desires me for his own, and I am impotent to deny anything in his presence. Therefore I will save myself from this misfortune."[52] In

as a scholar, which were by definition those of a private person, since they did not affect
the maintenance of public life in Provence" (p. 53). While this is true, we should
remember that Gassendi emphasized Peiresc's activities as a patron in the *Life*, which by
our definition means that Peiresc straddled and combined private and public life.
49 Gassendi to Naudé, March 1633, in *Opera omnia*, 6:33.
50 Pintard, *Le libertinage érudit*, pp. 127–8.
51 Gassendi to Naudé, July 1633, in *Opera omnia*, 7:57.
52 Ibid. Gassendi had returned to Digne in 1631 to perform his clerical duties. He remained
there until 1641. According to Sortais, Gassendi's Parisian friends urged him to return
to Paris during this period. The "great man" referred to here, according to Sortais, is
probably Henri-Louis Habert, lord of Montmor, who offered the philosopher room

other words, with Luillier, as with Peiresc, Gassendi was able to maintain his personal liberty, while another, more typical patronage relationship would have resulted in servitude and "perpetual society." Gassendi's attitude here seems to confirm Robert Mandrou's contention that patronage, while protecting the savant from the repressive institutions of the French church and state, also exposed the thinker to continual dangers. Mandrou argues that intellectuals were led, so to speak, "to put themselves under the direct and personal protection of those whose task it would be, if they were to expose themselves completely, to burn down and destroy them. Theirs was a double duplicity, or ambiguity, which shows us the degree of oppression that was felt and experienced by these men."[53] Bloch, a modern interpreter of Gassendi's thought, echoes this theme when he argues that Gassendi was an ideological materialist who disguised the heterodoxy of his thought with a hypocritical theological conformity.[54] Bloch's analysis, in turn, exaggerates the argument of Pintard, who softens his critique of Gassendi in later works by suggesting that he was not an intellectual heretic but rather a prudent and cautious individual who conformed to the political necessities of his time.[55]

Pintard has described in great detail the meetings of the freethinkers at the assembly or cabinet of the brothers Dupuy, which Gassendi and Luillier attended regularly while they were in Paris. The expanding intellectual horizons of the early seventeenth century necessitated a more communal and collective approach to learning, particularly in science, philology, and history.[56] Within the confines of the cabinet, all opinions were welcomed and all ideas were tolerated: "Each man, therefore, proposed his opinion

and board and a pension of three thousand livres (Sortais, *La philosophie moderne*, p. 9). This was a great deal of money at the time, when an income of six thousand livres was enough to support a "gentleman" (Robert Mandrou, *Introduction to Modern France, 1500–1640*, trans. R. E. Hallmark [New York, 1976]). It appears that money alone was not a determining factor in Gassendi's career as a client. For more on Montmor, with whom Gassendi spent the last years of his life, see Howard Jones, *Pierre Gassendi, 1592–1655* (Nieuwkoop, 1981), pp. 74–8.

53 Robert Mandrou, *From Humanism to Science*, trans. Brian Pearce (Harmondsworth, 1978), p. 218.

54 Olivier René Bloch, *La philosophie de Gassendi* (The Hague, 1971).

55 See Joy, *Gassendi The Atomist*, p. 17, for an excellent historiographical account of the differences between Bloch and Pintard's interpretations of Gassendi.

56 Pintard, *Le libertinage érudit*, p. 87.

in complete independence."[57] It is likely that in these meetings Gassendi found the ideal Epicurean community. The friendships he formed with three other intellectuals – Gabriel Naudé, Elié Diodati, and François de la Mothe le Vayer – produced an even more intimate group, which the members called the Tetrade. Pintard believes that this association also, like the one with Luillier, was a cover for all sorts of freethinking.[58] During this time Gassendi wrote to a friend, "I love the liberty of philosophizing greatly.... For shame! not to risk themselves, who pride themselves as philosophers, and who repute themselves men; and who, if they are not supported by the word of authority, fear, hesitate, waver, and fall down."[59] Nevertheless, he admits to another friend, subjects like mathematics, poetry, history, rhetoric, and alchemy are often studied in private and "never publicly venerated or mentioned."[60]

In fact, Gassendi was willing to court a certain amount of danger in pursuing his intellectual contacts. Although the Thirty Years War had just reached its most critical phase, with France openly entering the hostilities on the side of the Protestants, Gassendi continued to correspond with his German philosopher friends. In a very interesting letter to the Flemish astronomer and mathematician Godefroid Wendelin, written in 1636, Gassendi urged him to persevere with their correspondence:

Act, therefore, let us delay nothing since it may not be that that which we fear greatly, will come to pass, that our letters, if they were intercepted, would convict us of high treason. Certainly this would be an innocent arrow shot at us, since we do not mix ourselves in public affairs or plans. Those things which we consider are separate from the affairs of the republic and they have nothing in common with these kind of

---

57 Ibid., p. 30.
58 Ibid., p. 128. Naudé was a doctor, but, more importantly, librarian first to Francesco Cardinal Barberini and then to Mazarin. La Mothe le Vayer was tutor to the duke of Orleans and Louis XIV and a prolific author. For more on these *libertins érudits*, see J. S. Spink, *French Free-Thought from Gassendi to Voltaire* (London, 1960), pp. 17–22; Richard H. Popkin, *The History of Scepticism* (New York, 1964), pp. 89–112; and Robert Mandrou, *From Humanism to Science, 1480–1700*, trans. Brian Pearce (Harmondsworth, 1973), pp. 183–98. Naudé and La Mothe le Vayer had very active public careers. Whether they shared Gassendi's attitudes toward patronage remains to be investigated. Diodati was a Swiss Protestant diplomat who was instrumental in the dissemination of the works of Galileo. See Pintard, *Le libertinage érudit*, pp. 129–31.
59 Gassendi to Fienus, June 1629, in *Opera omnia*, 6:17.
60 Gassendi to Golius, March 1630, in ibid., p. 32.

common difficulties. It is the heavens, it is the earth, it is the nature of things toward which we direct our contemplation.... For which reason, we can securely traverse opposing battlefields while worshiping the world, as Diogenes once did.[61]

Seen from the perspective of this letter, Gassendi's desire to absent himself from public life shows an acute awareness of political realities and the dangers of too active a political life. In the 1630s, Gassendi also had very much in mind the example of Galileo, another example of a scholar who had wandered too far into the public battlefield and too far away from the world, sinking into the morass of ecclesiastical politics. Some scholars have argued that Gassendi did not complete his early critique of Aristotle, and delayed the publication of his work on Epicurus, out of fear of ecclesiastical persecution.[62] Indeed, he wrote to a friend in 1630 about his work on Aristotle, "Liberty in these things for me is somewhat greater than the condition of the current time bears.... Therefore I will do another kind of thing, and I will await a better chance, which, unless it smiles on me, I will blind myself to the Muses. I have another commentary on an ancient philosopher, but I will certainly hold it for myself and my friends."[63] In 1624, the Sorbonne, with the contrivance of the Parlement of Paris, had driven two anti-Aristotelians out of Paris, and attacks on Aristotle "were forbidden on pain of death."[64]

Nevertheless, if Gassendi lived in constant fear of the Church authorities, it is odd that he chose to write on a figure as controversial as Epicurus. In fact, ecclesiastical censure in France was far more limited than in the rest of Catholic Europe, because the Gallican liberties of the French church exempted the country from the jurisdiction of the Inquisition, and only the theological faculty of the University of Paris could declare beliefs heretical. Gassendi was certainly cautious, but there were many reasons why he did not put his work out into the public arena. He was both sincere

---

61 Gassendi to Wendelin, June 1636, in ibid., pp. 90–1.
62 Henri Berr, *Du scepticisme de Gassendi*, trans. B. Rochot (Paris, 1960), argues that Gassendi ceased his attack on Aristotle because of the persecution of the anti-Aristotelians in 1624, while Bernard Rochot, *Les travaux de Gassendi sur Épicure et sur l'atomisne, 1619–1658* (Paris, 1944), p. 17, argues that Gassendi was never in any danger from the established authorities because of his work on Epicurus.
63 Gassendi to Schickard, September 1630, in *Opera omnia*, 6:36.
64 J. S. Spink, *French Free-Thought*, pp. 89–90.

in his respect for the Church and its injunctions – a sentiment repeated many times in both his private correspondence and published works – and truly happy to disseminate his ideas only within the circle of his friends.[65]

Ultimately, Gassendi's actions suggest that Mandrou's characterization of French repression depicts twentieth-century totalitarianism, rather than seventeenth-century absolutism, and that Bloch's accusation of theological duplicity and hypocrisy overlooks the fact that in the seventeenth century caution and conformity could coexist with sincere religiosity. Thus, Gassendi's fear of the "perpetual servitude" involved in a patronage relationship was not so much the fear of suffering persecution, if his protectors were to discover his true opinions, as fear that the duties incumbent upon a client would cheat him of the time, repose, and freedom he needed for his studies.

However, the greater Gassendi's renown, the more his freedom was endangered. As Orest Ranum has pointed out, the great sought clients whose glory would increase their own.[66] Gassendi found himself in a very precarious position in 1637, because Peiresc had just died, and Gassendi had returned to Provence to perform his clerical duties. The governor of Provence was Louis-Emmanuel de Valois, count of Alais, closely connected to the royal family. In 1638, Gassendi was "invited to wait upon" Valois, and thus he entered the nobleman's service until the latter's death in 1653.

The nature of this new patronage relationship is indicated in the dedication to Valois at the beginning of Gassendi's biography of Peiresc: "In a word, you were so far ravished, with the admiration of his [Peiresc's] virtues, as to have a principal hand in persuading me to write his Life, and were for this cause willing to dispense with my attendance upon you, that I might in this my retirement, the sooner accomplish that work." *Noblesse oblige!*[67] In Gassendi's first letter to Valois in 1639, he wrote, "You bind me willingly, and I become yours by right."[68]

65 On the position of the Inquisition in France and the Gallican Liberties of the French church, as well as Gassendi's reaction to Galileo's condemnation, see my article "French Catholic Reaction to the Condemnation of Galileo."
66 Ranum, *Artisans of Glory*, pp. 31–2.
67 Gassendus, *Mirrour of true Nobility*, dedication to Valois, unnumbered.
68 Gassendi to Valois, August 1639, in *Opera omnia*, 6:95.

Gassendi's ties with Valois exposed him to many of the inconveniences of the patronage system he had longed to avoid. From the very beginning of their relationship, Valois sought to gain offices and rewards for Gassendi, and the savant could only accede to his wishes. Between 1639 and 1641, Valois attempted to have Gassendi elected as an Agent General of the French clergy, a national post which normally led to the episcopate. To this end, Valois enlisted the aid of Chavigny, the secretary of state, and finally the king himself. Gassendi thanked Valois for these efforts in the following way: "Hence it is not marvelous if I feel myself weighed down by so much benevolence, but also oppressed (which is hard for a soul acknowledging gratitude), and I foresee no time sufficient to give you the thanks I owe you."[69] This kind of rhetoric is familiar in the context of seventeenth-century patronage but differs sharply from the words that Gassendi used to describe his state of mind after the appointment was blocked: "[I]f you might know with what great joy I have been inspired by my burden having been transferred, because not for all the gold of Croesus would I wish to have been afflicted with these burdens. The voice of the Assembly [of the Clergy] has released me like the rose from the thorns."[70]

This theme continues in later letters: "Nothing could be more sweet, than not to have displeased you, because I took the initiative of claiming the tranquil life for myself.... I esteem especially quiet of soul and freedom for study."[71] Clearly, Gassendi regarded his service to Valois as servitude, and the rewards of patronage as a prison which would destroy tranquillity and freedom. After this affair Gassendi returned to Paris, where he remained until 1648. He told Valois,

When I returned, it was marvelous how much all my friends congratulated me that I was returned to myself and the muses.... Already it is

---

69 Gassendi to Valois, October 1639, in ibid., p. 97.
70 Gassendi to Valois, March 1641, in ibid., p. 104.
71 Gassendi to Valois, April 1641, in ibid., p. 105. This office was worth four thousand livres, which were to be split between Gassendi and the other contender for the post, who actually took on the duties of the office. See Sortais, *La philosophie moderne*, pp. 9–10. If Gassendi actually received this sum for the five-year term that the Agent General served, he would certainly have been free from financial need during this period. However, there is some indication that Louis XIII intervened in this election and replaced Gassendi and his rival with another candidate. See Mousnier, *Institutions of France*, pp. 367–8.

profitable, that I am able to enjoy peaceably the delightful association with old friends, but pleasure will be augmented when I may see you. Whether this may be before the end of the year, I do not dare to promise.[72]

If this sounds like the great escape, I believe that is precisely what it was. Although Gassendi remained a client of Valois for many years and attended him sporadically, he tried to keep his distance, living with Luillier and serving Valois mostly by advising him and instructing him in Epicurean philosophy through letters. (Valois, although an agent of the state, was not incensed but, rather, intrigued by these somewhat unorthodox ideas.)

Nevertheless, Gassendi was not able to avoid more public involvement. In 1645, Cardinal Alphonse de Richelieu, the cardinal-minister's brother and archbishop of Lyon, who had known Gassendi from the time when Richelieu had been archbishop of Aix in 1626–8, helped procure a professorship of mathematics at the Collège Royal for his client, although this new patron reassured Gassendi that he would not actually have to teach at the Collège but could continue to pursue his scholarly activities without distraction.[73] In 1653, Gassendi came under the protection of Henri-Louis Habert, lord of Montmor, with whom he lived until Gassendi's death in 1655. Montmor was the patron of a proto-scientific academy which evolved into the Royal Academy in the next decade.

It was also between 1641 and 1655 that Gassendi wrote the "Ethics" of the *Syntagma philosophicum* and rewrote the "Physics." His relationship with Valois, in particular, who relied on Gassendi for news of the court while the latter was in Paris, and for advice in governing Provence, brought him face-to-face with the necessity of finding a formula to mediate between the ethics of political abstention promoted by Epicurus and the reality of public involvement required by the social structure of seventeenth-century France. This formula evolved from an expanded definition of tranquillity, which Gassendi increasingly interpreted not only as a state possible for the wise man who withdraws from affairs but

72 Gassendi to Valois, March 1641, in *Opera omnia* 6:104.
73 Joy, *Gassendi The Atomist*, pp. 195–6. Joy mentions that this office was worth only six hundred livres.

also as the state characteristic of the politically and publicly active wise man.

We can see Gassendi beginning to develop his ideas in the letters written to Valois in the 1640s. He advised Valois to "maintain the placid state of the soul constantly, among these public storms."[74] He told him that it is possible to "obtain leisure in business [*otium in negotio*], tranquillity in the midst of the storm, serenity in a state of very agitated affairs," if one fixes on a goal, anticipates problems, and does not fluctuate.[75] No doubt this advice was meant for himself as well as for his patron, for men "may not ignore those things born to us by law and condition, so it is not permitted to be happy at every time."[76] Moreover, the calm or tranquil soul is virtuous: "Therefore, virtue is certainly immobile, but . . . it ought not to be quiescent. . . . Indeed virtue will be similar to the governor, who sits quietly in the stern of the ship holding the rudder, nor does he [pursue transitory things] as young men do, but he does many things greater and better."[77]

All of this advice must have seemed slightly mysterious to Valois, but it becomes clear when read in the context of the "Ethics." This work revolves around the discussion of the Epicurean notion of pleasure as the highest good. Epicurus had distinguished two different kinds of pleasure. The first is the pleasure of motion, or the active pleasure we associate with activities such as eating, drinking, and copulating. The higher kind of pleasure is the pleasure of rest, which is the same thing as tranquillity of soul, or the absence of agitation produced from either internal or external causes. The best way to obtain the pleasure of rest or tranquillity is to recognize what is natural and necessary, not to fear the gods, and to understand the physical nature of the universe.

In antiquity, Epicurus had been accused of advocating a kind of pleasure that was akin to death or to a stupefied state and that therefore was no kind of pleasure at all. Gassendi defended Epicurus against this charge and at the same time further developed Epicurus' concept of the pleasure of rest. He wrote, "Epicurus

74 Gassendi to Valois, July 1646, in *Opera omnia*, 6:252.
75 Gassendi to Valois, January 1644, in ibid., p. 214.
76 Gassendi to Valois, June 1641, in ibid., p. 110.
77 Gassendi to Valois, September 1641, in ibid., p. 113.

does not wish tranquillity and lack of pain to be like pure torpor, but rather he wished it to be a state during which the actions of life are accomplished peacefully and delightfully."[78] Moreover, Gassendi claimed, "the pleasure that Epicurus places in settled rest, he never understood as a lazy life or a rest like that of a drone or a worm" (p. 687). Indeed, according to Gassendi, this kind of pleasure is equally operative whether one leads an active or contemplative life, at least for the wise man who correctly understands that pleasure is tranquillity.

Thus, tranquillity or pleasure, rather than divorcing a man from affairs, becomes a tool for success in the world: "[T]he soul is called tranquil, not only when one lives in leisure, but also, especially, when one undertakes great and excellent things, without internal agitation and with an even temperament" (p. 717). A wise man will use the faculty of prudence – here Gassendi rebaptized the Epicurean calculus of pleasure and pain – to judge his own abilities in any matter, to weigh the capacities of others, and to utilize his past experience to help current enterprises (pp. 745 ff). In short, he will be prepared for any contingency: and if, by some chance, he does fail, he will not be destroyed, because he always retains his inner tranquillity and the knowledge that he did the best he could.

Tranquillity is a "continuing state," and it is very "natural" and very "lasting," but other pleasures are fluctuating and inconstant. Tranquillity "continues in an uninterrupted course, unless it is interrupted and perishes due to our own defects" (p. 715). Gassendi drew an analogy between the tranquil motion of a boat, which is smooth and placid not only when the boat is becalmed but "especially when it is blown by a favorable wind," and the tranquillity of the soul, which retains its calm whether one is at leisure or active and toiling (p. 718). He used the same analogy in the letter to Valois when the governor, who possesses virtue, is described as "quietly" holding the rudder of the ship.

Thus, Gassendi believed he had found a prescription for a life of happy calm within the turmoil of public affairs. Does this mean that he had abandoned the Epicurean ideal of withdrawal from the world into a life of private happiness spent with friends of like

78 Gassendi, in *Opera omnia*, 2:716. Further citations will be given in parentheses in the text.

mind? By no means. Also in the "Ethics" he asked "Do you see any courtier, anyone full of honors, and the managing of affairs, who is not disgusted by this kind of life? Do these men not envy the quiet of those whom they observe resting in a tranquil port, having escaped from the troubled sea?" (p. 763). Maintaining tranquillity in the midst of affairs remains second best to living a life of quiet retirement from the world: It is better to be at port than at sea. But Gassendi was a pragmatist, and his own experiences as a client in the social world of the seventeenth century had taught him that "we may not change fate . . . , and it is better to soften the harshness by our own consent, rather than to exacerbate it by fruitless opposition" (p. 769).

If we are enslaved, "the wise man is greater than to be compelled by an alien imperium. . . . Certainly the body can be put in chains, but the soul is too noble and free to be enslaved" (p. 771). Thus, the freedom to philosophize and the life of tranquil pleasure, which are guaranteed by a life of private withdrawal from the state, continue to exist as a state of mind in the midst of the troubles of the public arena.

10

Robert Boyle on Epicurean atheism and atomism

J. J. MACINTOSH

In Boyle's published works there are many references to Epicurean atomism[1] and atheism.[2] He refused to pronounce publicly on
the first but was willing to pronounce decidedly on the second. He
did not, however, publish any sustained piece directly on atheism,
though he seems to have continued to work on the topic of
atheism throughout his life. His firm belief in Christianity began
early – he underwent a conversion from unthinking to committed

Like every other worker on the Boyle manuscripts, I owe a debt of gratitude to the
Librarian and staff of the Royal Society Library. I am also indebted to the Royal Society for
permission to quote from the Boyle Papers. I am grateful, too, to Margaret J. Osler for a
number of helpful suggestions, both editorial and substantial, concerning this chapter.
1 Epicurus is one of a half dozen thinkers to whom Boyle constantly refers, the others
being Aristotle, Descartes, Gassendi, Paracelsus, and van Helmont, each of whom
garners considerably more references than other philosophical writers. (There are also
numerous references to Moses, Saint Paul, and Solomon, as well as the tracts written
specifically against Linus, Hobbes, and More.) Other seventeenth-century writers agreed
about the importance of Epicurus. Thomas Stanley, for example, in his *History of
Philosophy*, 2nd ed. (London, 1687), devotes more space to Epicurus than to any other
single philosopher. The history deals only with classical philosophers, and Epicurus gets
100 pages, which is about one-tenth of the three volumes. For comparison, Plato receives
41 pages and Aristotle just over 42. Considerable space is also devoted to the Pythagoreans, including Empedocles (90 pages) and the Skeptics (64).
2 See especially *The Usefulness of Experimental Natural Philosophy*, 1.4, "... a Requisite
Digression...," in Thomas Birch, ed., *The Works of the Honourable Robert Boyle*, 6 vols.
(London, 1772), 2:36 (hereafter otherwise unassigned numbers will be references by
volume and page number to this edition of the *Works*), and *Considerations about the
Reconcileableness of Reason and Religion*, 4:157. As well, there are relevant remarks and
passages in such works as *The Christian Virtuoso I* and *II*; *Of the High Veneration Man's
Intellect Owes to God*; *A free Inquiry into the vulgarly received Notion of Nature*, etc., but the
sustained work remained unpublished. In his *Life* of Boyle Birch mentions a variety of
such unpublished pieces known to him (1:ccxxxvi–vii).

Christianity at the age of thirteen – and continued until his death. In his will he left a sum of money to fund an annual series of lectures against atheism, the intent being to refute infidels (with a caution against internecine quibbling), spread the word, and soothe serious doubts. The eleventh sheet of his will reads in part,

Whereas I have an intention to Settle in my Life time the Sume of Fifty pounds per Annum for ever or att Least for a Considerable Number of yeares to be for an Annual Salary for some Learned divine or Preaching Minister from time to time to be Elected and Resident within the City of London or Circuite of the Bills of Mortality, who shall be enjoyned to performe the Offices following (*viz*ᵗ)

1 To Preach Eight Sermons in the yeare for proveing the Christian Religion agᵗ notorious Infidels (*viz*ᵗ) Atheists, Theists, Pagans, Jews and Mahometans, not descending lower to any Controversies that are among Christians themselves, . . .

2 To be Assisting to all companies and incourageing of them in any Undertakings for Propagating the Christian Religion to Forreigne Parts.

3 To be ready to satisfy such real Scruples as any may have concerning those Matters and to Answer such new Objections or Difficulties as may be started to wᶜʰ good Answers have not yet been made.[3]

In this chapter I shall argue that an examination of the details of his unpublished arguments against Epicurean atheism – which are interesting enough in themselves – suggests a possible reason for his failure to publish on this topic. Since Boyle constantly runs atheism and atomism together in his discussions, there is also the question "Was Boyle an atomist?" to be considered. Some hold that he was, early in life, but that later he either changed his mind or decided not to tell anyone that he was. If he wasn't, another question arises: *Why* wasn't he? Hadn't Gassendi made it clear to all that atomism was the thinking experimentalist's hypothesis of choice?[4] And why, whether or not Boyle was an atomist, did he never publish anything on the topic, save his constant disclaimers to the effect that he hadn't publicly committed himself in this area but might one day?

3 Quoted in R. E. W. Maddison, *The Life of the Honourable Robert Boyle* (London: Taylor & Francis, 1969), p. 274.

4 Sennert, Basso, and Magnenus, all of whom Boyle mentions in his published work, also thought of themselves as promoting atomism of one variety or another.

The notion of *atomism*, as a subspecies of Boyle's more generic notion of *corpuscularianism*, became clearer as the century went on. To anticipate slightly: The mature Boyle associated atomism with the Epicureans (and hence, by a further association, with atheism).[5] He also wanted to include people such as Descartes, Gassendi, and himself under a common banner, fighting the goodly fight against both Paracelsans and Peripatetics. Thus he was led to see anyone who had a doctrine of *minima naturalia*, who opposed occult qualities, and who explained change in terms of the interaction of material particles or quasi particles, as being on the same side, and was, in consequence, also led to play down such otherwise important differences as that between the plenists and the vacuists. Boyle was, to use his own term, a corpuscularian, uncommitted (publicly at least) about atomism as such but sufficiently wary of Epicurean atomism to need to make it clear that his views were distinct from it.[6]

Was the young Boyle an atomist? Was the mature Boyle a closet atomist? I cannot find, in either case, that he was. Boyle's language in the early fragment "Of ye Atomicall Philosophy" is certainly loose, and he does indeed talk of "atomical effluvia," "steemes of Atomes," and so forth,[7] but, even then his language is full of his later habitual caution. In 1661 he wrote,

Perhaps you will wonder ... that in almost every one of the following essays I should speak so doubtingly, and use so often, *perhaps, it seems, it is not improbable* and such other expressions, as argue a diffidence of the truth of the opinions I incline to, and that I should be so shy of laying down principles, and sometimes of so much as venturing at explications. But I must freely confess to you ... that having met with many things,

5 This was a common association at the time. Later Newton was to remark to David Gregory, "The philosophy of Epicurus and Lucretius is true and old, but was wrongly interpreted by the ancients as atheism." (Memoranda by David Gregory, 5, 6, 7 May 1694, in H. W. Turnbull, ed., *The Correspondence of Isaac Newton*, 7 vols. (Cambridge: Cambridge University Press, for the Royal Society, 1961), 3:338.

6 Even after Gassendi had made Epicurus intellectually palatable, there were social factors at work which made it difficult for Boyle to accept anything labeled Epicureanism. It was, for example, along with the associated views of Hobbes (or what were widely believed to be the views of Hobbes), the claimed doctrine of the "wits" that Boyle so disliked, who sat about in coffeehouses making jokes about religion. On this, see further David Berman, *A History of Atheism in Britain* (London: Croom Helm, 1988), Chaps. 1–2, and Michael R. G. Spiller, *"Concerning Natural Experimental Philosophie": Meric Casaubon and the Royal Society* (The Hague: Nijhoff, 1980), Chap. 5, "Epicurus and the New Philosophy."

7 Boyle Papers (hereafter *BP*), Royal Society, London, 26:162–75.

of which I could give myself no one probable cause, and some things, of which several causes may be assigned so differing, as not to agree in anything, unless in their being all of them probable enough; I have often found such difficulties in searching into the causes and manner of things, and I am so sensible of my own disability to surmount those difficulties, that I dare speak confidently and positively of very few things, except of matters of fact. And when I venture to deliver any thing, by way of opinion, I should, if it were not for mere shame, speak yet more diffidently than I have been wont to do.[8]

This habitual and lifelong caution[9] is to be found in full force in the "Atomicall Philosophy"[10] where Boyle speaks of the atomic hypothesis as being "very probable"[11] but does not really go farther. Most of the fourteen-page fragment is taken up with a discussion of odors: The young Boyle was very interested in the fact that the "steames of effluvia" which give rise to odors can last so long and with the fact that such apparently minute causal interactions can have such lingering effects.

I haue obseru'd with wonder [he wrote] that *such small creatures as* Partridge by but transiently passing by a field [perhaps] many houres perhaps the day before should in so short a time by a bare touch of the grasse or ground or stubble leave matter for so lasting an Effluviũ that the day after a Setter should find fresh remaines of it, & manifestly discouer that he does so.[12]

In this piece he also mentions his microscopic observations on cheese mites, which "seeme but . . . mouing Atome[s]," being so small themselves that "the parts that make the haire vpon the legs" must be "vnimaginably little . . . & how much more subtle must

---

8 *Certain Physiological Essays, and other Tracts* (1661), "Proemial Essay," 1:307.

9 Compare the "for ever or att Least for a Considerable Numbet of yeares" of his will.

10 Some time ago, R. H. Kargon suggested, in his *Atomism in England from Hariot to Newton* (Oxford, 1966, pp. 95–6), that the note added to the first page of the "Atomicall Philophy," "These Papers are without fayle to be burn't," "reflects Boyle's hesitancy to link himself publicly with a philosophy so tainted with atheism." This, as M. A. Stewart has pointed out, is a "romantic fiction" (*Selected Philosophical Papers of Robert Boyle* [Manchester: Manchester University Press, 1979], p. xxx). This note, like the note on BP, 26:163v, "Papers to be rifl'd & burn'd," is clearly a late addition.

11 BP, 26:163.

12 BP, 26:171. In my quotations from the manuscripts, insertions are placed within asterisks, deletions within square brackets; in shorter quotations I have simply given the final version.

be the animall spirits that run to & fro in nerues suitable to such
little legs."[13] Consequently each of the "multitude of Atomes
[which] must concurre to constitute the seuerall parts externall and
internall necessary to make out this little Engine" will also be
unimaginably little.[14] Thus Boyle certainly uses the language of
atomism in this piece, but it is by no means clear that he is
thinking of anything other than what he, here and later, also calls
minima naturalia.

It is worth noting that he includes among the atomists people as
diverse as Gassendi, Magnenus, Kenelm Digby, and "Des Cartes
& his disciples":[15]

The Atomicall Philosophy invented or brought into request[16] by
Democritus, Leucippus, Epicurus, & their Contemporaries, tho since the
invndation of Barbarians & Barbarisme expell'd out of the Roman world
all [the] but the casually escaping Peripateticke Philosophy, it haue been
either wholly ignor'd in the European Schooles or mention'd there but
as an exploded Systeme of Absurdities yet in our lesse partiall & more

---

13 BP, 26:169.
14 Boyle means this quite literally. Imagination is a function of the brain, and therefore
many things are either too small or too big to be imaginable, though they can be
conceived by that "higher faculty," the understanding: "a[n] *single* Atom[, whilst
such,] is too small to be ye object of any sense by ye Epicureans Confession or *rather*
Assertion. And I thinke it may be added by a just Consequence /that 'tis/ also too little
to haue an Image fram'd of it in the Imagination, since Experience informs us, that we
can not Imagine any thing a Thousand times less than any that we euer saw, and
therefore from ye Epicureans belieff of an Atome we may necessitate him to admit, that
there may be Conceptions in the mind fram'd by a higher faculty than ye Imagination,
which may, for Distinctions sake, be cald pure Intellection" (BP, 6:307–8). For the
same point concerning largeness, see, e.g., 2:21. The point at issue is discussed in my
"Perception and Imagination in Descartes, Boyle and Hooke," *Canadian Journal of
Philosophy*, 13 (1983), 327–52.
15 Boyle always thought highly of Gassendi, as he did of Descartes: The numerous
references to them attest to the care with which he read them, and his defense of
Descartes against More is obviously sincere: "Though not confining myself to any sect,
I do not profess myself to be of the Cartesian: yet I cannot but have too much value for
so great a wit as the founder of it, and too good an opinion of his sincerity in asserting
the existence of a deity, to approve so severe a censure, as the doctor is pleased to give of
him" (3:597–8). He also continued to think highly of Magnenus, making reference to *de
Manna* at various places in the printed works (see, e.g., 2:106, 3:320, 4:63, 5:68), though
presumably the reference here is to Magnenus's *Democritus reviviscens sive de atomis*,
published in 1646. References to "my noble friend Sir *Kenelm Digby*" (in 1663, 2:102)
drop off subsequently, but earlier his respect for Digby seems clear. Oldenburg felt, in
March 1660, that writing to Boyle of his "dislike of Mr. Digby . . . may be a mistake,
perhaps." (A. R. Hall and M. B. Hall, *The Correspondence of Henry Oldenburg: vol. 1,
1641–1662*, 11 vols. (Madison: University of Wisconsin Press, 1965), p. 358.)
16 "Invented or brought into request" by Democritus et al. because "the devising of the

inquisitiue times it is so luckyly reuiu'd & so skillfully celebrated in diuers parts of Europe by the learned pens of Gassendus, Magnenus, Des Cartes & his disciples our deseruedly famous Countryman Sr Kenelme Digby & many other writers especially those that handle magneticall & electricall operations that it is now growne too considerable to be any longer laugh't at, & considerable enough to deserue a serious enquiry.[17]

But, to put it mildly, it is not clear that all of the writers mentioned are atomists in the sense in which the later Boyle understood the term.[18] It seems that even at this early stage (presumably around 1650) what Boyle had in mind was not

atomical hypothesis commonly ascribed to *Leucippus* and his disciple *Democritus* is by learned men attributed to one Moschus a Phenician" (*Sceptical Chemist*, 1:497–8). Cf. Sextus Empiricus, *Against the Mathematicians*, I.363, and Strabo, *Geography*, XVI.2.24 (trans. H. L. Jones, 8 vols. [London: Heinemann, 1960], 7:271): "If one must believe Poseidonius, the ancient dogma about atoms originated with Mochus, a Sidonian, born before the Trojan times." Cudworth (*The True Intellectual System of the Universe* . . . , 1.1.9–10, 12–13) looks sympathetically on the suggestion that "this *Moschus* was no other than the Celebrated *Moses* of the *Jews*, with whose Successors the Jewish Philosophers, Priests and Prophets, *Pythagoras* conversed at *Sidon*." He adds, "Some Phantastick Atomists perhaps would here catch at this, to make their Philosophy to stand by Divine right, as owing its Original to Revelation; whereas Philosophy being not a Matter of Faith but Reason, Men ought not to affect (as I conceive) to derive its Pedigree from Revelation, and by that very pretence seek to impose it Tyrannically upon the minds of Men, which God hath here purposely left Free to the use of their own Faculties, that so finding out Truth by them, they might enjoy that Pleasure and Satisfaction which arises from thence."

17 BP, 26:162.

18 Digby, for example, remarks, "By which word *Atome*; no body will imagine we intend to expresse a perfect indiuisible, but onely, the least sort of naturall bodies." (Sir Kenelm Digby, *Two Treatises, in the one of which, The Natvre of Bodies; in the other, The Nature of Mans Sovle; is looked into: in way of discovery, of the Immortality of Reasonable Sovles* [Paris, 1744, reprints, New York: Garland, 1978], p. 38.) Like Descartes, Digby was a plenist; unlike Descartes he felt that *minima naturalia* could be discovered by experience: "He then that is desirous to satisfy himselfe in this particular; may putt himselfe in a darke roome, through which the sunne sendeth his beames by a cranie or litle hole in the wall; and he will discouer a multitude of little atomes flying about in that litle streame of light; which his eye can not discerne, when he is enuironed on all sides with a full light" (ibid., p. 53). Digby, we may agree, was at best an atypical atomist. (Boyle attributes to the Epicureans the doctrine that "all things are compos'd of Atomes so extreamly minute, that ye least Mote wch ye sun beams make visible to us, may consist of thousands, if not myriads, of them" [BP, 6:306]. Lucretius had suggested not that the dust motes were atoms but that their movement provided us with indirect evidence of atoms and their motion [*De rerum natura*, II.114–41].) Digby was, however, in full agreement with the rest of the seventeenth-century corpuscularians on one matter: "that all operations among bodies, are either locall motion, or such as follow out of locall motion" (*Two Treatises*, p. 36).

atomism but what he was later to call "corpuscularianism."[19] There is more to be said on this topic than I have space for here but, particularly in view of the unclear notion of atomism revealed in the early manuscripts, I remain unconvinced that Boyle ever was an atomist in the sense in which the mature Boyle would have understood that term. Of course the young Boyle may have thought he was an atomist – but that is a different matter.

What we have, in the manuscripts and in passing in the printed works, is a writer who is determinedly noncommital about the truth of atomism but who thinks of himself as having a number of arguments against the atheism that he believes to be associated with Epicurean atomism. Some of his arguments in this area occur in the printed works, but by and large they are to be found in the manuscripts, and in this essay those are the arguments on which I shall concentrate.[20]

This brings us back to the question of why Boyle left his work on atheism unpublished, despite his obvious – one might almost

19 In the *Origine of Forms and Qualities*, "Proemial Discourse," 3:5, Boyle mentions "that philosophy, which I find I have been much imitated in calling Corpuscularian." Leibniz, writing in 1669 ("The Confession of Nature against Atheists," in *Philosophische Schriften*, ed. C. I. Gerhardt, 7 vols. [Berlin, 1875–90], 4:106, trans. in L. E. Loemker, *Gottfried Wilhelm Leibniz, Philosophical Papers and Letters* [Chicago: University of Chicago Press, 1956], p. 169), gives a similar (and similarly heterogeneous) list of writers but does not call them atomists, preferring instead Boyle's own term: "At the beginning I readily admitted that we must agree with those contemporary philosophers who have revived Democritus and Epicurus and whom Robert Boyle aptly calls corpuscular philosophers, such as Galileo, Bacon, Gassendi, Descartes, Hobbes, and Digby, that, in explaining corporeal phenomena, we must not unnecessarily resort to God or to any other incorporeal thing, form or quality (*Nec Deus intersit, nisi dignus vindice nodus inciderit* [Horace, *Ars poetica*, 191]), but that, so far as can be done, everything should be derived from the nature of body and its primary qualities – magnitude, figure, and motion." It might be noted that Galileo, Gassendi, and Digby, as opposed to Descartes, accepted the notion of frigorific atoms, while the mature Boyle felt that there were conceptual difficulties with accepting such a notion, on the one hand, and experimental difficulties with rejecting it in favor of a kinetic theory on the other. (See "Of the Positive or Privative Nature of Cold," 3:752–3.) On the diversity of the atomists generally, see Lynn S. Joy, *Gassendi the Atomist* (Cambridge: Cambridge University Press, 1987), Chap. 1, and Christoph Meinel, "Early Seventeenth-Century Atomism: Theory, Epistemology, and the Insufficiency of Experiment," *Isis, 79* (1988), 68–103.

20 I shall be referring mainly to the early pages of Volume 2 of the Boyle Papers, but similar or related arguments are to be found elsewhere in the manuscripts, particularly in Volumes 6 and 7. I am grateful to M. A. Stewart and Michael Hunter for suggestions concerning this manuscript material.

say obsessive – interest in the subject. "The Honourable Robert Boyle, Esq., that profound Philosopher, accomplished Humanist, and excellent Divine, I had almost sayd Lay-Bishop," wrote Aubrey.[21] Well, why was the lay bishop so reluctant to put his views on atheism forward?

No doubt there were a number of factors. When we consider the general chaos that preceded and indeed accompanied a Boyle publication,[22] we might wonder whether a better question would be not Why did he fail to publish on these topics? but How did he ever succeed in publishing anything at all? This, I think, is not satisfactory. It might be the whole story, but there are other factors that, at any rate, look as if they might be relevant.

In 1678 Cudworth described

four several Forms of Atheism; First, the *Hylopathian or Anaximandrian*, that derives all things from Dead and Stupid Matter in the way of *Qualities and Forms*, Generable and Corruptible: Secondly, the *Atomical* or *Democritical*, which doth the same thing in the way of *Atoms* and *Figures*: Thirdly, the *Cosmoplastick* or *Stoical Atheism*, which supposes one *Plastick* and *Methodical* but *Senseless Nature*, to preside over the whole Corporeal Universe; and Lastly, the *Hylozoick* or *Statonical*, that attributes to all Matter, as such, a certain *Living* and *Energetick Nature*, but devoid of all Animality, Sense and Consciousness.[23]

Of these, with an occasional sideswipe at the Peripatics, the one that Boyle is chiefly interested in combating is the Epicurean, that is, the Atomical atheist. Often indeed, his arguments against atheism are simply ad hominem attacks on the views of Epicurus himself. He does not seriously consider the possibility that the actual claims of Epicurus might be patched up or strengthened. In part, at least, this is because he felt that the absence of a God who could do all that Boyle felt was necessary to ground an acceptable worldview was so central to the Epicurean position that, were such a deity to be added, we would no longer have any reason to think of the system as being that of Epicurus, or even an extension of it. On matters such as the constitution of the universe, its

21 John Aubrey, *Brief Lives*, "Boyle."
22 See J. F. Fulton, *A Bibliography of the Honourable Robert Boyle* (Oxford: Oxford University Press, 1961), for a variety of details on this topic.
23 Ralph Cudworth, *The True Intellectual System of the Universe: Wherein, All the Reason and Philosophy of Atheism Is Confuted, and Its Impossibility Demonstrated* (London, 1678), I.3.30, 134–5.

creation, and the place of its creator in a scientific worldview, Boyle sees himself as balanced between Epicurus and Descartes, and it is worth quoting him at length to see what he took the correct position to be:

Though I agree with our Epicureans, in thinking it probable, that the world is made up of an innumerable multitude of singly insensible corpuscles, endowed with their own sizes, shapes, and motions; and though I agree with the Cartesians, in believing (as I find that *Anaxagoras*[24] did of old) that matter hath not its motion from itself, but originally from God: yet in this I differ both from *Epicurus* and *Des Cartes*, that whereas the former of them plainly denies, that the world was made by any deity (for deities he owned,) and the latter of them, for aught I can find in his writings . . ., thought, that God having once put matter into motion, and established the laws of that motion, needed not more particularly interpose for the production of things corporeal, nor even of plants or animals, which, according to him, are but engines: I do not at all believe, that either these Cartesian laws of motion, or the Epicurean casual concourse of atoms, could bring mere matter into so orderly and well-contrived a fabrick as this world.[25] And therefore I think, that the wise author of nature did not only put matter into motion, but, when he resolved to make the world, did so regulate and guide the motions of the small parts of the universal matter, as to reduce the greater systems of them into the order they were to continue in; and did more particularly contrive some portions of that matter into seminal rudiments or principles lodged in convenient receptacles, (and, as it were, wombs,) and others into the bodies of plants and animals: one main part of whose contrivance did, as I apprehend, consist in this, that some of their organs were so framed, that supposing the fabrick of the

24 Boyle's footnote reads, "*Aristotle* speaking of *Anaxagoras*, in the first chapter of the last book of his Physicks, hath this passage: *Dicit* (Anaxagoras) *cum omnia simul essent, atque* quiescerent *tempore infinito, mentem* movisse *ac segregasses.*" (Aristotle, *Physics*, VIII.1.250^{b}24–6, trans. R. P. Hardie and R. K. Gaye [Oxford: Oxford University Press, 1930]: "Anaxagoras . . . says that all things were together and at rest for an infinite period of time, and that then Mind introduced motion and separated them.")

25 There are two issues here: How does Descartes think the world did come about, and how does he think it could have come about? Boyle is right to suggest that Descartes thought the world could have come about in this way, but Descartes says explicitly that this is not the way it did come about: In fact, God created it as a going concern, complete with "not just the seeds of plants but the plants themselves," etc.: *Principles of Philosophy*, III.45, trans. R. Stoothoff, in *The Philosophical Writings of Descartes*, trans. J. Cottingham, R. Stoothoff, and D. Murdoch (Cambridge: Cambridge University Press, 1985), p. 256 (Ch. Adam and P. Tannery, *Oeuvres de Descartes*, 11 vols. [Paris: Vrin 1964–76], 8A.100), and cf. *Discourse on Method*, pt. 5, in *Philosophical Writings*, pp. 133–4, Adam and Tannery, 6:45.

greater bodies of the universe, and the laws he had established in
nature, some juicy and spirituous parts of these living creatures must be
fit to be turned into prolific seeds, whereby they may have a power, by
generating their like, to propogate their species. So that, according to
my apprehension, it was at the beginning necessary, that an intelligent
and wise agent should contrive the universal matter into the world,
(and especially some portions of it into seminal organs and principles,)
and settle the laws, according to which the motions and actions of its
parts upon one another should be regulated: without which interposition of the world's architect, however moving matter may (for I see not
in the matter any certainty) be conceived to be able, after numberless
occursions of its insensible parts, to cast it self into such grand conventions and convolutions as the Cartesians call vortices, and (as I remember) *Epicurus*[26] speaks of under the name of προσκρίσεις καὶ δινήσεις
[aggregations and whirlings] yet I think it utterly improbable, that
brute and unguided, though moving matter should ever convene into
such admirable structures, as the bodies of perfect animals. But the
world being once framed, and the course of nature established, the
naturalist (except in some few cases where God or incorporeal agents
interpose,) has recourse to the first cause but for its general and ordinary support and influence, whereby it preserves matter and motion
from annihilation or desition; and in explicating particular phaenomena
considers only the size, shape, motion, (or want of it,) texture, and the
resulting qualities and attributes of the small particles of matter.[27]

Before moving on to Boyle's arguments against the Epicureans,
it might be as well to notice explicitly that Boyle clearly thought
well of Epicurus (as he did, also, for example, of Aristotle). He
reminds us that many of Epicurus's opinions are "innocent"[28] and
deserve embracing and takes care to make explicit the fact that
Epicurus himself was not an atheist:

For in his Epistle to Menoeceus, after he had said, that *really there are
Gods, since ye notice we haue of them is evident, but that they are not such as*

---

26 Boyle's note: "Epicurus in his epistle to Pythocles." (Epicurus to Pythocles, 90, see
Cyril Bailey, *Epicurus: The Extant Remains* (Oxford: Oxford University Press, 1926),
p. 61.)
27 *Origine of Forms and Qualities*, "An Examen of the Origin and Doctrine of Substantial
Forms," 3:48.
28 5:428. Not all agreed with Boyle. Casaubon, writing in 1667, thought Epicurus a
"detestable monster" in view of his "lewd doctrine, notorious stupiditie, and grosse
ignorance." See Meric Casaubon, "On Learning," Bodleian MS. Rawlinson D.36.1
(1667), 22 (extracts reprinted as App. 2 of Spiller, *Casaubon*).

men are wont to thinke them, he adds that memorable and excellent saying, that therefore *the Impious man is He, not who takes away ye Gods of the Multitude, but who adapts ye Opinions of the Multitude to the Gods*. Wch Sentence (by ye way) is very unfavourable to those, that in spite of such ancient Writers, as, *Cicero, Seneca & Diogenes Laertius*, will needs make Epicurus to haue but dissemblingly, & for fear, acknowledg'd any Gods; since the greatest danger of ye imputation of Atheism, was from the Vulgar, whose Deities we see he openly reiected.[29]

Elsewhere Boyle remarks,

In the controversy between Aristotle and Democritus . . . I take not upon me to determine here anything about the Truth of the Opinions, but only about the goodnes of the Argumentations. And thô I fear that neither has opin'd well, yet Democritus seems to have Philosophis'd the better of the two.[30]

Boyle's reason for concentrating on Epicurean atheists is simple: They were the modern enemy. He writes, "Our moderne Atheists . . . are *Somatists*, who admit no substance but Body; and relye wholly upon the Epicurean principles & Hypothesis."[31] The standard antiatheistic arguments will not prevail, Boyle believes, against "resolved" Epicurean atheists:

This was not ye only motiue I had to pass over in silence divers Arguments wont to be urged agst Atheists, for some I pretermitted, because as they are usually proposd, they seem to me not to be so warily proposd as I could wish, & as I thinke some of them may be by one that were not confin'd to brevity; and others of them I omitted, not because they are not considerable (wch. I desire you to take notice of, that my silence may not injure them) but because I thought that of how great force soever they may be against other sorts of Atheists they are improper to be urged against the Epicureans with whom alone we have to do *in this Paper*, who, allowing nothing substantial but Bodyes, reject many things & admit so very few, of those, that other sects of Philosophers and almost all mankind besides assent to, that he y^t will dispute with them upon their owne Principles must lay aside divers of ye brightest & sharpest weapons that haue been employed by ye Champions of a Deity, and will find [t]his Taske much harder to be well performd than perhaps any man that has not tryed will imagine. For, to omit other things, they reject not only God, but all Incorporeal Beings & Final

29 BP, 2:81. The reference is to Epicurus to Menoeceus, 123, in Bailey, *Epicurus*, p. 83.
30 BP, 9:106r.   31 BP, 2:3.

Causes except in some actions of Liveing Bodyes; and tho they seem to
deify Nature, yet in my Opinion, they must according to their Principles
make Nature it selfe but an Effect of Chance. . . .

And indeed some of Gods Attributes & his general Providence, and also
natures actings with designe in what she dos, are things supposed or
implyed in so many of mens Reasonings & Conclusions, that there are
[very often] *many aplauded* Discourses that would [not] be maimed or
weake, if all these were left, or taken, out. And therefore you need not
thinke it strange, that I never pretended to convert resolved Atheists.
For, besides ye difficulty of treating clearly and cogently of such abstruse
subjects as are many yt relate to Atheism; the Will and Affections haue so
great an influence upon some mens Understandings, that 'tis almost
as difficult to make them *beleiue*, as to make ym *Loue*, against their Will.
And it must be a very dazzleing Light, yt makes an impression upon
those that obstinately shut their Eyes against it. 'Tis not by Gods
ordinary workes, but by his Extraordinary Power, that such men must
be reclaimd to an acknowledgement of his [Essence] *Existence*. For
they y^t would find ye Truth, especially in matters of Religion, *must be*
diligent Inquirers after it, and *may be* strict Examiners of it, but *must not be*
resolued Enemies to it. For to such, if to any, God is a Sun, that is not to
be discover'd but by [its] *his* owne Light.[32]

Thus there are both formal and psychological reasons to expect
the determinedly unconvinced to remain so. Elsewhere, Boyle
expands upon the psychological aspect:

The Vitious Affections and Habits, and the depraued frame of Mind,
to be met with in most Atheists, dos very much indispose them to be
convinc'd of the otherwise sufficient Proofes of a Deity.

The sinful Lusts, unruly Passions, and corrupt Interests of such vitious
Persons as Atheists are commonly obserued to be, cannot but haue a
great stroke in their Disinclinations, or the Judgements they pass of the
truth or falseness of things, and the force of Arguments that are employd
to proue them: ffor so predominant an Affection as Selfe Loue is wont to
be, especially in men, that thinke themselues born for themselues, and
acknowledge no Superior Being, cannot but make a man very indispos'd
to approue Arguments that would establish a Doctrine, wch he foresees
will highly condemn him, both as to his Opinions and his Practises. And
ye natural Connexion he discerns betwixt ye Conviction of a Deity, and
ye condemning that course of Life, wch ye Deity must needs abhor, and
will severely punish, makes it little less than morally impossible for
an already degenerate man to consent to part with Atheism, whilst he

32 BP, 2:63–4.

resolues to stick to his Vices; many of wch naturally flow from it, and cannot be plausibly excus'd, nor quietly enjoyed, without it.[33]

To some extent then, Boyle is not trying to offer arguments which will convert, or even convince, the uncommitted.[34] What is his strategy? He has, negatively, a series of arguments to show that Epicurean atheism will not, and indeed cannot, work. Then, positively, he has an ingenious three-stage maneuver to arrive at the nature of God. First he opts for a fairly standard design argument, to show that there is a deity of some sort or other. (In fact Boyle also has in this context another, and much more interesting proof. Of this, more later.) Having done that, the question will be, he thinks, what sort of instituted religion ought one to embrace? That is the function of miracles: They show us which of the instituted religions one ought to embrace.[35] Boyle was fortunate enough to be brought up in the Christian faith and to discover that the miracles which God has given to humanity point, in fact, to that faith as the correct one. Finally, having found the right instituted religion, we may then turn to the writings and traditions of that religion to discover the genuine revelations that God has given to us, and these will allow us to know what his properties are.

It is important to bear this interesting three-stage maneuver in mind when reading Boyle's arguments against the atheists, for

33 BP, 6:301.
34 Boyle's stance here is not unusual. For similar contemporary positions, see Berman, *Atheism*, Chaps. 1–2.
35 Boyle's two favorite miracles are (1) the miracle of Pentecost, which he takes to be both empirically well attested and consonant with the dignity of the deity, and – for those who might be inclined to think miracles rare or a thing of the past – the miracle of the attaching of the human soul to the human fetus: "And since also there are [every day,] in the world *every day,* many Myriads of human Engines (if I may so call them) produc'd in the wombs of their Mothers, wth. an admirable Organization & Sensitive life, before the superveneing of the rational Soul: Since, I say, these things are so, and that there is no meerly Physical or Mechanical Agent that can make an intimate Union between two such differing Beings; but this Union as well as most of the Laws of it, depends upon the free or arbitrary institution of God, that is pleas'd to unite two such differing substances; and this particular Soul rather than any other, to this particular Body; we may argue thence that the Providence of God doth in a peculiar mãner extend[s] to[o], and imploy[s] itself about, Mankind, and every individual person compris'd in it; and dos for the sake of Mankind almost every moment in the day; work Physical Miracles, (if I may so [call] stile them;) that is, form Animals of such a Compounded nature, as the Mechanical powers, or Laws of matter & motion, would not wth. out a peculiar interposition of God, be able to produce. (BP, 2:62)

otherwise it will seem at various stages as if he is quite unsophisti-
catedly begging the question, when in fact the argument, though
it may be weak in other ways, is innocent of that fallacy.

What are Boyle's arguments against the Epicureans? First of all,
we should note that he takes the position of the Epicurean atomist
very strictly. In this context Boyle doesn't really canvass the
notion that one might be a kind of Epicurean atomist without
subscribing, for example, to the Epicurean theory of what he calls
the declination of the atoms.[36] This is an interesting and unusual
lack of philosophical subtlety on Boyle's part, but I think it
doesn't really harm his position, for the main thrust of the
argument would not be affected if he were to be more charitable to
Epicurus and the Epicureans. Still, it is not clear that he realizes
this, and the fact that he does not feel able to be more charitable in
argument may reveal something about his confidence in his own
position.[37]

To begin, Boyle reminds us that God is a primary and transcendent being, so first, some incomprehensibility is only to be
expected, and second, as a consequence, the notion of God may
be subject to "specious, and perhaps knotty, objections,"[38] but
finally, and importantly, the "same or other insuperable ones,
may be urg'd against some opinions the Atheist must embrace, if
he reject ye Being of a Deity."[39]

It is in the spelling out what is involved in this final point that
we find most of Boyle's anti-Epicurean arguments. Boyle takes
the Epicureans to hold, centrally, that the world is composed of
atoms that are sempiternal (though, importantly for Boyle's argument, the world as currently constituted had a beginning); that the
the atoms in question are indestructible; that they come in an

---

36 It has often been pointed out that the "swerve" is an ad hoc device in the Epicurean
   system of atoms falling in Euclidean space. It seems, however, that it would be
   unnecessary in a positively curved space such as we believe ourselves in fact to inhabit.

37 As a boy Boyle was certainly troubled by religious doubt, which drove him at one stage
   to contemplate suicide. By the time he was twenty-one, however, he had learned to live
   with these doubts: "But . . . neuer . . . did these fleeting Clouds [of doubt], cease now &
   then to darken/ obscure/ the clearest serenity of his quiet: which made him often say
   that infections of this Nature were such a Disease to his Faith as the Tooth-ach is to the
   Body; for tho it be not mortall, 'tis very troublesome." (Quoted in Maddison, *Life of
   Boyle*, p. 35.)

38 BP, 2:1. Boyle deals with this point further, but I shall not, in this chapter, be able to
   follow him in this.

39 BP, 2:1.

indefinitely large variety of shapes and sizes but not of speeds; that they have a preferred direction, though with an unexplained declination which stops them from traveling forever in parallel lines; and that every atom has the power of self-motion. In this last case, incidentally, Boyle is being more than usually unfair. For one of the big changes that the seventeenth century brought in was the common acceptance of inertial laws: What that meant in practice was that changes no longer required an explanation; only changes of state required that.[40] But Epicurus was writing from within a framework whose dynamics required explanation for change *simpliciter*. However, there was no real reason for requiring such a view of a seventeenth-century Epicurean, anymore than there would have been for requiring it of Boyle himself.[41] With this as background, we may look briefly at Boyle's objections to the Epicurean point of view.

Epicurus (and by extension the modern "Somatists") must admit, Boyle argues,

(1) that every particular atom is self-existent;[42]
(2) that each atom has existed from eternity – this follows, says Boyle, somewhat unconvincingly,[43] from (1);
(3) that every atom is incorruptible and indestructible;
(4) that each atom must have several *"Emanatiue"* attributes

40 Boyle makes this very point, indeed, against the Epicurean account of the behavior of atoms: "One of the most fundamentall Customes or Laws of Nature, . . . is *that every undivided Body will alwayes continue in that state wherein it is, unless it be put* out of it by some Externall force" (BP, 2:6). However, the implications of inertial laws took some time to be realized, and Kant still found it worthwhile to make this point explicitly a century later (*Critique of Pure Reason*, A207n = B252n).

41 Interestingly, in the already-mentioned "[Of ye Atomicall Philosophy]" Boyle defends Epicurus and Democritus against what he considers the unfair charge that their atoms must be "Mathematicall points" and notes that it is enough if they are *minima naturalia*: "[B]y Atoms the Assertors of them /vnderstand/ not indiuisible or Mathematicall points which are so void of quantity that the subtle razor of Imagination it selfe cannot dissect them but minima Naturalia or the smallest particles of bodyes which they call Atomes not because they cannot be suppos'd to be diuided into yet smaller parts (for they allow them . . . both quantity & figure as wee shall see anon) but because tho they may be further diuided by Imagination yet they cannot by Nature, which not being able in her resolutions of Naturall bodyes to procceed ad infinitū) must necessarily stop somewhere & haue some bodyes which shee can possibly no further . . . subdiuide & which therefore may be justly termed Atomes. (BP, 26:162–3)

42 The arguments that follow are culled from the first thirty pages of BP, vol. 2.

43 Unconvincingly, that is, in an intellectual context that does not blink at Heisenberg uncertainty and the consequent possibility of virtual particles becoming real. But it would have seemed like a plausible move to Boyle's contemporaries.

("By wch are meant such as flow immediately from it's own nature wthout the intervention of any cause or Agent").[44] Examples are its determinate bulk or size, and its particular figure. We need, that is, an explanation for a given atom's having, not size in general, but just this precise size which it has, and so for the other qualities.

Boyle suggests that we might add, as emanative qualities, "Gravity or *weight* (for Epicurus calls it βαρος) or the Internall power that every Atome is affirm'd to haue of moueing it selfe continually."[45] He notes, further, that the Epicurean hypothesis cannot give a good account of the recoiling of an atom when hit. For, "according to the Epicureans, both the motion of an Atome, and its tendency downwards, are coñaturall and Essentiall to it" ("as indeed they must say so," Boyle adds, "because they can assigne no cause of either"[46]): Why, then, are these tendencies lost when it hits another body?

Moreover, each atom must "be in a sort Independent from all other Beings." It acts "rather *conjunctly* than *dependently*" in aggregates. "Nor . . . is there any other reason then their own nature to be giuen of ye Impenetrability of Atomes."[47]

So far then, we have the following: The Epicurean, by denying that there is a god, cannot avoid believing things that

confound his understanding; since the most abstruse and perplexing Attributes, such as selfe-existence, Eternity, selfe-motion, &c. must necessarily belong either to God or to matter; and if he will not ascribe them to the Deity, he must doe it to the despicablest Atome. And sure tis less inconvenient in Philosophy to admit *one* Being to be endowed wth. Properties that we cannot perfectly comprehend, then to allow *many millions* of Beings, each of them endowed wth. some *such* Attributes and Propertyes; and wth. some *others* of wch. no reason can possibly be giuen, how it came by them. For in effect the Epicurean makes his Atomes so many little gods: And we can noe more giue our understandings satisfaction about the perplexing Difficulties that incumber the notion of an eternally Existent Being, when tis affirm'd of an Atome, then we can when tis ascrib'd to God.[48]

And there is as much trouble understanding how an atom could put itself in motion as there is in understanding how God could do it. Still more problems arise when we consider atoms acting

44 BP, 2:4.　45 BP, 2:5.　46 BP, 2:5.　47 BP, 2:6.　48 BP, 2.7.

together. For (1) they must be given both a motion downward and their declination, neither of which is proved by any argument a priori. (2) Nor is any reason offered for the fact that every atom (and they "differ exceedingly in size as well as figure"[49]) just happens to be too small to affect any of our senses. (3) Their movement is implausibly fast: as fast as a thought, says Epicurus. (4) Also implausibly, they all move with equal swiftness: "For how should any one of them excite or moderate it's motion to make it just of the same velocity wth. that of other Atomes, whose degree of velocity it knows nothing of, nor has any thing to doe with?"[50]

We have, as noticed earlier, (5) the problem of transferring motion and (6) the problem of cohesion;[51] "ffor to ascribe this to a Law of nature would be improper for *Them* to doe, who must acknowledge all nature subsequent to y^e. Primary affections of

49 BP, 2:8.
50 BP, 2:9. This seems to be mere caviling. Boyle is being interestingly, unnecessarily, and uncharacteristically ungenerous to his hypothetical opponent. It is a perfectly reasonable point against Epicurus, but there is no reason whatsoever to suppose that any of Boyle's Epicurean contemporaries need be committed to this particular Epicurean doctrine, even if they accepted the rest: It is certainly not entailed by the other doctrines mentioned here. Neither Epicurus nor his commentators explain very clearly why the atoms move with equal speeds, nor why their speed is "inconceivably" fast. (The relevant sections are in the Letter to Herodotus, in Bailey, *Epicurus*, §§ 61, 46^b, p. 37.) A possible reason for Epicurus's position may be seen by considering two of Aristotle's arguments against the void (in *Physics*, VI.215a24–216a7). Aristotle points out that, given certain plausible assumptions (such as that, for a given impelling force, a projectile's velocity is proportional to the density of the medium through which it travels), velocity in a vacuum would be infinite. That runs counter to the empirical evidence. So if we accept a void there must be a physically arbitrary limiting velocity (one of Boyle's "emanative" properties). But there is no further reason for such a limiting velocity not to be the same for all particles. Thus they would all have the same speed *in vacuo*. Aristotle notes further that we are still left with the implausibility that there will be a medium, possibly hypothetical, through which projectiles will move against resistance, with the same speed at which they move through the unresisting void; this might explain why, for Epicurus, the atoms traverse "every comprehensible distance in an inconceivably short time": That is sufficient to make it unlikely that there is a real medium in which they could move with the same velocity as they have *in vacuo* and also comes as close as a vacuist can to allowing Aristotle's point that velocity in a vacuum should be infinite. Given his dislike of emanative properties that are not God-given and his liking for theoretical parsimony, Boyle ought in fact to have *approved* of this feature of atoms, given the context in which Epicurus wrote!
51 This problem (why don't things, including corpuscles and their hypothetical hooks, fall apart?) was a problem for the seventeenth-century corpuscularians, as Leibniz pointed out explicitly in the above-mentioned "Confession of Nature against Atheists," a point repeated in the *Nouveaux essais* with a back reference to the 'Confession of Nature." (*Nouveaux essais sur l'entendement humain*, ed. André Robinet and Heinrich Schepers

Atomes, to be the product or effect of chance."[52] (7) We should notice, too, that the "casually concurring aggregate of Atomes" is in fact not "a rude heap or Chaos, but a most curious, as well as most vast *Automaton*,"[53] and it is most unlikely that it would have come about by chance. As Dryden put it, in words reminiscent of Boyle's own, "No Atoms casually together hurl'd/Could e're produce so beautifull a world."[54]

At this point in the development of his thesis Boyle almost finds a nice, if old, argument, but he unfortunately wanders off, en route, into one of his usual puzzlements about infinity:[55]

If wee deny *God* and assert *matter*, to be eternal, the understanding is stil distress'd & confounded to conceiue, how an Atome had lasted as many Ages, or yeares, or any other determinate measure of Time, a thousand yeares agoe, as it has done to day; (since at both those Dates, it has lasted infinite Ages or yeares;) and yt supposeing such measures to be applyed to the Duration of an Atome, it may be said to haue lasted as many yeares as Months, and as many dayes as houres, thô in a yeare there be twelue Months, and an houre be but the twenty fourth part of a day.[56]

[Berlin: Akademie-Verlag, 1962], pp. 60 and 223.) The note on *cohesion* in the excellent English translation of the *Nouveaux essais* by Peter Remnant and Jonathan Bennett gives useful cross-references to other discussions of this topic in the work of Leibniz and other seventeenth-century scientists. (*New Essays on Human Understanding* [Cambridge: Cambridge University Press, 1981], p. xxxiv.)

52 BP, 2:10   53 BP, 2:10

54 John, Dryden "To My Honored Friend, Sr Robert Howard, on His Excellent Poems," lines 31–2 (*The Poems of John Dryden*, ed. James Kinsley [Oxford: Clarendon Press, 1958], 1:13). At BP, 1:16 and 1:18, we have: "And when they tell us, that by the Casual Justlings of Atoms, they shuffled themselves into this goodly and orderly Fabrick of the World; and by other lucky Concourses of smaller Numbers *there were* made at first certain Wombs, as it were, in the Earth, whence sprang not only other Animals, but Men themselves, Rational Souls and all: when I say, they assent to such Fancys as these, that are not only destitute of proofs, but are so fable-like, that many parts of *Ovids Metapmorphosis* are far less so; few men, if any, shew themselves more credulous. &c." And at BP, 2:14r he writes, "[T]here remaines an insuperable difficulty to conceiue how a confusd rabble of Atomes could casually [range] *iustle* themselues *at ye beginning* into so beautifull, orderly, & curious a Frame as the World is." The point is repeated at various places in the printed works, e.g., 1:445, 3:48, 5:428.

55 Boyle, who often though unnecessarily tells us that he was no mathematician, is constantly interested in infinity, particularly, as here, in the fact that some proper subsets of denumerable infinities are themselves denumerable infinities. He is also repetitively excited by the fact that the diagonal and sides of a square are incommensurable. Like his near contemporary Spinoza, Boyle was not a thinker for whom a mathematical example soon lost its novelty. For a more positive view of Boyle's mathematical ability see M. Boas, *Robert Boyle and Seventeenth-Century Chemistry* (1958; reprint, New York: Kraus Reprint, 1968), pp. 34–6.

56 BP, 2:11.

The argument he has just missed is, of course, the one that is most familiar to modern readers from Kant's antinomies but that is also to be found in Saint Thomas and, earlier still, in Algazel: that if the world were infinite in past time, then any given moment or event would be the completion of an infinite series of finite parts with a lower limit to their size: but infinite series comprised of such parts cannot be completed, Q.E.D.[57]

Boyle now proceeds to make some points for which he claims no originality but that he thinks are telling against the materialist. Matter, by itself, will not only not explain motion but, more important, it will not explain four important empirical features of the world: the lawlike behavior of matter; the occurrence of spontaneous motion (self-moving powers); the existence of seminal principles (the ability of organisms to breed, and indeed to breed true); and the occurrence of thought.

We cannot say simply that "things proceed in nature as we see them to doe, because they need but follow the course that things were put into at first," for, as we have already noticed, "there remains an insuperable difficulty to conceiue how a confusd rabble of Atomes could casually [range] *iustle* themselues *at ye beginning* into so beautiful, orderly, & curious a Frame as the World is."[58]

---

57 Algazel, quoted in Averroes, *Destructio destructionum philosophiae Algazelis*, I (IX, 9rb; 10rb), apparently finds the argument plausible; Kant thinks it unanswerable given that "the sensible world" is a thing in itself (*Critique of Pure Reason*, A507 = B535); Aquinas (*Summa theologiae*, 1a.46.2, ad 6) points out the fallacy.

58 BP, 2:14r. The point is, of course, that in Boyle's preferred alternative story, God, as we have already seen, (1) creates matter; (2) puts it into motion; (3) establishes its laws; (4) guides its direction in such a way that, following the laws (which he sustains), our present world will ensue; (5) adds seminal principles; and (6) is always on hand to give out human souls, as required. Boyle's friend John Locke pointed out that even if we thought that incorporeal souls were not required, since matter itself might think, it would still be necessary to have a deity to superadd the power of thinking to matter, since it is no part of matter's essence that it can think. For Locke, what it amounts to is this. Thinking, or at least, human sense perception, requires one of two miracles: either the conjoining and interacting of a corporeal and an incorporeal substance, or the superaddition of thought to brute matter. In either case a deity is required. (John Locke, *Essay Concerning Human Understanding*, ed. Peter H. Nidditch [Oxford: Clarendon Press, 1975], 4.3.6.) Newton, incidentally, agrees with Boyle's main point: "The *Vis inertiae* is a passive Principle by which Bodies persist in their Motion or Rest, receive Motion in proportion to the Force impressing it, and resist as much as they are resisted. By this Principle alone there never could have been any Motion in the World. Some other Principle was necessary for putting Bodies into Motion, and now that they are in Motion, some other Principle is necessary for conserving the Motion," (*Opticks*, Q. 31).

Since the Epicureans allow the (present) world to have had a beginning, there are various seminal problems, such as the propagation of species and the formation of the original animals.

It is straining probability, Boyle thinks, to argue that atoms could by chance produce animals, and even if, by chance, an animal were produced, it is most unlikely (a) that it would have generative organs; (b) that they would be in working order; (c) that chance would have produced a contemporaneous mate for it, which also just happened to have working, appropriate, and matching generative organs. Moreover, it is noteworthy that in the history of the world no one has ever observed animals being formed in this manner.[59]

But the most important omission is this. The Epicurean hypothesis is not as universal as is pretended. For, leaving aside the problem about the origin of the world and of animals, this theory cannot account for the rational soul:

For I haue never yet seen proposd any mechanical Contrivances to make out how meer matter can be brought to frame Notions of Universals, and of things yt are not ye Objects of any of our senses;[60] and how it can affirm, deny, doubt, and suspend its Assent: and yet less, how it can

---

59 In Part I, Essay 4, of *The Usefulness of Natural Philosophy*, ("Containing a requisite Digression concerning those, that would exclude the Deity from intermeddling with Matter,") Boyle refers with approval to Lactantius's attack in *Divinae institutiones* (II.11), on the Lucretian account of the generation of animals (Chap. 12 in J.-P. Migne, *Patrologia latina*, vol. 6). The attack in Book III of the same work is also very much in line with Boyle's views. Boyle was well read in Lactantius, and is gentle with him, as he is with Saint Augustine, for being witty (and wrong) about the Antipodes, though he had a lifelong dislike of "Wit Profanely or Wantonly employ'd" (Maddison, *Life of Boyle*, p. 21). And in connection with the attack on Epicurus, he quotes, without giving the source, a passage from Lactantius's *The Wrath of God* (*De ira dei*, Chap. 10; in Migne, *Patrologia latina*, 7:106): Tanta ergo qui videat, et talia, potest existimare nullo effecta [RB: affecta] esse consilio, nulla providentia, nulla ratione divina, sed ex atomis subtilibus exiguis concreta esse tanta miracula? Nonne prodigio simile est, aut natum esse hominem qui haec diceret, [RB: ut Leucippum,] aut extitisse qui crederet, ut Democritum qui auditor eius fuit, vel Epicurum in quem vanitas omnis de Leucippi fonte profluxit. (Who, seeing things so great and so remarkable [the back reference is to the earth and the star-studded heavens], can believe that such wonderful things were produced by no plan, no providence, no divine reason, but by the joining together of small, minute atoms? Isn't it amazing that someone should have been born who said such things, or that there existed those who believed him? Such were Democritus, who was his pupil, and Epicurus, all of whose folly flowed from the font of Leucippus.)

60 These are straightforwardly Aristotelian points, generally assented to by a wide variety of Scholastic philosophers, though Boyle would not, perhaps, have relished explicitly admitting that he had such allies.

frame and connect long chaines of just Consequences about abstracted subjects; and can make solid Ratiocinations about surd Numbers, Incommensureable Lines, and other abstruse things, I know yt matter & motion may perform much, and more than vulgar Philosophers will easily admit or imagine. But I cannot find y^t ye mechanical Powers of matter can reach so far, as to produce ye Powers and Operations of ye Humane Intellect.[61]

Boyle remarks that he doesn't expect to see a mechanical engine that can demonstrate the principles of mechanics or demonstrate the incommensurableness of the side and diagonal of a square. All this is important for him in this context, because he thinks, first, that if the soul has faculties or operations not deducible from matter, then it is incorporeal and hence may survive its separation from the human body, but second, and more interesting, that if the soul is incorporeal, then, since (being incorporeal) it could not have been created by matter, it must have been created by a deity. (The existence of free will, he argues, provides a further reason for believing in the incorporeality of the soul.)

This is the second proof of the existence of a deity, mentioned earlier in this chapter, and provides the natural-theology lead-in to the subsequent moves, via miracles and revelation, to a full-blown faith.

Boyle has other and interesting arguments, including his own version of the ten-leaky-buckets fallacy (the version, in fact, taken up by Alvin Plantinga in our own day[62]), and a splendid if wild

---

61 BP, 2:15. Cf. Boyle's *Christian Virtuoso*, pt. 1 (1690), 5.517: "The real philosophy ... teacheth us to form true and distinct notions of the body, and the mind; and thereby manifests so great a difference in their essential attributes, that the same thing cannot be both. This it makes out more distinctly, by enumerating several faculties and functions of the rational soul; such as to understand, and that so, as to form conceptions of abstracted things, of universals, of immaterial spirits, and even of that infinitely perfect one, God himself: and also to conceive, and demonstrate, that there are incommensurable lines, and surd numbers; to make ratiocinations, and both cogent and concatenated inferences about these things; to express their intellectual notions, *pro re nota*, by words or instituted signs to other men; to exercise freewill ... ; and to make reflections on its own acts, both of intellect and will. For these and the like prerogatives, that are peculiar to the human mind, and superior to anything, that belongs to the outward senses, or to the imagination itself, manifest that the rational soul is a being of an higher order than corporeal.

62 One leaky bucket won't hold water, but perhaps ten might; one unsound argument won't prove God's existence, but perhaps several might. For Plantinga's use of the argument, see "Reason and Belief in God," in Alvin Plantinga and Nicholas Wolterstorff, eds., *Faith and Rationality* (Notre Dame, Ind.: University of Notre Dame Press, 1983), pp. 16–93, esp. p. 80.

version of Ockham's razor, noticed in passing earlier – but space does not allow me to go into these.

We have enough, however, to see what his preferred strategy would have been had he published: (1) identify the enemy as the Epicureans; (2) show that their system is incomplete and insufficient; (3) point out that it has at least as many difficulties as the deity system (indeed, it has many of the same difficulties); (4) use Ockham's razor to get from this position to the deity hypothesis, if that hypothesis is tenable; (5) show that it is tenable by showing its need or peculiar fitness to account (a) for the world generally, or (b) specifically for the existence of the incorporeal soul; (6) given that, use our knowledge of real or claimed miracles to discover that the Christian religion is the one that a believer in God ought to embrace; and finally (7) use the revelations of the Christian religion to discover the nature of God and our duties to God.

Why didn't he publish? I suggest that, in his own mind, he had never quite managed to convince himself that, to borrow Margaret Osler's happy phrase, Gassendi had "baptized" Epicurus.[63] Indeed, he may have felt, as I suggested earlier, that Epicurus was unbaptizable, much as Aristotle proved to be for Saint Thomas. On the other hand, he could hardly, at some level of consciousness, have failed to notice that one could be an atomist without being an old-fashioned Epicurean atomist. But his arguments against the somatic atheist were arguments against someone who was an old-fashioned Epicurean atomist. So he couldn't publish on atomism without taking account of the fact that Epicurus could be, and should be, brought up to date. And he couldn't do that without undercutting his arguments against the somatic atheist. Similarly, he couldn't publish his arguments against atheism without updating his arguments to take account of the fact that there were atomists who weren't, in the old sense, Epicurean atomists. Now in fact his arguments would, I think, have worked as well against the new materialists as against the old.[64] But then, much

---

63 M. J. Osler, "Baptizing Epicurean atomism: Pierre Gassendi on the Immortality of the Soul," in Margaret J. Osler and Paul Lawrence Farber, eds., *Religion, Science, and World View* (Cambridge: Cambridge University Press, 1985), pp. 163–83.

64 There is much more to be said here, including the fact that the notion of "chance" was undergoing transition in the seventeenth century (see, e.g., Ian Hacking, *The Emergence of Probability* [Cambridge: Cambridge University Press, 1975]), but Boyle's arguments in general rely on the way in which the atheist's explanations cut out a level before the

of the detail of the manuscript writings would have been inappropriate. So I conclude, finally, that a possible main, or anyway contributing, reason for not publishing was the need to recast his arguments. Typically, when Boyle published late in life on a subject he had thought about earlier he employed the same arguments and considerations he had thought of earlier. Here he couldn't do that, and that disincentive may have been, along with his habitual disinclination to publish, just sufficient to prevent him from going to press.[65]

believer's, and this is a general truth about the difference between atheists' explanatory systems and those of believers, not something that depends on particular details. Additionally, Boyle is quite right to point out to us that the atomism of his time, and *a fortiori*, the atomism of Epicurus, could not even begin to explain sentient life.

65 But, as Boyle would have said, all of this upon the by: I assert all of it hypothetically, and none of it absolutely.

11

Stoic and Epicurean doctrines in Newton's system of the world

B. J. T. DOBBS

INTRODUCTION

In the last edition of his *Opticks* Isaac Newton had this to say about the ultimate particles of matter. "It seems probable to me, that God in the Beginning form'd Matter in solid, massy, hard, impenetrable, moveable Particles, of such Sizes and Figures, and with such other Properties, and in such Proportion to Space, as most conduced to the End for which he formed them."[1] Newton also observed that the primitive particles never wear out or break, and their permanence guarantees that nature will be "lasting," with the "Changes of corporeal Things . . . placed only in the various Separations and new Associations and Motions of these permanent Particles."[2]

Those phrases seem at first glance to place Newton firmly in the Epicurean camp – to align him fully with the doctrines of the ancient atomists, who argued for indestructible solid bodies, or atoms, that might vary in weight, shape, and size.[3] But that simple view of Newton's position must rapidly be modified on four counts: the first, religious and theological; another, scientific; the next, historical; the last, philosophical. Although the four categories overlap to a certain extent, one may treat them separately for purposes of analysis, and in this chapter I shall argue that Newton accepted only a very incomplete and partial version of

---

1 Isaac Newton, *Opticks, or A Treatise of the Reflections, Refractions, Inflections & Colours of Light* (New York: Dover, 1952), p. 400.
2 Ibid.
3 Norman Wentworth DeWitt, *Epicurus and His Philosophy* (Minneapolis: University of Minnesota Press, 1954), pp. 155–70.

Epicureanism. Probably the most significant modifications of
ancient atomism effected by Newton in his system of the world
derived from Stoicism, and it will be profitable later to explore the
tension and synthesis in Newton's thought between discreteness
and continuity in the natural world: in other words, between the
atomistic ideas of discrete particles in Epicureanism and the tonic
continuum pervading the whole world in the Stoic *pneuma*.

NEWTON'S RELIGIOUS AND THEOLOGICAL OBJECTIONS TO EPICUREANISM

In the passage from Newton's *Opticks* already quoted, one
may see immediately one of his religious and theological objections to Epicureanism: the issue of creation. Epicurus had insisted
that matter was "uncreatable."[4] To Epicurus that was probably
the logical counterpart of the indestructibility of matter, and the
Epicurean doctrine was nothing more or less than an early philosophical statement of what came, in the nineteenth century, to be
known as the law of the conservation of matter. Matter can be
neither created nor destroyed. Although that nineteenth-century
law has now, in the twentieth century, been subsumed under
broader principles of conservation, it was part of the foundation of
classical physics and chemistry. Why, then, should Newton have
objected to it?

Epicureanism had, of course, been revived in the context of
Christian Europe in the seventeenth century, at a time when the
notion of an eternal material universe was totally unacceptable.
The first chapter of Genesis declared that "in the beginning God
created the heavens and the earth," and part of the Christianization
of ancient atomism had been to insist that in the beginning God
created the atoms.[5] Newton's own conservation law was thus
incomplete. Matter could not be destroyed, but it had been created
by divine fiat in the beginning.

Yet another religious/theological problem for Newton focused
on the motion of the atomic particles. The atoms are always in
motion, according to Epicurus, with both linear and vibratory

4 Ibid., p. 156.
5 Genesis 1.1; Thomas Franklin Mayo, *Epicurus in England (1650–1725)* (Dallas: Southwest Press, 1934), pp. 30–1, 67, 110, 136, 140; Robert Hugh Kargon, *Atomism in England from Hariot to Newton* (Oxford: Clarendon Press, 1966).

motions. In addition, the atoms are capable of swerving spontaneously, veering ever so slightly from linear motion.[6] But what the atoms do not have, in the Epicurean system, is any sort of divine guidance. The atomists had been called "atheists" even in antiquity because of that fact, and those who revived the corpuscular philosophy in the seventeenth century, not wishing to be labeled atheists themselves, insisted that God had given the particles their motion in the beginning, when he created them.[7] All that resulted, then, was due not to random corpuscular activity but rather to that initial divine guidance. That doctrine was correct as far as it went, Newton thought, but it simply did not go far enough.

Following the lead of the Cambridge Platonists, Newton came to realize that initial divine guidance, given only at the time of creation, was insufficient; this notion would lead inevitably to deism. In order to maintain the Christian view of a providential God who continues to guide and maintain his creation, one must incorporate into the system of the physical world some provision for continuous divine guidance of the atoms.[8]

NEWTON'S SCIENTIFIC OBJECTIONS TO EPICUREANISM

On the scientific side, Newton had problems with Epicurean explanations of cohesion. The question of cohesion had always plagued theories of discrete particles, atomism having been criticized even in antiquity on this point. The cohesion of living forms, in particular, seems intuitively to be qualitatively different from anything that the random, mechanical motion of small

6 DeWitt, *Epicurus*, p. 157.

7 B. J. T. Dobbs, *The Foundations of Newton's Alchemy, or "The Hunting of the Greene Lyon"* (Cambridge: Cambridge University Press, 1975), pp. 100–5; Henry Guerlac, "Theological Voluntarism and Biological Analogies in Newton's Physical Thought," *Journal of the History of Ideas*, 44 (1983), 219–29; idem, "Newton et Epicure," in Henry Guerlac, *Essays and Papers in the History of Modern Science* (Baltimore: Johns Hopkins University Press, 1977), pp. 82–106; Kargon, *Atomism*, pp. 64, 67–8, 87–9; Margaret J. Osler, "Descartes and Charleton on Nature and God," *Journal of the History of Ideas*, 40 (1979), 445–56.

8 Mayo, *Epicurus in England*, pp. 128–43; Robert H. Hurlbutt III, *Hume, Newton, and the Design Argument*, rev. ed. (Lincoln: University of Nebraska Press, 1985), pp. 3–132; Jacob Viner, *The Role of Providence in the Social Order: An Essay in Intellectual History. Jayne Lectures for 1966*, Memoirs of the American Philosophical Society, no. 90 (Philadelphia: American Philosophical Society, 1972), pp. 8–9; Keith Hutchinson, "Supernaturalism and the Mechanical Philosophy," *History of Science*, 21 (1983), 297–333.

particles of matter might produce. Nor does atomism explain even mechanical cohesion in inert materials very well, for it requires the elaboration of ad hoc, unverifiable hypotheses about the geometric configurations of the atoms or else speculation about their quiescence under certain circumstances. In the various forms in which corpuscularianism was revived in the seventeenth century, the problems remained, and variants of ancient answers were redeployed.[9]

The most comprehensive answer to the problem of cohesion in antiquity had been given by the Stoics. The Stoics postulated a continuous material medium, the tension and activity of which molded the cosmos into a living whole and the various parts of the cosmic animal into coherent bodies as well. Compounded of air and a creative fire, the Stoic *pneuma* was related to the concept of the "breath of life" that was thought to escape from a living body at the time of death and allow the formerly coherent body to disintegrate into its disparate parts. Although always material, the *pneuma* becomes finer and more active as one ascends the scale of being, and the proportion of the more corporeal air in the *pneuma* decreases as the proportion of the less corporeal fire increases. Thus the deity, literally omnipresent in the universe, is the hottest, most tense, and most creative form of the cosmic *pneuma*, or aether: pure fire, or nearly so. The cosmos permeated and shaped by the *pneuma* is not only living; it is rational and orderly and under the benevolent, providential care of the deity. Though the Stoics were determinists, their deity was immanent and active in the cosmos, and one of their most telling arguments against the atomists was that the order, beauty, symmetry, and purpose to be seen in the world could never have come from the random, mechanical action of small particles of matter. Only a providential God could produce and maintain such lovely, meaningful forms.[10]

The original writings of the Stoics were mostly lost but not

---

9 Lancelot Law Whyte, *Essay on Atomism: From Democritus to 1960* (1961; reprint, New York: Harper & Row [Harper Torchbooks] 1963); E. C. Millington, "Theories of Cohesion in the Seventeenth Century," *Annals of Science*, 5, (1941–7), 253–69; Kargon, *Atomism*, pp. 43–54; A. A. Long, *Hellenistic Philosophy: Stoics, Epicureans, Sceptics* (London: Duckworth, 1974), pp. 152–6; B. J. T. Dobbs, "Newton's Alchemy and His Theory of Matter," *Isis*, 73 (1982), 511–28.

10 Shmuel Sambursky, *Physics of the Stoics* (1959; reprint, London: Hutchison, 1971); Stephen Toulmin and June Goodfield, *The Architecture of Matter* (1962; reprint, New

before ideas of *pneuma* and *spiritus* came to pervade medical doctrine, alchemical theory, and indeed the general culture of late antiquity with form-giving spirits, or souls, and active, guiding, vital principles. Alchemy, in particular, carried much Stoicism into the seventeenth century, and, as is now fairly well known, Newton studied alchemical literature exhaustively. In the end Newton chose to modify Epicureanism by adding to the atoms and the void certain "forces" or "active principles" that served, in the microcosmic part of his system, to account for cohesion and life. Newton thus drew upon certain aspects of Stoic thought to create a new and better atomism.[11]

NEWTON'S OWN INTELLECTUAL DEVELOPMENT

However, when I turn to Newton's own intellectual development, I must offer a historian's *apologia*. Up to this point I have argued that Newton found problems in Epicurean doctrines, both from the Christian and from the scientific point of view. He did do those things, but I fear I have made the story sound too simple, as if Newton sat down one day with the Epicurean texts and consciously and rationally decided how those texts had to be modified. History is rather messier than that, and historians always want to make philosophical purities more complex by

York: Harper & Row [Harper Torchbooks], 1966), pp. 92–108; Mary B. Hesse, *Forces and Fields: The Concept of Action at a Distance in the History of Physics* (1962; reprint, Westport, Conn.: Greenwood Press, 1970), pp. 74–9; John M. Rist, ed., *The Stoics* (Berkeley and Los Angeles: University of California Press, 1978), esp. Robert M. Todd, "Monism and Immanence: The Foundations of Stoic Physics," pp. 137–60, and Michael Lapidge, "Stoic Cosmology," pp. 161–85.

11 Mary Anne Atwood, *Hermetic Philosophy and Alchemy: A Suggestive Inquiry into "The Hermetic Mystery" with a Dissertation on the More Celebrated of the Alchemical Philosophers*, rev. ed. (New York: Julian Press, 1960); D. P. Walker, *Spiritual and Demonic Magic from Ficino to Campanella* (1969; reprint, Notre Dame, Ind.: University of Notre Dame Press, 1975); idem, *The Ancient Theology: Studies in Christian Platonism from the Fifteenth to the Eighteenth Century* (London: Duckworth, 1972); idem, "Medical Spirits: Four Lectures," Boston Colloquium for the Philosophy of Science, 27 October–19 November 1981; Rosaleen Love, "Some Sources of Herman Boerhaave's Concept of Fire,"*Ambix*, 19 (1972), 157–74; idem, "Herman Boerhaave and the Element–Instrument Concept of Fire," *Annals of Science*, 31 (1974), 547–59. See also B. J. T. Dobbs, *The Janus Faces of Genius: The Role of Alchemy in Newton's Thought* (Cambridge University Press, forthcoming).

insisting on that ragged and often obscure thing called the "historical record." In Newton's case the historical record is extraordinarily complete and really must be utilized if one is to grasp the form his eclecticism took.

Newton did indeed struggle with the ancient rivalries of Epicureanism and Stoicism and produce a sort of eclectic synthesis of them that was a better match for the natural world than either ancient philosophy had been. But he came to that success rather late and in an oddly piecemeal and perhaps illogical fashion. In fact, Newton was neither Stoic nor Epicurean at first; he was, rather, a Cartesian for about twenty years.

I will not drag the reader in great detail through the manuscript record of Newton's thought that sustains this label, but from the 1660s, in his student notebook, then in various documents of the 1670s, until he actually began work on the *Principia* in 1684, one finds Newton to have been a rather wholehearted disciple of Descartes.[12] For purposes of the analyses relevant to this essay, one may now isolate two aspects of the Cartesianism that appears in those early papers of Newton's.

On the one hand, Descartes was not an atomist in the classical sense. A believer in the corpuscular structure of matter he was, certainly, but Descartes did not hold that the particles were uncuttable. Cartesian corpuscles suffered attrition, were ground down by friction into other sorts of matter.[13] Similarly, in all of Newton's early systems, particles of gross matter were attenuated into the finer particles of aether, while, conversely, aethereal particles condensed to form all the grosser sorts of matter.[14] Newton did not talk at all in these early papers about permanent

12 J. E. McGuire and Martin Tamny, *Certain Philosophical Questions: Newton's Trinity Notebook* (Cambridge: Cambridge University Press, 1983); Isaac Newton, "Of Natures obvious laws & processes in vegetation," Dibner Library of the History of Science and Technology, Smithsonian Institution, Washington, D.C., Dibner MSS 1031B; idem, "An Hypothesis explaining the Properties of Light, discoursed of in my severall Papers," in H. W. Turnbull, J. P. Scott, A. R. Hall, and Laura Tilling, eds, *The Correspondence of Isaac Newton*, 7 vols. (Cambridge: Cambridge University Press for the Royal Society, 1959–77), 1:362–82 (Newton to Oldenberg, 7 December 1675).

13 J. F. Scott, *The Scientific Work of René Descartes (1596–1650)* (New York: Garland, 1987), pp. 67–8, 160–1, 167–81.

14 McGuire and Tamny, *Certain Philosophical Questions* pp. 362–5, 426–31; Newton, "Of Natures obvious laws & processes in vegetation," fols. 3v–4r; Newton to Oldenberg, 7 December 1675, in Newton, *Correspondence*, 1:365–6. Newton's speculative systems are discussed in detail in Dobbs, *Janus Faces*, Chap. 4.

primitive particles that never wear out or break. Although one knows that Newton was introduced quite early to the Epicureanism of Gassendi, which came to him by way of Charleton's epitome, Newton made no commitment to atomism until later.[15] Interestingly enough, Newton's own copy of the *De rerum natura* of Lucretius, though showing signs of concentrated study, could not have been purchased before 1686, and that year may very well have been the time when Newton finally became an atomist.[16]

On the other hand, returning to Newton's Cartesianism, Descartes's system of the world was a plenistic one, completely filled with invisible swirls of aethereal vortices that served in the Cartesian system to explain the action of gravity, among other things, in a mechanical way.[17] Descartes was determined to exclude all occult principles from his philosophy, as is well known, and that included the apparent "attraction" of gravity, as well as the shaping spirits, souls, forms, and active principles of Peripatetics, Stoics, and alchemists. Descartes's physics was impact physics: Motion could be transferred from one body to another only by mechanical impact or pressure. Hence the postulate of a gravitational aether that *pushed* heavy bodies toward the earth, that *pushed* the planets around the sun, in much the same way that chips of wood may be swirled about the center in a whirlpool of water. Newton fully accepted that Cartesian explanation of gravity also, and one may find in Newton's early papers several speculative systems established upon Cartesian principles, in which gravity is explained by the mechanical impact or pressure of the gravitational aether.[18]

Had Newton's intellectual development ceased in 1683, there would have been no *Principia* in 1687, and perhaps no Scientific Revolution either. But in 1684 and 1685 a series of events forced Newton to break definitively with his Cartesian heritage and to

15 A. Rupert Hall, "Sir Isaac Newton's Note-book, 1661–65," *Cambridge Historical Journal*, 9 (1948), 239–50; Richard S. Westfall, "The Foundations of Newton's Philosophy of Nature," *British Journal for the History of Science*, 1 (1962–3), 171–82; idem, *Force in Newton's Physics: The Science of Dynamics in the Seventeenth Century* (New York: American Elsevier, 1971), pp. 324–6; idem, *Never at Rest: A Biography of Isaac Newton* (Cambridge: Cambridge University Press, 1980), pp. 89–97.

16 John Harrison, *The Library of Isaac Newton* (Cambridge: Cambridge University Press, 1978), p. 183, item 990: Titus Lucretius Carus, *De rerum natura libri VI* (Cambridge, 1686), now Trinity College, Cambridge, NQ.9.73.

17 Scott, *Scientific Work of Descartes*, pp. 167–75.

18 See especially the references in n. 14, this chapter.

consider several issues anew.[19] Chief among his new concerns was the question, Plenum or void? Is the universe full of the sort of imperceptible mechanical aether postulated by Descartes, or is there only empty space between the particles of matter, as Epicurus had claimed?

The visit of the astronomer Edmund Halley to Newton in August of 1684 redirected Newton's thoughts to celestial dynamics. Newton had made some inconclusive studies on the topic in the 1660s but had since then been immersed in other types of work. A brief exchange with Robert Hooke in 1679/80 had suggested a new way of analyzing planetary motion to Newton, but by the time of Halley's visit in 1684 Newton had mislaid his papers on the subject.[20] Stimulated by Halley's visit, however, Newton undertook his calculations again and began the work that led to his scientific masterpiece of 1687, *The Mathematical Principles of Natural Philosophy*. Only in the course of writing that work did Newton confront the problems that inhered in all his earlier aethereal gravitational systems that had been formulated on Cartesian principles.[21]

Theorem 1 of Newton's first tract on motion was a mathematical demonstration of a general area law for bodies revolving about an immovable center of force, a demonstration that remained substantially unchanged through all versions of the preliminary papers on motion and then became Proposition I, Theorem I, of Book I of the *Principia*: "The areas which revolving bodies describe by radii drawn to an immovable center of force do lie in the same immovable planes, and are proportional to the times in which they are described."[22] Earlier in the seventeenth century, Johann Kepler had determined that just such a

19 B. J. T. Dobbs, "Newton's Rejection of the Mechanical Aether: Empirical Difficulties and Guiding Assumptions," in Arthur Donovan, Larry Laudan, and Rachel Laudan, eds., *Scrutinizing Science: Empirical Studies of Scientific Change* (Dordrecht: Kluwer Academic, 1988), pp. 69–83.

20 John Herivel, *The Background to Newton's "Principia": A Study of Newton's Dynamical Researches in the Years 1664–84* (Oxford: Clarendon Press, 1965); Derek T. Whiteside, "Before the *Principia*: The Maturing of Newton's Thoughts on Dynamical Astronomy," *Journal for the History of Astronomy*, 1 (1970), 5–19; Westfall, *Never at Rest*, pp. 381–3, 402–4.

21 Whiteside, "Before the *Principia*."

22 Isaac Newton, *Sir Isaac Newton's "Mathematical Principles of Natural Philosophy" and His "System of the World"*, trans. Andrew Motte (1729), rev. ed., ed. Florian Cajori, 2 vols. (Berkeley and Los Angeles: University of California Press, 1966), 1:40–1; idem, *Isaac*

law of equal areas in equal times applied to the motion of planets about the sun, but Kepler could not really demonstrate the law mathematically in a general form.[23] The area law for planets in elliptical orbits that Kepler had determined but could not demonstrate mathematically was subsumed in Newton's general demonstration.

The problem, then, that Newton encountered here at the earliest stage of his writing is simple and obvious, in retrospect, but was one that proved insurmountable within the context of Cartesian philosophy. The observed area law of Kepler should not fit so closely with the exact area law derived mathematically by Newton, if the heavens are filled with a material aethereal medium such as Descartes had postulated.[24] The orbital speed of the planet varies from moment to moment; the planet moves more rapidly when nearer the sun, less rapidly in the parts of the orbit more distant from the sun. Unless the medium, the aether, is somehow disposed to move with exactly the same variable speed that the planetary body exhibits, the planet should encounter enough resistance from the medium to cause an observed deviation from the mathematical prediction, just as projectiles in the resisting terrestrial atmosphere are observed to deviate from mathematical prediction.[25]

Newton's "Philosophiae naturalis principia mathematica": The Third Edition (1726) with Variant Readings, ed. Alexandre Koyré and I. Bernard Cohen, 2 vols. (Cambridge, Mass.: Harvard University Press, 1972), 1:88–90; idem, *The Mathematical Papers of Isaac Newton*, ed. Derek T. Whiteside, 8 vols. (Cambridge: Cambridge University Press, 1967–80), 6:35–7, 539–42; Westfall, *Never at Rest*, pp. 411–14; Herivel, *Background*, pp. 258–9.

23 Owen Gingerich, "Johannes Kepler," in *Dictionary of Scientific Biography*, New York: Scribner, 1970–80), 7:289–312; J. L. E. Dreyer, *A History of Astronomy from Thales to Kepler*, 2nd rev. ed. (New York: Dover, 1953), pp. 387–8; Gerald Holton, *Thematic Origins of Scientific Thought, Kepler to Einstein* (Cambridge, Mass.: Harvard University Press, 1973), pp. 74–6, 79–80; Curtis Wilson, "Kepler's Derivation of the Elliptical Path," *Isis*, 59 (1968), 5–25; idem, "How Did Kepler Discover His First Two Laws?", *Scientific American*, 226:3 (1972), 93–106.

24 For a more extended discussion of this point, see Dobbs, "Newton's Rejection of the Mechanical Aether."

25 Isaac Newton, *De gravitatione*, in idem, *Unpublished Scientific Papers*, ed. and trans. A. Rupert Hall and Marie Boas Hall (Cambridge: Cambridge University Press, 1962), pp. 90–156, esp. pp. 112–13, 146–7. The manuscript of this work (University Library, Cambridge, Portsmouth Collection MS. Add. 4003) must date from late 1684 or early 1684/85, although it has in the past been seriously misdated by other scholars to the 1660s or 1670s. Its importance in the foundation of classical mechanics is attested by recent French and German editions: Isaac Newton, *De la gravitation ou les fondements de la*

Newton quickly realized that the close fit between observed law and mathematical law in effect precluded the mechanical gravitational aether of Descartes. In Newton's own words,

If there were any aerial or aetherial space of such a kind that it yielded without any resistance to the motion of comets or any other projectiles [such as planets] I should believe that it was utterly void. For it is impossible that a corporeal fluid [such as Descartes's aether] should not impede the motion of bodies passing through it, assuming that ... it is not disposed to move at the same speed as the body.[26]

Newton's realization that no form of the hypothetical gravitational aether of mechanical philosophy could be reconciled with actual celestial motions was evidently rather a shock to him, and it forced him to make a break with Cartesian philosophy.

Newton thus found himself in a difficult situation, in the *Principia* and post-*Principia* years, with respect to a causal principle for gravitation. He had been forced to make a dramatic break with the orthodox mechanical philosophy of his day, the philosophy that was generally understood, among advanced thinkers at the time, to be the most promising method of approaching the study of the natural world. He did not reject the entire system of mechanical thought, but he did question one of its most basic assumptions: that motion could be transferred only by the impact or pressure of one material body on another.[27]

The focus must now turn to Newton's search for a different sort of causal principle for gravitation during the ensuing twenty-five years. It seems one must recognize that Newton was dismayed by the fact that a combination of mathematics, observation, and experimentation had forced him to abandon the mechanical causal principle he had accepted for so long. It has often been assumed that he was content to be an early version of a philosophical positivist and to let his mathematical principles stand alone, but a close scrutiny of the surviving evidence from Newton's papers

mécanique classique, ed. and trans. Marie-Françoise Biarnais (Paris: Belles Lettres, 1985); Isaac Newton, *Über die Gravitation ... Texte zu den philosophischen Grundlagen der klassischen Mechanik*, translated into German and edited by Gernot Böhme (Frankfurt/M.: Klostermann, 1988). The German edition contains a facsimile reproduction of the original manuscript.

26 Newton, "De gravitatione", in *Unpublished Scientific Papers*, pp. 112–13, 146–7, my brackets.

27 Dobbs, "Newton's Rejection of the Mechanical Aether."

will not sustain that interpretation.[28] Quite the contrary. Newton had always accepted the Renaissance view of history as a decline from an original golden age, a time in which there had existed an original pure knowledge of things both natural and super-natural, an ancient wisdom subsequently lost or garbled through human sin and error and through temporal decay. By the time of the writing of the *Principia*, in the mid-1680s, Newton had already been engaged for a long time in attempts to restore the original truths once known to mankind by decoding obscure alchemical texts and by searching ancient records for the original pure religion. So when the (for him, modern) mechanical explanation of gravity failed, it was only natural that he should turn to ancient sources in an attempt to recapture the true explanation of gravity once known to the wise ancients. What he found there in antiquity that he thought to be correct he then incorporated into his cosmic system.[29]

Since the question of plenum versus void was of most immediate concern, Newton may first have begun his serious study of ancient thought on that issue by reading the atomists. As already noted, Newton's copy of Lucretius was published in 1686, so it could not have been purchased before that year. On the other hand, Newton had known Charleton's version of Gassendi's work on Epicurus since the 1660s. (There is no evidence that Newton ever read Gassendi directly, except for his *Institutio astronomica* in a 1682 edition.[30]) But the events of the 1680s had thoroughly shaken Newton's faith in Cartesian principles, so he was ready to take a fresh and unprejudiced view of ancient atomism. In all

28 B. J. T. Dobbs, "Newton's Alchemy and his 'Active Principle' of Gravitation," in P. B. Scheuer and G. Debrock, eds., *Newton's Scientific and Philosophical Legacy* (Dordrecht: Kluwer Academic Publishers, 1988), pp. 55–80; idem, *Janus Faces*, Chap. 6.

29 Dobbs, "Newton's Alchemy and His 'Active Principle' of Gravitation."

30 McGuire and Tamny, *Certain Philosophical Questions* pp. 20–1, 318–19: The editors of Newton's undergraduate notebook have documented Newton's use of Charleton's *Physiologia Epicuro-Gassendo-Charltoniana* of 1654 but found no evidence of Newton's use of Gassendi's *Syntagma philosophicum* of 1658. Although they do not rule out Newton's possible familiarity with Gassendi's *Animadversiones in decimum librum Diogenis Laertii* of 1649, they incline to the position that Newton read the Epicurean fragments directly as they were recorded in Diogenes Laertius, *De vitis dogmatis et apophthegmatis eorum qui philosophia claruerunt, Libri X*. Newton's copy of Diogenes Laertius (London, 1664) survives, with signs of Newton's dog earing: Harrison, *Library*, p. 133, item 519. The only work of Gassendi's recorded by Harrison as having been in Newton's library is Gassendi's *Institutio astronomica* (Amsterdam, 1682); see Harrison, *Library*, p. 147, item 650.

probability, it was at this time that Newton rejected the Cartesian postulate of corpuscles that could be ground down to become different sorts of matter and opted for the indestructible atoms of the ancients.

Did Newton also accept the void of ancient atomism? Here the evidence is more ambiguous. He was later to pin his argument for interstitial spaces empty of ordinary matter to certain experimental facts: that light can be transmitted through "pellucid" bodies without being stifled and lost, as it surely would be if it hit a truly solid particle, and also that magnetic and gravitational forces could pass through even very dense and opaque bodies "without any diminution." Newton called his interstitial spaces "pores," but they were not really void spaces, since they were permeated by the forces of cohesion and fermentation operating in micromatter, as well as by the cosmic force of gravitation.[31]

With respect to a possible cosmic void existing between the heavenly bodies, one must take into account Newton's letter to Bentley in 1692/93: "Tis unconceivable that inanimate brute matter should (without ye mediation of something else wch is not material) operate upon & affect other matter without mutual contact."[32] The context of the discussion was some remarks of Newton to Bentley concerning the force of gravity, and the statement Newton made there denying the possibility of "action at a distance" was perfectly general: He found it "so great an absurdity" that he believed "no man who has in philosophical matters any competent faculty of thinking can ever fall into it."[33] How, then, did Newton propose to explain the action of gravity? The sun reaches out somehow, across millions of miles, to hold the planets in their orbits. Once Newton had cleared the heavens of the mechanical aethereal whirlpools of Descartes, there seemed to be nothing there in those vast spaces. At least there were no material substances there, but Newton hinted to Bentley that perhaps the force of gravity was mediated by something else that was not material.

If there were no material aether in the heavens, then what was there? God, of course. Newton often stated his strong belief in the

31 Newton, Opticks, pp. 266–9, 401.
32 Newton to Bentley, 25 February 1692/93, in Newton, Correspondence, 3:253.
33 Ibid., p. 254.

literal omnipresence of God, and by the 1690s he also began to make at least private statements to the effect that gravity had its foundation only in the will of God – that is, that there was no material cause for it. By 1713 he was willing to hint in public that such was his belief, in the General Scholium that he added to the second edition of the *Principia*, published in that year.[34]

Henry More and other contemporaries of Newton were already treating the deity as an incorporeal yet three-dimensional being whose immensity constituted infinite three-dimensional space, and it is possible to see Newton's ideas on this subject as the "fruition of a long tradition," a tradition extending from Aristotle through Newton, in which Aristotle's finite plenum was slowly and by painful steps converted into the void, infinite, three-dimensional framework of the physical world required by classical physics. Newton's God-filled space was the penultimate development in the process by which concepts of space were developed by attributing to space properties derived from the deity. After Newton's time, the properties remained with the space, while the deity disappeared from consideration.[35]

But there were theological subtleties to Newton's view that deserve consideration. Simply to describe God as omnipresent – present where there is no body and also present where there is body – is to give a priority to body and the void that is theologically unacceptable. The creator must be conceived as prior to his creation, and Newton's conception surely owed something to the theological construct of God "as the ground of all being," in which God is the place of the world but the world is not his place. Newton sometimes used the Hebrew word *māqôm* (place) as an expression for God's omnipresence, and in the General Scholium he cited many of the Old Testament texts upon which Jewish theologies of space were based. "In him are all things contained and moved," Newton said, citing the passages from Kings, Psalms, Job, and Jeremiah that the rabbis and Jewish philosophers had pondered. In addition, he cited texts from "Moses" in Deuteronomy, from John, and from Acts, thus bringing in the "sacred

34 Dobbs, "Newton's Alchemy," and the references cited there.
35 Edward Grant, *Much Ado about Nothing: Theories of Space and vacuum from the Middle Ages to the Scientific Revolution* (Cambridge: Cambridge University Press, 1981), esp. pp. 182–264.

writers" from the most ancient Hebrew authority through Christ himself and earliest Christian antiquity.[36]

Were one to stop there, one would have to conclude that Newton's ideas were dependent only upon the received texts of the Judeo-Christian tradition. But although Newton's conception of God's omnipresence was undoubtedly richer than that required for a conception of physical space, enriched as it was by a profound appreciation of the religious awe of God's immediacy, ubiquity, and plenitude that the ancients had expressed, the sacred writers had given Newton no assistance in the matter of gravity. It was one thing to let them speak for him and reinforce his own sense of God's presence everywhere. It was quite another thing to ask them how God's omnipresence subsumed the mathematical laws of gravity he had found. On that point the sacred writers were silent.[37]

But perhaps the ancient natural philosophers were not entirely so. Newton's search for ancient opinions on gravity led him into all varieties of ancient thought, many of which are beyond the scope of this chapter. Those he found most satisfactory, however, were certain pre-Socratics and certain Stoics. In a draft from the 1690s Newton himself called attention to "the Stoics" as having had the right ideas:

Those ancients who more rightly held unimpaired the mystical philosophy as Thales and the Stoics, taught that a certain infinite spirit pervades all space *into infinity*, and contains and vivifies the entire world. And this spirit was their supreme divinity, according to the Poet cited by the Apostle. In him we live and move and have our being.[38]

Then, in a footnote to the General Scholium, Newton cited not only Thales but Pythagoras and Anaxagoras as well, as their ideas had been reported in Cicero's *De natura deorum*, two passages from

36 Brian P. Copenhaver, "Jewish Theologies of Space in the Scientific Revolution: Henry More, Joseph Raphson, Isaac Newton and Their Predecessors," *Annals of Science*, 37 (1980), 489–548; Newton, *Principia*: Motte–Cajori, 2:545; Koyré–Cohen, 2:762. Newton's citations were to 1 Kings 8:27; Psalms 139.7–9; Job 22.12–24; Jeremiah 23.23–4; Deuteronomy 4.39 and 10.14: John 14.2; Acts 17.27–8.

37 For a more extended discussion of this point, see the references in n. 28, this chapter.

38 Isaac Newton, draft Scholium, University Library, Cambridge, MS. 3965.12, fol. 269, as quoted in J. E. McGuire and P. M. Rattansi, "Newton and the 'Pipes of Pan,'" *Notes and Records of the Royal Society of London*, 21 (1966), 108–43, on p. 120. The original Latin of this manuscript may be found in Paolo Cassini, "Newton: The Classical Scholia," *History of Science*, 22 (1984), 1–58.

Virgil, one from Philo, and one from Aratus.[39] The concept of the deity as an infinite mind, soul, or spirit penetrating and pervading all things was the common factor in all of Newton's citations. One may take a single example to illustrate the point, that of Aratus. Aratus, whose name is now not widely known, was a Stoic poet whose didactic poem on the heavens opened with a hymn to Zeus as the all-pervasive providential deity.

> The sky is our song
> and we begin with Zeus; for man cannot speak
> without giving him names: the streets are detailed
> with the presence of Zeus, the forums are filled,
> and the sea and its harbors are flooded with Zeus,
> and in Him we move and have all our being.[40]

Aratus studied under Zeno the Stoic in Athens around 227 B.C., and, though this prologue to his astronomical work is thought to be his own, it is similar in some ways to the Stoic Cleanthes', "Hymn to Zeus."[41] The famous Pauline statement on God as the ground of all being, in Acts 17.27–8, was probably drawn from the prologue of Aratus, for the poem was immensely popular in Hellenistic and Roman circles, and Newton seemed to indicate his knowledge of that filiation in his draft comment of the 1690s, when he mentioned "the Poet cited by the Apostle." Aratus is also known to have influenced Virgil.[42]

But perhaps even more fundamental to Newton's use of Stoic ideas was the work of Philo and his successors. Newton's reference, in the *Principia*, was to Philo's *Allegorical Interpretation* of the Mosaic account of creation. In the passage Newton cited Philo argued that Moses called the mind "heaven," and Philo also said that "world" (cosmos), "in the case of mind," meant "all incorporeal things." Those are almost certainly the notions that attracted Newton's attention, for what one finds in Philo is a

39 Newton, *Principia*: Motte–Cajori, 2:545; Koyré–Cohen, 2:762.

40 Aratus, *Phaenomena*, lines 1–9 (trans. Stanley Frank Lombardo, 1976). Cf. Stanley Frank Lombardo, "Aratus' *Phaenomena*: An Introduction and Translation," Ph.D. diss., University of Texas at Austin, 1976, p. 102; *Sky Signs: Aratus' Phaenomena*, ed. and trans. Stanley Lombardo (Berkeley: North Atlantic, n.d.), p. 1.

41 Lombardo, "Aratus' *Phaenomena*," esp. pp. vii, 1–3, 30–2, 54–7, 102–3; *Sky Signs*, unpaginated Introduction, p. 1, and unpaginated n. 7; Cleanthes, *Hymn to Zeus*, in T. F. Higham and C. M. Bowra, eds., *The Oxford Book of Verse in Translation* (Oxford: Clarendon Press, 1938), pp. 533–5, no. 483 (trans. Michael Balkwill).

42 Jean Martin, *Histoire du texte des "Phénomènes" d'Aratos* (Paris: Klincksieck, 1956).

Platonizing Stoicism in which the traditionally material Stoic deity has been dematerialized – made "incorporeal."[43]

The more orthodox of the ancient Stoics had certainly held everything to be material, including the *pneuma* or aether as deity, probably in conscious reaction to the philosophical inadequacies of Aristotle's unmoved mover. The Stoics maintained the four traditional elements of earth, water, air, and fire, supposed to constitute the entire universe. These elements were ranked in order of ascending tonicity or tensional activity and divinity, and the cosmic *pneuma* that filled the heavens and permeated the entire universe was thought to be composed of the hottest possible combination of fire and air, the most tense, active, and divine substance in the world – in short, the Stoic deity.[44]

But the cosmic *pneuma*, as active principle, permeated and blended completely with body (conceived as passive principle), and the tensional powers of the *pneuma* held the body together. Stoic insistence upon the corporeality of the cosmic divine *pneuma* had always made the Stoic theory of total blending somewhat problematic. Are corporeal *pneuma* and corporeal body to fuse completely, and so become something that is neither God nor matter? Are two corporeal bodies to occupy the same space at the same time?[45] The original Stoic arguments on blending seem to have more or less avoided those issues by treating the *pneuma*, in its motion through matter, as a shaping, cohesive force, but another solution to the difficulty was to create a Platonizing Stoicism in which the corporeal (but active and divine) Stoic *pneuma* was made spiritual and incorporeal but still mingled and

43 Philo, *Allegorical Interpretation of Genesis II., III (Legum allegoria)* (Loeb Classical Library, 1962), 1:1. David Winston, *Logos and Mystical Theology in Philo of Alexandria* (Cincinnati: Hebrew Union College Press, 1985), pp. 9–12, provides a recent summary of modern opinion on Philo Judaeus. See also Samuel Sandmel, *Philo of Alexandria: An Introduction* (New York: Oxford University Press, 1979), and Max Pulver, "The Experience of the *pneuma* in Philo," trans. Ralph Manheim, in *Spirit and Nature: Papers from the Eranos Yearbooks*, ed. Joseph Campbell (Princeton: Princeton University Press, 1982), pp. 107–21, originally published in German in *Eranos-Jahrbuch*, 13 (1945). Newton owned a 1640 edition of Philo: Philo, *Judaeus, Omnia quae extant opera . . .* (Paris, 1640): see Harrison, *Library*, p. 216, item 1300, now missing.

44 David E. Hahm, *The Origins of Stoic Cosmology* (n.p.: Ohio State Universtiy Press, 1977), pp. 3–28, esp. pp. 14–15; H. A. K. Hunt, *A Physical Interpretation of the Universe: The Doctrines of Zeno the Stoic* (Melbourne: Melbourne University Press, 1976).

45 Hunt, *Physical Interpretation*, pp. 47, 50; Robert B. Todd, *Alexander of Aphrodisias on Stoic Physics: A Study of the "De mixtione"* (Leiden: Brill, 1976); Paul Hager, "Chrysippus' Theory of *pneuma*," *Prudentia*, 14 (1982), 97–108.

blended with every body. That was the solution chosen by Philo, by many of the church fathers, and by the prominent sixteenth-century Renaissance Stoic Justus Lipsius.[46] It was also the solution Newton needed.

The tenor of Philonic Stoicism was to shift from a materialistic monism toward a Platonic dualism, within which there existed a distinction between spirit and matter that was quite foreign to most of the ancient Stoics. Under the dispensation of a Platonizing Stoicism, with its incorporeal deity as the active principle, Newton was able to use the Stoic concept of the deity as a tensional force, binding the parts of the cosmos together, penetrating and mingling with all bodies, and that without the inconveniences attendant upon the corporeal *pneuma* or aether.[47]

In the matter of gravity, Newton had recognized that no material, mechanical cause would serve and had been forced to make a break with corporeal causality. The evidence he had in hand denied the presence of the corporeal aether. Yet gravity acted, and it seemed to act as if it penetrated to the very centers of bodies. Only spirit could penetrate in that way without constituting a frictional drag by acting on the surfaces of bodies and/or on the surfaces of their internal parts. In the context of Stoic thought, as interpreted by Philo and Lipsius, the all-pervasive spirit was the active principle, acting everywhere to penetrate and bind the passive principle of matter. The conceptualization of gravity as active principle, as subsumed by the literal omnipresence of God, as a spiritual force binding all together, was to serve Newton for many years. Stoic the idea was, certainly, but Newton used it in its Platonizing version, in which the deity was wholly immaterial, noncorporeal, yet all-pervasive.

NEWTON'S PHILOSOPHICAL ECLECTICISM:
CONCLUSION

In 1706, in what was then Query 23 of the Latin *Optice*, and again in Query 31 of the next English *Opticks* in 1717/18,

46 Sandmel, *Philo*; M. Spanneut, *Le Stoïcisme des pères de l'Église: de Clément de Rome à Clément d'Alexandrie* (Paris: Seuil, 1957); Léontine Zanta, *La renaissance du stoïcisme au XVIᵉ siècle* (Paris: Champion, 1914); Jason Lewis Saunders, *Justus Lipsius: The Philosophy of Renaissance Stoicism* (New York: Liberal Arts, 1955).

47 For a more extended discussion of this point, see Dobbs, "Newton's Alchemy and His 'Active Principle' of Gravitation."

Newton openly designated gravity as an active principle in his most general statement on the forces associated with matter. He described the particles of matter, as one has already seen, in Epicurean terms as indestructible. But then he went on to say, "It seems to me farther, that those Particles have not only a *Vis inertiae*, accompanied with such passive Laws of Motion as naturally result from that Force, but also that they are moved by certain active Principles, such as is that of Gravity, and that which causes Fermentation, and the Cohesion of Bodies".[48] Within the context of Stoic doctrines, the divinity of Newton's active principles now seems perfectly obvious. Defined against the passive principles of matter, they are the eternal cause, *efficiens*, mind, reason, force: the divine rational principle of the pre-Socratics, Philo, and the Stoics.[49]

One is now in a position finally to evaluate Newton's position regarding the existence of a cosmic void. Did Newton ever accept that portion of Epicureanism? Clearly, the answer must be negative. Newton had filled the Epicurean void with the all-pervasive divine substance of Stoicism. Newton had resolved the ancient tension between atomicity and the continuum, between Epicureanism and Stoicism, by incorporating elements of both into his system of the world.

48 Newton, *Opticks*, pp. 400–1.
49 See esp. Saunders, *Lipsius*.

12

Locke, Willis, and the seventeenth-century Epicurean soul

JOHN P. WRIGHT

In recent years, historians of early modern thought have raised questions concerning the relation of the philosophy of John Locke to the seventeenth-century Epicurean revival instituted by Pierre Gassendi.[1] While many have seen a strong element of Cartesianism in Locke's thought,[2] some of these historians have argued that the main thrust of his philosophy should be understood in the context of this seventeenth-century Epicureanism. One of the

I am grateful for comments on an earlier draft of this paper that was presented to a symposium of the Canadian Philosophical Association in Quebec City in May of 1989 by François Duchesneau and Tom Lennon and for comments made in correspondence by John Yolton.

1 For a discussion of the secondary literature, see John R. Milton, "Locke and Gassendi: A Reappraisal," *Oxford Studies in the History of Philosophy*, no. 2 (Oxford: Clarendon Press), forthcoming. Milton convincingly challenges the claim that Gassendi had a major influence on the development of Locke's philosophy. However, Milton points out that there are two references to Gassendi's philosophical views in Locke's Notebooks, both of which concern Gassendi's views about "the nature of human and animal souls" (see n. 12, this chapter). See also Fred and Emily Michael, "The Theory of Ideas in Gassendi and Locke," *Journal of the History of Ideas*, 51 (1990) 379–99; Richard W. Kroll, "The Question of Locke's Relation to Gassendi, "*Journal of the History of Ideas*, 44, (1984), 339–59; David Fate Norton, "The Myth of 'British Empiricism,'" *History of European Ideas*, 1 (1981), 331–44; Howard Jones, "Gassendi and Locke on Ideas," in R. J. Schoeck, ed., *Acta conventus Neo Latini bononiensis: Proceedings of the Fourth International Congress of Neo Latin Studies* (Binghamton: Medieval and Renaissance Texts and Studies, 1985), pp. 51–9.

Locke's comments on the soul and the "essence of man" have been discussed in the context of Gassendism by François Duchesneau in *L'empiricisme de Locke* (The Hague: Nijhoff, 1973), Chap. 3, esp. pp. 96–108. The views expressed in what follows seem to me to be close to those of Duchesneau, though differing in emphasis. I stress the speculative side of Locke's thought, as well as the Gassendist conception of the soul as principle of life.

2 For example, James Gibson, *Locke's Theory of Knowledge and Its Historical Relations* (Cambridge: Cambridge University Press, 1917), esp. Chap. 9.

most central, and at the same time puzzling concepts in Locke's *Essay Concerning Human Understanding* is the concept of soul. While Locke never gave a definitive account of the soul, he did present two hypotheses relating to the "essence" of the soul which were important in contemporary disputes and later philosophical developments.[3] In this chapter I shall suggest that these hypotheses represent Locke's own modifications of a seventeenth-century Epicurean account of the soul. Moreover, I shall argue that two major themes concerning the soul in the *Essay* have roots in this philosophy.

The best-known of Locke's hypotheses was put forward in Book 4 of the *Essay*. Throughout much of the *Essay* Locke wrote about the soul from a Cartesian perspective – apparently the perspective of much of his audience. For example, in the chapter of the *Essay* in which he discussed personal identity, Locke wrote that he was "taking, as we ordinarily now do, ... the Soul of a Man, for an immaterial Substance, independent from Matter." As such it is conceivable that it can be united to different bodies at different times.[4] In Chapter 23 of Book 2 of the *Essay* he discussed "our *Idea* of our Soul, as an immaterial Spirit"; so considered, it is "a Substance that thinks, and has a power of exciting Motion in Body, by Will, or Thought" (2.23.22). These are what Locke called the "primary Ideas" which are "peculiar to Spirit." In discussing the soul in this way he was considering the idea of soul "as contradistinguished" from that of body (2.23.17–18). However, in a famous passage in Chapter 3 of Book 4 of the *Essay*, Locke put forward the hypothesis of "the Soul's Materiality." According to this hypothesis, God may have "given to some Systems of Matter fitly disposed a power to perceive or think" (4.3.6). Locke maintained that it was impossible for us to decide with certainty between this hypothesis and that of the Cartesians.

However, right at the beginning of Book 2 of the *Essay*, Locke did unequivocally reject the Cartesian view that thought is the essence of the soul. He wrote that thought or "the perception of *Ideas* [is] ... to the Soul, what motion is to the Body, not its

3 For the history of Locke's hypotheses that the soul may be material see John W. Yolton, *Thinking Matter* (Minneapolis: University of Minnesota Press, 1983).
4 John Locke, *An Essay Concerning Human Understanding*, ed. P. H. Nidditch (Oxford: Clarendon Press, 1975); Bk. 2, Chap. 27, sec. 27. Further references to the *Essay* will be to book, chapter, and section numbers, cited in parentheses in text.

Essence, but one of its Operations" (2.1.10). He went on to insist, against the Cartesians, that the soul of a human being can go on existing in sleep even though there is no thought. But Locke never gave a positive account of the essence of the soul, nor did he tell his readers what it does when it is not thinking.[5] It is plausible that he thought of it as continuing to exist as a particular kind of material substance. In spite of Locke's attempt to be evenhanded in his treatment of soul and body, in Chapter 23 of Book 2, a clear asymmetry emerges in the course of his discussions elsewhere. In Chapter 4 of Book 2 he wrote that "*Solidity* . . . seems the *Idea* most intimately connected with and essential to Body"; he is unequivocal that it is the essence of body in his subsequent dispute with Bishop Stillingfleet.[6] When matter is not moving, it still continues to be solid and so remains extended and resists penetration by other bodies. Moreover, following Newton, Locke suggested that God has superadded universal attraction to bodies; thus an unmoving body might continue to attract other bodies. Locke's positive accounts of the nature of body and its operations when it is not moving stand in contrast to his merely negative comments on the essence of the soul.

There is, I believe, no reason to limit the operations of the Lockean soul to those ascribed to the soul by the Cartesians – that is, thinking and will – especially on the hypothesis that the soul is material. In Chapter 1 of Book 2 (section 12) of the *Essay*, Locke wrote ironically of the Cartesians, who "so liberally allow Life, without a thinking Soul to all other Animals" besides man and so "cannot then judge it impossible, or a contradiction, That the Body should live without the Soul." He went on to discuss how the soul of Castor can go off while he is sleeping and think in

5 In an article ("Locke on the Essence of the Soul," *Southern Journal of Philosophy*, 17 [1979], 455–64), Garth Kemerling has argued that on Locke's view "the essence of the soul is the power to think, which a substance may have whether or not it is actually thinking" (p. 459). But while, for reasons laid out in Kemerling's article, it is clear that the dispositional power of thought forms part of the "nominal essence" of the soul for Locke, it cannot be the real essence of the soul. Locke wrote that "the Essences of things are not conceived capable of any . . . variation" in degree (*Essay*, 2.19.4). However, it is clear that the power of thought, no less than actual thinking itself, has different degrees during the life history of an individual. Moreover, for reasons laid out in this chapter, we should not confine Locke's account of the "nominal essence" of the soul to the power of thought.

6 See *Essay*, 2.4.1, and *Mr. Locke's Second Reply to the Bishop of Worcester* in *The Works of John Locke*, 10 vols. (London, 1823), 4: 460.

the body of Pollux. I would suggest that, in spite of his use of this thought experiment to show the absurdity of the Cartesian view, which presupposes that Pollux can sleep "without a Soul" (2.1.12), there is no reason to ascribe to Locke himself the Cartesian view that life can continue without a soul. In section 18 of the same chapter Locke identified the view that the "human Soul" always thinks with the view "that a Man always thinks" – and in the new chapter on "Identity and Diversity" (added in the second edition of the *Essay*) Locke claimed that "the identity of the same *Man* consists . . . in nothing but a participation of the same continued Life . . . by constantly fleeting Particles of Matter, in succession vitally united to the same orgainzed Body" (2.27.6). When Locke stressed that it is "not the *Idea* of a thinking or rational Being alone, that makes the *Idea* of a *Man* in most Peoples Sense; but of a Body so and so shaped joined to it," he was allowing for a wider view of the essence of man than that of the Cartesians (2.27.8). In the conclusion to this chapter I shall argue that, in putting forward his hypothesis about the "essence of Man" in Book 3 of the *Essay*, Locke was modifying and extending a seventeenth-century Epicurean conception of the material soul as the principle of life.

A good source for the seventeenth – century Epicurean account of the soul is François Bernier's *Abrégé de la philosophie de Gassendi*. This book was in Locke's library, though his own knowledge of the Gassendist conception of the soul was probably well developed long before its publication in the mid-1670s.[7] (Locke had also become personally acquainted with Bernier during his sojourn in France [1675–8].[8]) Bernier introduced his discussion of the soul in the midst of his account of Gassendist physics, at the point at which he distinguished the animal creation from other physical bodies. Bernier wrote that he intended to use the term soul "for the principle of life and sentiment." He went on to argue that this soul which produces life and sense in bodies is material.[9]

7 John Harrison and Peter Laslett, eds., *The Library of John Locke*, 2nd ed., (Oxford: Clarendon Press, 1971), no. 283, p. 84.

8 Maurice Cranston, *John Locke, a Biography* (London: Longmans, 1957), p. 170.

9 François Bernier, *Abrégé de la philosophie de M. Gassendi* [1674–8], 2nd ed., 7 vols. (Lyon, 1684), vol. 5, Bk. VI, p. 461. My translation. In his "Question of Locke's Relation to Gassendi," p. 340, Richard Kroll suggests that Locke did not read Gassendi's *Institutio logica, et philosophiae Epicuri syntagma* and that Locke may well have imbibed

At the same time, it is important to note that Bernier carefully distinguished the soul from what he called the "mind" [*l'esprit*] – which was an immaterial substance.

The ideas of soul and mind which we find in Bernier are also to be found in a thinker who was certainly far more important in the formation of Locke's own thought, namely the English anatomist and physician Thomas Willis. Locke attended and transcribed the medical lectures which Willis gave as Sedleian Professor of Natural Philosophy at Oxford in 1663 or 1664.[10] Many of these lectures were later incorporated in a book entitled *Two Discourses Concerning the Soul of Brutes, Which is also that of the Vital and Sensitive Part of Man*, which Willis had first published in Latin in 1672.[11] While the views on soul and mind in the early lectures which we possess are fairly sketchy, it is reasonable to assume that Locke would have imbibed the systematic views which Willis set out in his later book.[12] I shall go on to examine these views in the main body of

Gassendi's views on method by reading a translation of a shorter work which constituted the section on Epicurus in Thomas Stanley, *The History of Philosophy*, 4 vols. (London, 1655–62). However, Locke does refer to Gassendi's major work, the *Syntagma philosophicum* (1658), in his Medical Commonplace Book of 1659–66.

Bernier listed a number of reasons for holding that the soul is "a sort of fine fire or little flame which, insofar as it is active [*en vigueur*] or remains lit constitutes the life of the animal" (p. 479). His last reason seems particularly interesting, in the light of Locke's own later discussion at the beginning of Book 2 of the *Essay*: "Finally the fifth [reason] is taken from the continual agitation of the phantasy, which results in the fact that the images of things never cease and *the animal always thinks* [*l'Animal pense incessament*] not only in waking but yet in sleep, as is shown by dreams" (p. 484, my italics). It seems clear that there were others besides orthodox Cartesians who held the view which Locke attacks at the beginning of Book 2 of his *Essay*.

10 Kenneth Dewhurst, *Thomas Willis's Oxford Lectures* (Oxford: Sandford Publications, 1980), pp. viii–ix, pp. 43–4.

11 Thomas Willis, *De anima brutorum* (London, 1672). I shall cite the English translation of 1683: Thomas Willis, *Two Discourses Concerning the Soul of Brutes, Which is also that of the Vital and Sensitive Part of Man*, trans. S. Pordage (London: 1683). This translation will be cited in the text as *Soul*, in parentheses.

12 See, for example, Dewhurst, *Oxford Lectures*, pp. 125–9. Locke did not have the 1672 edition of *De anima brutorum* in his library, though he did have the 1682 *Opera omnia* which included this work. Locke's interest in the Epicurean doctrine of the soul is indicated in a citation that John Milton (see n. 1, this chapter) gives from Locke's Notebooks: "Animam brutorum et sensitivam hominis esse materialem [The soul of brutes and the sensitive soul of man (is) material] v: Bacon Advancement 208, 209. Willis de Anima Brut: Gassendi Phys: Sec 3 1.1 c.11." Locke is quoting from a pamphlet entitled *A Philosophical Discourse of the Nature of Rational and Irrational Souls* (1695) by one Matthew Smith. Locke could, of course, have originally imbibed this view from any or all of the sources cited by Smith.

this essay and conclude by returning to a discussion of the relevance of these views for Locke's own account of the soul and his hypothesis about the essence of man.

WILLIS'S ACCOUNT OF THE CORPOREAL SOUL

At the beginning of his *Two Discourses*, Willis wrote that upon the "Hypothesis of the *Epicureans*, as it were its basis, the Philosophers of this latter Age have built all their doctrines of the Soul." Among these philosophers Willis included René Descartes and Kenelm Digby, as well as Pierre Gassendi. According to Willis, the common Epicurean basis of the doctrines of these philosophers lay in the belief that "the Soul (is) plainly Corporeal, and made out of a knitting together of subtil Atoms" – in particular, light, round atoms, like those which compose fire. Willis recognized that in naming all the modern doctrines of the soul "Epicurean" he was saying something which some might find paradoxical – for these doctrines were themselves in many ways "opposite" to one another (*Soul*, pp. 2–3).

Willis's claim that Descartes's philosophy of the soul is rooted in that of Epicurus may well puzzle us. For Descartes is commonly considered the philosopher who characterized the soul or mind as entirely immaterial and divorced it from the body. However, it is important to note that Willis is using the term "soul" (*anima*) in a traditional sense – not that which Descartes himself and his followers recommended.[13] For Willis, the soul is that "by which the Brutes as well as Men live, feel [and] move" (*Soul*, Preface, n.p. [1]).[14] Descartes ascribed life to a corporeal principle in the heart, and bodily motion to "animal spirits" flowing through the nerves to the muscles: The former is nothing but a fire "without light,"[15] and the latter consists of "extremely small bodies which

13 See Descartes, *Meditations on First Philosophy*, Fifth Set of Replies (to Gassendi), in *The Philosophical Writings of Descartes*, trans. and ed. John Cottingham et al., 2 vols. (Cambridge: Cambridge University Press, 1984), 2:248. (Henceforth this set will be cited in parentheses in text as *PWD*, with volume number and page.) See also Nicolas Malebranche, *The Search after Truth*, trans. Thomas M. Lennon and Paul J. Olscamp (Columbus: Ohio State University Press, 1980), pp. 491–8.

14 On page 40 he appeals to Book 9, Chapter 11, Section 3 of Gassendi's "Physics" to support his doctrine.

15 Descartes, *Discourse on the Method*, pt 5, in *PWD*, 2:134; cf. Descartes, *Passions of the Soul*, no. 8, in *PWD*, 1:332.

move very quickly, like the jets of flame that come from a torch."[16] In his Replies to Gassendi's Objections to his *Meditations*, Descartes even said that he ascribed what was generally called "sensation" as well as motion, "for the most part" to the body; he attributed "nothing belonging to them to the soul, apart from the element of thought alone" (PWD, 2:243). For Descartes, the primary functions of life, sensation, and motion are brought about by tiny, invisible particles such as those proposed by Epicurus, and if one follows Willis in considering the soul to be the principle which is responsible for these functions, then it is reasonable to say that the soul, for Descartes, is corporeal.

Willis, following Gassendi, described the human being as a "Two-soul'd Animal" (*Soul*, Preface, n.p. [1]). As Margaret Osler has recently stressed, Gassendi – no less than Descartes – considered the higher soul of man to be immaterial.[17] Willis concurred in this judgment, and for him too it was the primary ground for arguing for the immortality of the mind or higher soul in man.[18] In the case of both Gassendi and Willis, there was here a radical break with the classical Epicurean argument – presented by Lucretius in Book III of *De rerum natura* – that the *animus* as well as the *anima* is material, and that hence we should not fear death.[19]

In order to understand what is distinctive in the seventeenth-

---

16 Descartes, *Passions of the Soul*, no. 10, in PWD, 1:331–2.

17 Margaret J. Osler, "Baptising Epicurean Atomism: Pierre Gassendi on the Immortality of the Soul," in Margaret J. Osler and Paul L. Farber, eds., *Religion, Science, and Worldview: Essays in Honor of Richard S. Westfall* (Cambridge: Cambridge University Press, 1985), pp. 163–83.

18 He wrote that he agreed with "that famous Philosopher, *Peter Gassendus*, who . . . differencing the Mind of Man, as much as he could, from that other Sensitive Power has (as they say) divided the whole heaven between: Because when he had shewed this to be Corporeal, Extensive, and also Nascible or that may be born, and Corruptible, he saith that the other was an Incorporeal Substance, and therefore Immortal, which is Created mediately by God, and infused into the Body" (Willis, *Soul of Brutes*, p. 40).

19 In Book III of his poem, Lucretius had claimed to prove "the Mind material too, because it moves, / And shakes the limbs . . . / All which is done by Touch, and what e're touch, / Are Bodies, then the Mind and Soul are such" (T. Lucretius Carus, *The Epicurean Philosopher, His Six Books, De natura rerum, Done into English Verse . . .*, trans. Thomas Creech, 2nd ed. [Oxford, 1683], p. 74). (Hereafter I shall cite this edition in text as *Creech*.) The terms translated as "mind" and "soul" in this famous late seventeenth-century translation are (respectively) *animus* and *anima* (see Lucretius, *De rerum natura*, ed. and trans. W. H. D. Rouse [Cambridge, Mass: Harvard University Press], pp. 182–3). Lucretius goes on to argue, on the basis of their material nature and interdependence, that all sentience is destroyed at death.

century Epicurean doctrines of the soul of a thinker like Willis, we need to concentrate on what he has to say about the lower soul and on his description of how the functions of the soul are divided between the lower and the higher souls.

THE LOWER SOUL AS THE PRINCIPLE OF LIFE

The distance between Willis's use of the term "soul" and that of Descartes becomes clear when we consider Willis's general characterization of what the English translation calls his "Psychelogie or Doctrine of the Soul." Willis states that as soon as the soul begins to act "it performs chiefly these two offices; viz. First to frame the Body as it were its domicil or little house, and then that Body being wholly made, to render it apt and fitted to all the Uses necessary both to the Kind, and to the Individuum." According to Willis, the soul is composed of a subtle matter which, "rightly disposed," provides a genetic blueprint for the formation of the body (*Soul*, pp. 6–7). Moreover, it contains patterns both for the characteristic motions of the whole body, once formed, and for that of its individual organs. Willis retained a traditional conception of the soul[20] as a principle which creates and controls the life processes of the visible body. But, like Gassendi, he broke with the accepted views[21] in ascribing life to atoms structured in a certain way.

In stressing the substantial corporeal nature of the principle of life, sense, and motion in animals, Willis was adopting and transforming a traditional Epicurean theme. Lucretius had sought to convince his readers "that the *Soul* / Is *part*, and not the *Harmony* of the Whole" (Creech, p. 72). He opposed a doctrine which considered the soul to be a kind of epiphenomenon – fundamentally the doctrine presented by Simmias, in Plato's *Phadeo*, through a musical analogy.[22] Willis, in arguing for his own conception of the substantial soul, was rejecting the Scholas-

20 I have discussed this in an unpublished paper entitled "The Embodied Soul in Seventeenth-Century French Physiology."
21 On the traditional nature of the teaching at Oxford at the time of Willis and other *vertuosi*, see Hansruedi Isler, *Thomas Willis, 1621–1675* (New York: Hafner, 1968), p. 5.
22 Plato, *Phaedo*, 85e–86e.

tic doctrine that "the Soul of the Beast is an Incorporeal Substance, or Form" (*Soul*, p. 4). He argued against this Scholastic version of the doctrine of Simmias on the ground that the Scholastic view implies that the souls of all animate beings are incorruptible and that, if it were true, the world would become overpopulated with discarded souls – including those of "Fleas, Flys, and of other, more vile Insects" (p. 23). To avoid such a consequence, it is more reasonable to maintain that the souls of irrational animals and the lower soul of human beings are "Co-extended with the whole Body," divisible into parts – and hence quite mortal (p. 5).

Willis directly followed Lucretius in maintaining that the *anima* is spread throughout the body. Willis maintained that in the intact body different senses operate simultaneously, and, in so doing, "all the Exterior members Exercise the sense of feeling and motion." Moreover, the soul acts at the same time in producing the motions of the internal organs of the body. The soul, he argued, cannot operate in all these different organs of the body without being extended. In further support of his conclusion, Willis observed that when certain animals are cut into pieces the soul is retained in each and "exercises its Faculties . . . of Motion and Sense, in every one of the divided members . . . . So Worms, Eeles and Vipers, being cut into pieces, move themselves for a time, and being pricked will wrinkle themselves together" (p. 5).[23]

Willis divided the corporeal soul into two parts, "the Vital and Animal" (p. 22). The "vital" part of the soul is essentially a fire which lies in the blood. This fire is a "heap of most subtil Contiguous particles . . . existing in a swift motion"; these particles are constantly replaced by "Sulphureous" particles from food and "Nitrous" particles from the air. The continuous reaction of these particles supports the life of the body, which Willis

---

23 Such phenomena are easily accounted for through the two-soul doctrine of the seventeenth-century Epicureans. It is interesting to note the metaphysical difficulties of an eighteenth-century physiologist such as Robert Whytt, who cited similar examples – but who also argued for the unity of the higher and lower souls in both men and beasts. See my "Metaphysics and Physiology: Mind, Body and the Animal Economy in Eighteenth-Century Scotland," *Studies in the Philosophy of the Scottish Enlightenment*, Oxford Studies in the History of Philosophy, no. 1 (Oxford: Clarendon Press, 1990), pp. 251–301, esp. pp. 288–92.

conceived as an invisible flame which is identical to the vital heat (pp. 5–6; cf. pp. 22–3). He stressed that this vital part of the soul is "scarce sensible or knowing" (p. 55); its acts are carried out without any cognitive function of the animal itself – "as it were by a Law from the Creator" (p. 41).

THE SENSITIVE PART OF THE LOWER SOUL

The second part of the soul of brutes is sensitive and is composed of "the animal Liquor or Nervous Juyce, flowing gently within the Brain and its Appendexes" (*Soul*, p. 22). This juice, the "Animal Spirits," is "distilled" from the most subtle parts of the blood into the brain and cerebellum (p. 23). Willis also described the animal spirits as being like "Rays of Light" that are "sent from the Flame of the Blood" (p. 24). He explained how some of these spirits descend into the middle part of the brain, where they "are kept in great plenty, for the business of the Superior Soul." Others flow into the spinal marrow and nerves and blow up the passages in which they are contained. They then "enter into the nervous Fibres, planted in the Muscles, Membranes, and Viscera," and they turn these into "the proper and immediate Organs of the Sence and Motion." Unlike Descartes, Willis does not think that muscular motion occurs purely through hydraulic pressure. When the spirits are "shut up within the Muscles," they mix "with Sulphureous Particles from Blood" and thereby cause a kind of explosion which accounts for the motions of the muscles.[24]

While Willis stresses that the "Animal" part of the soul is spread out in this way in the body, he also makes it clear that perception itself and the coordinating functions take place in the brain. In particular Willis located perception and imagination in the "Corpus Collosum," which joins the two hemispheres of the brain:

24 This aspect of Willis's theory has been stressed by Georges Canguilhem in *La formation du concept du réflexe aux XVIIe et XVIIIe siècles*, 2nd ed. (Paris: Vrin, 1977), p. 62. According to Canguilhem, Willis, unlike Descartes, considered the animal spirits to be "a power which must be actualized." On Descartes's hydraulic view, the spirits are conceived as fully actualized. For the context of Willis's chemical theory, see the fine study by Robert G. Frank, Jr., *Harvey and the Oxford Physiologists* (Berkeley: and Los Angeles: University of California Press, 1980), esp. pp. 248–50.

The Images or Pictures of all sensible things, being sent or intromitted by the Passages of the Nerves, as it were by Pipes or strait holes, pass first of all thorow the streaked Bodyes, as it were an objective Glass, and then they are represented upon the Callous Body, as it were upon a white Wall; and *so induce a Perception, and a certain Imagination of the thing felt*: Which Images or Pictures there expressed, as often . . . import nothing besides the mere Knowledg of the Object. (p. 25, my italics)

It is here also that the locus of the complex responses of the whole animal to the external world is to be found. Willis notes that "if the sensible species . . . impressed on the Imagination, promises any Thing of Good or Evil," the animal spirits are reflected back and cause motions in the body for "embracing or removing" the object perceived. He went on to present a complex set of speculations about the location of various sensory and motor functions, based on his own magnificent work in comparative brain anatomy.[25]

There is no doubt that Willis ascribed to the lower corporeal soul both a kind of knowledge and – although he wavered somewhat on this – a kind of reasoning. At the beginning of his book he stressed that he opposed Descartes, and agreed with Gassendi, in maintaining that there is a kind of "sensitive" use of reason in brute animals (*Soul*, p. 4). In Chapter 6, which is entitled "Of the Science or Knowledge of Brutes," Willis stressed that while one cannot consider animals to have (as we do) a "Rational Soul," nevertheless they do seem to "choose Acts, which flow from Council or a certain Deliberation" (p. 32). His argument is directed toward showing that while animals sense, and make simple judgments through sensation, there is no reason to ascribe an immaterial soul to them. Matter itself is not "meerly passive," as some have maintained: It is active and self-moving (p. 33). Thus the activity which distinguishes a living, sensing body from a dead or sluggish one is really not, in principle, different from that which distinguishes a body in combustion from the same body before it has been kindled. But his strongest argument in support of the claim that matter can perceive and reason is taken from anatomy: When he considers the art and workmanship of the

25 Comparative anatomy played a major role in Willis's famous *Cerebri anatome* (London, 1664); the diagrams in this book were done by Christopher Wren, who worked closely with Willis. See Isler, *Thomas Willis*, esp. pp. 86, 102–3.

brain, how it is far more complex in its structure than a pipe organ, which is set up so that the wind that passes through it can produce magnificent harmonies, then he cannot see why the animal spirits passing through the myriad passages that have been constructed in the brain cannot produce the complex perceptions and behaviors of the animal. Combining his two arguments, Willis concludes that there is no reason to think that matter, as it is put together in a living body, cannot produce even the most complex actions of animals. In some of the "more perfect Brutes" we must recognize a capacity to vary the type of actions they perform and "of Composing them in themselves" (pp. 33–4).

Willis distinguished three levels of behavior in nonhuman animals. In the first place, there are actions done from a "natural Instinct," or a knowledge which is "born with them." Such actions are generally for self-preservation or the propagation of the species. They derive from an implanted disposition of nature. Willis stressed that in explaining such actions there is no reason to assume any degree of reason or freedom (pp. 34–5). The second level of "brute actions" is based on instinct combined with sensory experience. In describing actions of this kind, Willis gives a description of the development of what we now call a conditioned response:

A Dog being by a staff struck, or by the flinging of a stone, perceives the hurt received by the senses, and easily retains the Idea in his Memory, but the Instinct dictates to him that the like stroke may be shunned afterwards, wherefore, when he sees a staff held out before his eyes, or a stone taken up, fearing thence the like hurt, he hastily flies away. (p. 37)

While Willis writes of a conjoining of natural instinct with "acquired" notions of sense and says that the animal is led to "Propositions or Assumptions" of what is to be done, he still gives a psychophysiological explanation based on the formation of traces in the "Phantasie and the Memory" (p. 36). But there is also another level of acquired knowledge which Willis identifies in higher animals – for which he does not attempt any mechanical explanation. In explaining the cunning of foxes or the apparent sagacity of hunting dogs, Willis wrote of "a certain kind of Discourse or Ratiocination." For example, he accounted for the behavior of a fox that plays dead in order to catch chickens by the fact that the fox once discovered by accident that a chicken approached him while he lay exhausted on the ground. After this, according to Willis, the fox formed a habit of lying still in order to catch chickens. Willis even has the fox concluding his reasoning with the general proposition that "Foxes for the taking of their Prey, use again the same Wiles" – though he stressed that all the notions used in this reasoning are those "of the Sensitive Soul" (pp. 37–8).

THE RATIONAL SOUL IN MAN

Willis's chapter "Of the Science of Brutes" is immediately followed by one describing the relation of the corporeal soul with the rational soul in man. Willis's stated aim in the chapter, in spite of what he had just written about higher animals, was to show that the operations of the corporeal soul are in no way equal to those of the rational soul in man. But, as we shall see, he also ended up explaining just how he conceived of the relation of the two souls in human beings.

Willis carefully considered the three main cognitive operations – "simple Apprehension, Enunciation, and Discourse" – as they are carried out in the two souls. On Willis's account, the primary "apprehending" function of the rational soul, or mind, is abstraction. The mind begins by forming abstractions from sensible images apprehended by the lower soul, but when the mind reaches metaphysical thoughts, such as "God, Angels, It self, Infinity, Eternity," it goes beyond all sensible things and beholds what is "wholly immaterial." According to Willis, this proves that "the Substance or Nature of the Rational Soul" must itself be "Immaterial and Immortal" (*Soul*, pp. 38–9).

Second, Willis stressed the difference between the "enunciations" or judgments formed by the mind and those formed by the corporeal soul. He noted that a human being "not only beholds all enunciations conceived by the phantasie" but also can judge whether they are true or false. Unlike the images combined purely by the imagination, the judgments formed by the intellect are directed: "It calls [the imagination] away from these or those Conceptions, and directs it to others." The rational soul chooses to limit its conceptions in order to attend to the matter at hand and does not merely continuously pass from one notion to another. Moreover, it makes certain judgments that cannot be made purely by sense or imagination – for example, when, "by a reflected Action, it supposes it self to think" and thus knows its own existence (p. 39).

Finally, in noting the superiority of human reason, Willis stressed the complexity and order of the reasoning involved in various human sciences. He claimed that "the whole Encyclopaedia or Circle of Arts and Sciences" shows the human mind, "if not divine, at least to be a particle of Divine Breath, to wit, a Spiritual Substance, wonderfully. Intelligent, Immaterial and which therefore for the future is Immortal" (pp. 39–40).

THE RELATION OF THE TWO SOULS IN MAN

Finally let us consider what Willis says about the relation between the mind and sensitive soul in human beings. Before addressing directly the question of the connection between the functions of the lower and higher soul, Willis (following Gassendi) rejected the Scholastic view that the rational soul itself performs "not only the Offices of the Intellect and Discourse, but also the other Offices of Sense and Life." The absurdity of the Scholastic view is shown by the fact that the lower-soul functions are carried out in the human being long before the rational soul manifests itself. How, he asked, "should the Sensitive Soul of Man, which subsisting at first in Act, was material and extended, foregoing its Essence at the coming of the Rational Soul, degenerate into a mere Quality?" (p. 40). The material soul must remain a substantial, independent being.

Willis also cited the authority of Gassendi in support of his own positive account of the relation of the two souls. On this account, in acting on the corporeal soul the rational soul acts as an independent entity and thereby "effects the Form, and Acts of the humane Body" (p. 41). Although Willis held that the rational soul is immaterial, he did not hesitate to locate it in the body. He maintained that it only operates in the principal part of the brain – at the locus of the imagination – and there it exercises its functions. Willis stressed that it does not preside over the whole of the lower part of the soul – certainly not the natural functions such as digestion, which are carried out through a law of the Creator. The rational soul, in its own nature, is nothing but "a full and perfect power of understanding" – though its actions depend upon the brain and animal spirits (p. 42). While noting that the human intellect knows what it knows directly, Willis pointed out that when it dwells in the body, "darkness is poured on it." Thus, far from having direct and intuitive knowledge, it must attain its understanding "by reasoning, that is, successively, and proceeding as it were by degrees."

By following the Epicurean two-soul account of the human being, Willis formed a conception of the union of mind and corporeal part of man which sharply differed from that which is to be found in the philosophy of Descartes. Willis described the rational soul as being within the compass of the corporeal soul and claimed that the former, by "using as it were its Eyes, and other powers, understands" (p. 42). I think that his description of the rational soul as a perceiver and manipulator of the images of the corporeal soul is more than just an analogy, for Willis. Earlier on he described how "the Rational soul, as it were presiding, beholds the Images and Impressions represented by the Sensitive Soul, as in a looking Glass, and according to the Conceptions and notions drawn from thence, exercises the Acts of Reason, Judgment and Will" (p. 32). For Willis, the higher soul perceives the images of the lower soul and so operates on and reacts to an entity which is already thinking (in Descartes's sense). In his *Optics*, Descartes had warned his readers against conceiving of the image or picture in the brain as resembling the object from which it proceeds, "as if there were yet other eyes within our brain by which we perceive it." For Descartes, who claimed that the essence of the single soul is to think, the union of mind and body is not effected through a perception of a body by the mind. Rather, he held that there is a kind of "natural" connection whereby certain motions in the brain are originally connected with certain sensations of the mind.[26]

But, of course, there is a certain incommensurability between Descartes's view of the body and that of an Epicurean like Willis; for the latter did not hesitate to ascribe what Descartes called "thought" to the body itself. For Willis, as we have seen, the body has its own kinds of perceptions and desires. Indeed, Willis went

---

26 PWD, 1:167. Different interpretations of this theory are given by John Yolton, *Perceptual Acquaintance from Descartes to Reid* (Minneapolis: University of Minnesota Press, 1984), pp. 23–7, and Margaret Dauler Wilson, *Descartes* (London: Routledge & Kegan Paul, 1978), pp. 207–20.

on to describe the relationship between the two souls as one between two active beings. The proper relationship of the two souls is one in which the higher soul controls the lower one through its acts of will. The rational soul, or mind,

not only perceives, but [also] . . . governs and moderates, all Concupiscences, and Floods of Passions, that are wont to be moved also within the Phantasie; and so, as it approves these Affections, and rejects those, now excites others, now quiets them, or directs them to their right ends, the Rational Soul it self is said to exercise certain Acts of the Will. (p. 43)

However,

the Corporeal Soul does not so easily obey the Rational in all things. . . . She being busied about the Care of the Body, and apt by that pretext, its natural Inclination, and indulging Pleasures, most often grows deaf to Reason. . . . The lower Soul, growing weary of the yoak of the Other, if occasion serves, frees it self from its Bonds, affecting a Licence or Dominion. . . . (ibid.)

Willis went on to describe this relationship between the two souls as one of "intestine Strife." He clearly described the conflict as one between two perceiving beings with opposing desires.

CONCLUSION: JOHN LOCKE AND THE SEVENTEENTH-CENTURY EPICUREAN SOUL

In conclusion, let me return to a discussion of the relation of the views of Willis to those of Locke. At the beginning of this essay, I mentioned Locke's famous suggestion that the deity may have superadded perception or thought to "some systems of Matter fitly disposed."[27] Locke's language here directly echoes Willis's claim, which we considered earlier, that the corporeal soul is made "out of matter rightly disposed" (Soul, p. 6).[28] Moreover, in the Preface to the Reader in Souls of Brutes Willis had written that there is no reason to think that God is "not able to impress strength, Powers and Faculties to Matter, fitted to the offices of a Sensitive Life".[29] Like Willis, Locke argued for his suggestion

27 Locke, Essay, 4.3.6. See John W. Yolton, Thinking Matter, esp. Chap. 1.
28 The Latin reads; "ex materiâ rîtè dispositâ." De anima brutorum, in Thomas Willis medicinae doctoris . . . , Opera omnia (Geneva, 1680), p. 9.
29 This parallel between the views of Locke and Willis is stressed by Isler, Thomas Willis, p. 179. The Latin reads, "non potuisse vires & facultates materiae imprimere, vitae sensitivae muniis accommodas" (De anima brutorum, n.p.).

on the ground that God, being all-powerful, is quite capable of superadding thought to matter. Of course, Willis's hypothesis is only about the lower, corporeal soul; Locke's suggestion is that we extend the scope of this soul to include the highest functions of man.

There are a number of other Epicurean themes in Locke's writings about the soul which he may well have first encountered through the teaching of Willis. One of the major themes of Locke's *Essay Concerning Human Understanding* is that children must attain understanding by degrees, as they mature. As we have seen, this is an important theme in Willis's account of the rational or intellectual soul. Like Willis, Locke rejected the Cartesian view that animals are pure machines which lack any perception. He argued, for example, that one cannot explain the way birds learn tunes without postulating that they "have Perception, and retain Ideas in their Memories" (2.10.10). Like Willis also, Locke held that animals conceive only particulars and do not have the power to abstract (2.11.10).

However, I do not think that one can simply identify Locke's views about thought and sensation with those of Willis. In spite of the fact that Locke rejected Descartes's conception of the soul itself at the beginning of Book 2 of the *Essay*, he stressed what I take to be a major Cartesian theme[30] – namely, that all thought is conscious. Locke emphatically denied that any being can think without "perceiving that it does so" (2.1.19). Like Descartes,[31] Locke considered all thought to be inherently reflexive and both sensation and imagination to be modes of thought (2.19.1). Willis, on the other hand, distinguished between the perceptions of the higher and lower souls – maintaining that in performing its operations the latter does not "perceive it self to know or imagine" (*Soul*, p. 39).[32] It appears that when Locke ascribed

---

30 See, for example, Descartes's definition of thought in his Replies to the Second Set of Objections (PWD, 1:113). See also Robert McCrae, "Descartes' Definition of Thought," in R. J. Butler, ed., *Cartesian Studies* (New York: Barnes & Noble, 1972), pp. 55–70.

31 See Descartes's characterization, in the Second Meditation, of what it means to have a sensory perception in a restricted and precise sense (PWD, 1:19).

32 Willis's text here is not entirely unambiguous. This view was even more explicitly maintained in a book published in the same year as Willis's *Souls of Brutes* – namely, Ignace-Gaston Pardies, *Un discours de la connoissance des betes* (see *Oeuvres du R.P. Ignace-Gaston Pardies* [Lyon, 1725], p. 451).

sensation to animals, unlike Willis and Gassendi, he was committed to the claim that they possess some degree of reflective
thought.

Moreover, there are important methodological differences between Locke's study of the mind in the *Essay Concerning Human
Understanding* and the study of the soul in the writings of Willis
and Gassendi. Right at the beginning of the *Essay* Locke
announced that in following the "Historical, plain Method" he
would not "*at present*, meddle with the Physical Consideration
of the Mind" (1.1.2). While Locke did not entirely reject such a
consideration, it is clear that he did not mean it to be a part of
his project in the *Essay*. The close tie between physiology and
psychology which we find in the work of Willis is absent from
Locke's *Essay*. While there are suggestions about the physical
operations of the mind or soul throughout the book, they are
presented in a tentative way.[33]

Finally, let me turn to the second of Locke's hypotheses which I
mentioned at the beginning of this chapter. Throughout most of
the *Essay* Locke ascribed the higher functions of the human soul to
what he calls "man."[34] In the opening chapter of Book 2, which I
considered at the beginning of this chapter, Locke formulated his
central question as a question about "when a *man* begins to have
any *Ideas*" (1.2.23). In later chapters, such as those on power and
moral relations, Locke made the "man" the subject of free action
(2.21.8) and moral accountability (2.28.8). As I have already
noted, in a section that he added to the second edition of the *Essay*
Locke stressed that the idea of "man" is that of a living being
organized in a certain way. He wrote that what we mean by man
is "an Animal of a certain Form" (2.27.8). He also stressed that
in an animal "the Motion wherein Life consists" comes "from
within." In so doing he was ascribing what he elsewhere (2.21.2–
4; 2.23.28) called "active power" to life itself. It would appear that

33 For a list of references to Locke's physiological hypotheses regarding brain operation,
see John Yolton, *Locke: An Introduction* (Oxford: Blackwell, 1985), p. 111. For the
hypothesis of the physiological underpinnings of the association of ideas, see in
particular my "Association, Madness and the Measures of Probability in Locke and
Hume," in Christopher Fox, ed., *Psychology and Literature in the Eighteenth Century*
(New York: AMS Press, 1987), pp. 83–104.
34 See Darryl Fanick, "John Locke's Moral Person," M.A. thesis, University of Windsor,
1988, esp. pp. 32–3. Also, John Yolton, *Locke: An Introduction* (Oxford: Blackwell,
1985), Chap. 1.

by ascribing the higher functions to what he called "man," Locke was recommending that we conceive of them as more developed manifestations of a widespread principle underlying animal life.

Locke did not accept the two-soul doctrine of the seventeenth-century Epicureans. Like Descartes, he considered the principles of sense and intellect in man to be identical. However, Locke went farther than either Descartes or contemporary Epicureans in suggesting a unified conception of the human being. We have seen that Locke rejected the Cartesian view that thought is the essence of the soul. While Locke never explicitly suggested any account of what he called the "essence" of the soul as such, he did propose an account of the "real Essence" of man. In Book 3 of the *Essay* Locke defined the "real essence" of a substance as the "real constitution of any Thing, which is the foundation of all those Properties, that are combined in, and are constantly found to co-exist with the nominal essence" (2.6.6). While he denied that we have actual knowledge of the real essence which underlies the various different properties that are constantly combined in a human being, he gave a clear account of what such knowledge might be like:

Had we such a Knowledge of that Constitution of *Man*, from which his Faculties of Moving, Sensation, and Reasoning, and other Powers flow; and on which his so regular shape depends, . . . we should have a quite other *Idea* of his *Essence*, than what is now contained in our definition of that *Species* . . . : And our *Idea* of any individual *Man* would be as far different from what it now is, as is his, who knows all the Springs and Wheels, and other contrivances within, of the famous Clock at *Strasburg*, from that which a gazing Country-man has of it, who barely sees the motion of the Hand, and hears the Clock strike, and observes only some of the outward appearances. (3.6.3)

This passage gave Locke's readers an idea of the nature of the principle on which all the faculties of a human being probably depend. What he here called "the Constitution of *Man*" appears to be a principle of the same kind as the "corporeal soul" of Willis.[35] Locke is suggesting that his principle provides a genetic blueprint for the complete human being – for the intellect and will, as well as the more basic functions of sensation and life itself.

35 That is, as described at the beginning of the section of this chapter entitled "The Lower Soul as the Principle of Life."

It is important to stress that on Locke's two hypotheses neither the material soul nor the essence of man is considered to be a primary substance: that is to say, neither is a substance in the way that God, an atom, or an immaterial spirit is a substance (cf. *Essay*, 2.27.2). But the soul is also not the "form" of a substance in the way that it was for Aristotle. François Bernier had insisted that "the soul is a certain thing" and went on to describe it as a "very fine substance . . . with a disposition, or habit, or symmetry of parts within the overall mass of the body."[36] Similarly, on Locke's hypotheses the soul or real essence of man would be a certain structure of atoms within the mass of the body. This structure would provide the foundation for that particular natural kind which we call "man," or a human being.

36 *Abrégé de la philosophie de M. Gassendi*, pp. 466 and 478.

13

The Epicurean new way of ideas: Gassendi, Locke, and Berkeley

THOMAS M. LENNON

The background to seventeenth-century philosophy may be conveniently summarized under two headings. One is the recrudescence of skepticism, resulting from a variety of factors, among them the recovery of classical texts, the Reformation, the rise of national states, and the voyages of discovery. My second heading is another such factor in the rise of skepticism, the emergence of the New Science. The mechanico-mathematical account of the world that was developed in the seventeenth century not only undid the reigning Aristotelian physics but also systematically called into question its whole worldview and thus contributed to the *crise pyrrhonienne* of the period. Yet the New Science, by its very name, was prima facie incompatible with skepticism, and it is to this opposition that the fundamental philosophical problems of the period relate.

The best-known program developed to deal with the opposition was of course Descartes's. He famously argued that skepticism, consistently applied, refutes itself, and further, that we are thereby led to none other than the New Scientific view of the world. The method of doubt leads us to a metaphysically vindicated clear and distinct idea of extension, the principal attribute of bodies, on which all their other attributes depend. Now, the Cartesian "new way of ideas," as it was later called, was an exceedingly complex issue, and protracted debate on the issue produced several versions of it among Arnauld, Malebranche, Régis, and others. What they all had in common, however, was the conception of an idea as containing the intelligible or universal element of the material particulars it represented. This is why all

Cartesians subscribed to the principle that "what is contained in the clear and distinct idea of a thing can be truly affirmed of that thing," a principle to which the Port-Royal *Logic* ascribes the certainty and evidence of knowledge of natural things.[1] Furthermore, as Descartes argued at length with respect to the wax, in *Meditations* II, the idea differs both in kind and in source from the deliverances of the senses.

As set out in the *Meditations*, the Cartesian program was immediately subjected to criticism, at greatest length and most systematically by Pierre Gassendi. Previously Gassendi had argued the cause of skepticism against the Aristotelians; now he turned the skeptical attack on the new dogmatist who claimed to know the essence of things. In both campaigns Gassendi's principal ally is Epicurus, whose philosophy, while nonskeptical, nonetheless shows how the dogmatism of both Aristotle and Descartes can be avoided.[2] Thus was established the lifelong project of Gassendi's rehabilitation of Epicurus, under three rubrics. Morally, Epicurus was to be defended against the charge of having instituted a philosophy epitomized by Horace's phrase "de grege porci" and of having led the life of a pig himself. Metaphysically, Epicurean atomism was to be made theologically acceptable by construing atoms as created and also finite in number, thus requiring providence instead of chance to explain order in the world. And epistemologically, as an answer to skepticism and on behalf of the New Science, Epicurus's failure properly to cultivate mathematics was explained on the basis of its perceived irrelevance to morality and was to be remedied by a more mathematical account of atomism. As Dijksterhuis, Koyré, and many others have noted, Gassendi was not appreciably successful in this effort. Even so, he was able to perform, and thus give a mathematically accurate and fairly sophisticated account of, Galileo's unperformed experiment of dropping stones from the mast of a moving ship. Most important, however, by appealing to Epicurus, Gassendi was able to define the object of knowledge in such a way as to steer a via media between the dogmatism of Aristotle and Descartes, on the

---

1 Pt. 4, Chap. 6.
2 For the evidence of Gassendi's view of Epicurus in these terms, see Bernard Rochot, *Les travaux de Gassendi sur Epicure et sur l'atomisme, 1619–1658* (Paris, Vrin, 1944), esp. pp. 79–80.

one hand, and skepticism on the other. This seventeenth-century Epicurean via media is my main concern here, and a good place to begin is with Gassendi's theory of truth.

Gassendi, in the *Syntagma philosophicum*, distinguishes between two sorts of truth. One is called truth of existence (*exsistentiae*) or of being (*exstantiae*). The other is called truth of judgment (*Iudici*) or of statement (*Ennuntiationis*).[3] Truth of being has no opposite. This concept is the residue of the Parmenidean dialectic that is to be found among those atomists such as Democritus and Leucippus who denied all status to the void. The idea may be expressed by saying that reference of any sort to the nonexistent is impossible. The upshot is that, in this first sense, a thing is said to be true in that it is just what it is. Thus, says Gassendi, do we speak of "true gold" or a "true man" without meaning thereby that fool's gold is not true fool's gold or a painting of a man not a true painting. This is the sense of truth that Locke, among others, calls "metaphysical." Truth and falsehood, according to Locke, properly speaking belong only to propositions, either mental or verbal, and perhaps elliptically to words and ideas. But "both *Ideas* and Words, *may* be said to be true in a metaphysical Sense of the Word Truth; as all other Things, that any way exist, are said to be true; i.e. really to be such as they exist."[4]

Truth of statement, on the other hand, consists in a conformity ("in conformitate . . . consistat") between the statement or judgment and the thing judged. Here there may be discrepancy, and therefore falsehood. Gassendi quotes Sextus, who was quoting Epicurus: "[T]he true is that which is as it is said to be; the false is that which is not as it is said to be."[5] This, of course, is what Locke later calls "truth of propositions."

For Gassendi, truth of being is the more fundamental. Without it there would be no truth of statement, for the double reason that there would be nothing judged and also that there would be no

3 *Opera*, I.67a. There have been two editions of Gassendi's *Opera omnia*: Lyons, 1658, and Florence, 1727. The former is the standard edition, standardly referred to by volume, page, and column. Craig B. Brush has provided the most useful English translations: *The Selected Works of Pierre Gassendi* (New York: Johnson Reprint, 1972). I have departed from these translations, however, where I felt it appropriate to do so.

4 An *Essay Concerning Human Understanding*, ed. P. H. Nidditch (Oxford: Clarendon Press, 1979), 4.5.2; p. 574. Also, 2.32.1–2; p. 384.

5 *Opera*, I.68a. Gassendi also cites Aristotle, *De interpretatione*, 19a.33.

statements, all of which, both true and false, must have truth of being. In addition, to seek the truth of a thing is to seek whether it is, and what it is; hence, says Gassendi, "it seems that truth may be defined as that which really is ["Id, quod res est"] or rather, as that which really exists ["Id, quod res exsistit"].[6] Now, among things that are, some are manifest and some are hidden ["manifestae . . . occultae"]. Manifest things are known by themselves – for example, "the light of day or that it is day, or the external aspect of things ["externa . . . rerum facies"], which spontaneously strikes our eyes and discloses itself to sight without any obstructing veil between."[7] Of hidden things there are three sorts. The *totally* hidden are such as to be unable to fall within our comprehension: for example, whether the number of the stars is even or odd. The *naturally* hidden are such as to be known, as we would say, through inference. For example, while pores in the skin are not manifest, or known by themselves, they can still be known from the occurrence of sweat. Finally the *temporarily* hidden are those things which, though naturally manifest, are circumstantially hidden. Thus a fire may be hidden by a building or Constantinople by its distance from us. The totally hidden is unknown and unknowable, and the manifest is readily knowable, even if not known. Hence, the philosopher seeks the truth of the temporarily hidden and the naturally hidden, especially the naturally, since the temporarily hidden involves no great dispute.[8]

The important point for us to note is that coming to know the naturally hidden – the philosopher's principal object – and, if we could but know it, the totally hidden as well, is a matter of what in the literature has been variously called "transduction" or "transdiction."[9] This is a form of induction, which, although it moves from observables to unobservables, involves no new concepts: What is said at the macroscopic level is said univocally at the microscopic level. There is not the least suggestion here that the move to what is hidden would involve the transition from qual-

6 *Opera*, I.68a.   7 Ibid., I.68b.
8 G. Verbeke reminds us, and Gassendi himself points out, that both the three divisions and the examples of them come from Sextus Empiricus. The source is *Against the Logicians*, II.145–7.
9 See E. McMullin, *Newton on Matter and Activity* (Notre Dame, Ind.: University of Notre Dame Press, 1978), pp. 15–16.

ity to substance, or from phenomenon to noumenon, or from exemplar to exemplification, or from universal to particular. Gassendi's position is radically different on this point from the traditional skeptics', and Gassendi himself is aware of this difference. In discussing the instrument of judgment by which hidden truth is discovered, he raises the problem of the separator, that is, the "criterion." The skeptics are those who deny that there is a criterion, even that it is the senses.

They say that the appearance of things, or what things appear to be extrinsically, is one thing and that the truth of things, namely what things are in themselves, is another thing; and when they say that nothing can be known with certainty and that there is no criterion, they speak not of what things appear to be and of what is disclosed through the senses as if by some unique criterion, but of what things are in themselves, which is so hidden [*occultatum*] that no criterion can reveal it.[10]

The premise is that honey, for example, tastes sweet to most, but not all, of us. The conclusion is that we do not know the nature of honey. The linkage of the argument is *not* that because of this relativity of taste we do not know the real taste. That is, the skeptic is *not* arguing that we do not know the nature of honey because we have no way of adjudicating between the statements of those who say it is sweet and those who say it is bitter – for neither statement can be true. He is *not* arguing that we do not know the real taste of honey because we find it sweet and it may really be bitter, or conversely, and that we have no way of telling who is right – for in his view no one is ever right. We are going to be wrong no matter how it tastes, for how it tastes will never be a quality of the honey itself. Arguments from relativity of sense perception are designed precisely to show that what things appear to be depends on us or the circumstances. Nothing will be in its nature sweet; nothing will be in itself what it appears to be. For, "if honey were sweet in itself and according to its own nature ["in se, et se undum suam naturam"], it would also appear as such to all who have the power of tasting honey."[11]

Gassendi does not quite put it in these terms, but for him the thing as it is in itself is not just unknowable but inconceivable

10 *Opera*, I.70a.   11 Gassendi, *Disquisitio metaphysica*, I.1–2; *Opera*, III.286b.

or ineffable. If we could know it, then, in Nelson Goodman's language, we would have a description of it apart from our ways of describing it – we would have the thing, apart from all versions of it.[12] Or in Berkeley's language, if you prefer, we would have a conception of the thing, apart from our ways of conceiving it.[13] This is the point on which Gassendi attacks Descartes, in his *Objections* to the *Meditations*, when he rhetorically inquires as to what is known about the wax, when it is alleged to be known by the mind alone (*inspectio mentis*), and apart from its sensible, imaginable qualities of color, scent, and shape. Gassendi contends that in fact such an object is inconceivable.[14]

The critical difference between Gassendi and the skeptics is reflected in their theory of signs. Having drawn the Aristotelian distinction between the "universal," or "necessary," sign (τεκμήρειον) and the "particular," or "probable," sign (for which Aristotle had no name but which Quintilian simply called *signum*), Gassendi further drew Sextus's distinction within necessary signs between those that are "indicative" (*indicativa*) and those that are "telling" (*commonefactiva*).[15] A telling sign pertains to what is temporarily hidden and poses no problem: It is a reminder of its usually perceived accompaniment. Thus smoke is a telling sign of fire, lactation of pregnancy, and dawn of the rising sun. These, we might say, are straightforwardly inductive signs, although they are nonetheless necessary, since they involve movement from a

12 Nelson Goodman, *Ways of Worldmaking* (Indianapolis: Hackett, 1978), esp. pp. 2–4.
13 Berkeley, *Principles of Human Knowledge* par. 5, in *The Works of George Berkeley, Bishop of Cloyne* (London: Nelson, 1948–57), 2:43.
14 *Oeuvres de Descartes*, 12 vols., ed. Ch. Adam and P. Tannery, rev. ed. (Paris: Vrin / C.N.R.S., 1964–76) 7:271–2. *The Philosophical Works of Descartes*, trans. E. S. Haldane and G. R. T. Ross (Cambridge: Cambridge University Press, 1911), p. 147.
15 *Opera*, I,81a. Lynn Sumida Joy, *Gassendi the Atomist: Advocate of History in an Age of Science* (Cambridge: Cambridge University Press, 1987), pp. 169–71, points out that earlier in the *Animadversiones* Gassendi had picked up on Epicurus's language to distinguish the necessarily true (τεκμήρειον) from the non-necessarily true (σήμειον) conclusions of inferences and that Gassendi characterized the inference to the void as an instance of the latter. This raises the question as to whether the inference to atoms is not of the same sort. If the existence of atoms is a non-necessary conclusion, I have no explanation for Gassendi's apparent shift of position to the *Syntagma*. In either case, however, the point of this paragraph seems to me to remain intact. Indeed, Joy points out that the distinction in the *Animadversiones* is part of an attempt to defend Epicurus against Sextus.

universal claim (e.g., "All smoke is a sign of fire"). An indicative sign, however, pertains to things that are naturally hidden, "not because it indicates a thing in such a way that the thing can ever be perceived and the sign can be visibly linked to the thing itself, so that it could be argued that where the sign is the thing is too, but on the contrary, because it is of such a nature that it could not exist unless the thing exists, and therefore whenever it exists, the thing also exists."[16] Now, the skeptics accept the telling sign as a sign of how not to get burned. But they reject the indicative sign, and the reason they do so, it seems to me, is that it typically purports to lead us from quality to substance, or from exemplar to exemplification, or more generally from appearance to reality. Gassendi, on the other hand, unequivocally accepts the indicative sign, and the example of it he gives shows why. The example is his preferred one of things naturally hidden, namely, pores in the skin. The argument is that just as at the macrolevel there must be passages, in order for one body to pass through another, so at the microlevel of the skin there must be pores to allow us to transpire sweat. Indicative signs are admissible because they rely only on transduction.

Gassendi thus undercuts the opposition between skeptic and dogmatist by radically reconstruing the proper object of knowledge. The dogmatist and the skeptic both agree that if we know, then what we know is different from what it appears to be through the senses and is therefore known independently of the senses. Both understand the upper half of the Divided Line as expressing what knowledge would be. But for Gassendi, what is naturally hidden is knowable in the very same terms as that which is manifest. (Indeed, even the totally hidden, which can never be known, is at least describable in these terms: It makes sense to speak of the number of the stars as even or odd. Moreover, at least one of those who disagree on it has a true belief on the matter.) This is because the object of knowledge is the object of empirical science, which is to say, for Gassendi, of mechanism. Thus Gassendi's atomism, as an account of a mechanistic world, is not an ontology designed to compete with Cartesian dualism to become the replacement for Aristotelian Scholasticism's hylomorphism.

16 *Opera*, I.81a.

Locke later gets it just right when, following Boyle, he goes no farther than to call it the "corpuscularian hypothesis."[17]

There are interesting and significant developments of this hypothesis among later Gassendists, such as François Bernier and Gilles de Launay. But the most significant version of it, of course, was advanced by Locke himself, for it is at the basis, not only of his distinction[18] between primary and secondary qualities but also of his distinction between real and nominal essences, and thus of his whole theory of kinds. Not much attention is required here to the first distinction: No chapter of the *Essay Concerning Human Understanding* is better known than the eighth chapter of Book 2, where Locke argues that by assuming the corpuscularian hypoth-

17 This gives a rather different picture of Gassendi from that suggested by R. H. Popkin. Popkin allows that Gassendi "accomplished one of the more important revolutions of modern times, the separation of science from metaphysics." In his view, Gassendi's science gives us knowledge of appearances; of things in themselves we have no knowledge, since in metaphysics we can only be skeptics. "Scientific explanation, which for Gassendi is in terms of an atomic theory, accounts for our experience of sense qualities, but does not tell us anything about the nature of things-in-themselves, except how they appear in relation to us.... Gassendi advocated total skepticism about the world beyond appearance." Popkin, *The History of Scepticism from Erasmus to Spinoza* (Berkeley and Los Angeles: University of California Press, 1979), p. 143. On my view, Gassendi's science explains some appearances by giving knowledge of at least some things that are hidden. But of other hidden things, even of the same sort, we are ignorant. With respect to the latter, my Gassendi is a temporizing, pragmatic skeptic as opposed to Popkin's permanent, theoretical skeptic. Finally, my Gassendi would be a kind of dogmatist in metaphysics; he does not advocate that we suspend judgment in such questions (pyrrhonian skepticism), nor does he claim that we can know we are ignorant of the truth in such matters (Academic skepticism); rather, his position would be that we know that there is nothing to be known. This also distinguishes my interpretation from the partial phenomenalism Ralph Walker attributes to Gassendi. According to Walker, Gassendi employs a coherence analysis of the qualitative aspects of ordinary experience, which are the appearances of things, but remains a skeptic about the real natures of things. See Walker, "Gassendi and Scepticism," in *The Skeptical Tradition* (Berkeley and Los Angeles: University of California Press, 1983), esp. pp. 323–4. I am grateful to Lisa Sarasohn for leading me to clarify my interpretations of these questions. For a review of the main interpretations of Gassendi, including his relation to skepticism, see Joy, *Gassendi*, pp. 12–19. Joy makes Gassendi's rejection of skepticism very clear, especially by emphasizing his distinction between mathematical and physical concepts as a rebuttal of skeptical challenges to atomism – a distinction that additionally vindicated the role of sense perception. See Chap. 7, esp. p. 154.

18 Or perhaps better: distinctions. J. J. MacIntosh has shown that for Locke, as for others who either draw a primary – secondary distinction or have it ascribed to them, several distinctions may be involved. I daresay that however many are picked out in Locke, the corpuscularian hypothesis underlies them all. MacIntosh, "Primary and Secondary Qualities," *Studia Leibnitiana*, 8:1 (1976), 88–104, esp. p. 100.

esis – namely, that bodies act only by impulse – and what is minimally required of the world for that hypothesis to be true, we can account for the world as we experience it. That is, if we assume that the world consists of objects having primary qualities of solidity, mobility, and so forth, at both the perceptible macroscopic level and at the imperceptible microscopic level, then it can be seen how we have two kinds of ideas: ideas of primary qualities, which qualities are real, inseparable, and resemble their ideas; and ideas of secondary qualities, like color and scent, which qualities are only imputed, are not inseparable, and do not resemble their ideas. I take none of this to be very controversial at this stage of Locke scholarship.

More controversial is the second distinction, between nominal and real essences, in terms of which Locke discusses kinds. He says repeatedly that species or essences are abstract ideas with words annexed to them.[19] This of course is consonant with Locke's nominalism, according to which "all things, that exist, being Particulars,"[20] there cannot be extra-mental universals. The upshot is that species or essences, as a kind of universal, are "the Workmanship of the Understanding, since it is the Understanding that abstracts and makes those general *Ideas*."[21] These are the essences Locke calls *nominal*. Concerning *real* essences, which are not the workmanship of the understanding, Locke thinks there are two opinions. "The one is of those, who using the Word *Essence*, for they know not what, suppose a certain number of those Essences, according to which, all natural things are made, and wherein they do exactly everyone of them partake, and so become of this or that *Species*."[22] This is the Scholastic view that essences are substantial forms. For Locke it comes to grief on two grounds: (1) There are productions such as monsters that do not fit essences thus conceived as "forms or moulds," and (2) such unknowable essences would be useless in our actual classifications.[23]

"The other, and more rational Opinion, is of those, who look

19 See *Essay*, 4.4.17; 3.3.12; 3.3.15; 3.1.2.　　20 Ibid., 3.3.1; p. 409.
21 Ibid., 3.3.12; p. 415.　　22 Ibid., 3.3.17; p. 418.
23 See also ibid., 3.6.5; p. 441: "[T]o talk of specifik Differences in Nature, without reference to general *Ideas* and Names, is to talk unintelligibly." Also, 3.6.14–25. Whether, in thus rejecting this first opinion, Locke thus rejects every opinion that there are real natural kinds is perhaps another question. M. B. Bolton has strongly argued that Locke admits real natural kinds that are not substantial forms.

on all natural Things to have a real, but unknown Constitution of their insensible Parts, from which flow these sensible Qualities, which serve us to distinguish them one from another, according as we have Occasion to rank them into sorts, under common Denominations."[24] In this sense the real essence of a thing just is that thing in its microscopic structure, so that strictly speaking, there are as many different kinds as there are things. This is why Locke says that at a given time all of a thing's qualities are essential to it, although at another time it could be without any of them. But insofar as the shapes, for example, of the microscopic particles of two things differ, we may have different ideas of their secondary qualities and may classify them as belonging to different kinds. (In the case of animate things, we probably would do so, according to Locke.) In any case, a curious upshot is that Locke nominally – but only nominally – departs from the mechanistic tradition, since for him the microscopic structure which is the real essence of a thing is its secondary qualities. A more important result is that the real essence of a thing is open, at least in principle, to empirical investigation of the most straightforward sort. With either microscopes, or, as Locke suggests, with "microscopical" eyes, we could perceive the real essence of a thing. Under such conditions, says Locke, our ideas of secondary qualities would change or cease altogether; we would have ideas of primary qualities at the microscopic level of exactly the same sort that we now have at the macroscopic level. The distinction that Locke draws between the improper, Scholastic sense of "real essence " and his own, proper sense is thus the same distinction Gassendi had drawn between the skeptics' improper object of knowledge and his own proper one.

Unlike his primary–secondary quality distinction, Locke's real–nominal essence distinction is a matter of exegetical controversy and has been since the seventeenth century. Most notably, Bishop Stillingfleet was unable to make it out. A quick account of his perplexity might be to point out that from Stillingfleet's perspective – which was more than somewhat, if far from entirely, Aristotelian – Locke had confused the primary and secondary senses of substance.[25] He had confused Dobbin, which is neither present in nor predicable of anything else, with horseness,

---

24 Locke, *Essay*, 3.3.17; p. 418.   25 See *Categories*, Chaps. 2, 5.

which though not present in anything else is predicable of individuals such as Dobbin.

It is difficult, in one sense, to distinguish Locke and Stillingfleet on substance, since they both use much the same language in stating and defending their views.[26] Locke himself claimed to be at a loss to make out the distinction.[27] Whether this is a ploy on Locke's part, or, as I think is more likely, whether Locke, from his point of view, really did not understand an important part of Stillingfleet's position, there is an important distinction, and it is one that is clearly indicated by Stillingfleet: Our knowledge of substance, for him, must have a nonempirical source. Sense and reflection, the experience that Locke takes to be the only source of ideas, are insufficient for a proper idea of substance, which in fact is had from *reason*. In some sense Locke of course recognizes substance, according to Stillingfleet, but he "has discarded [it] out of the reasonable part of the world."

Stillingfleet claimed that "the general Idea is not made from the simple *Ideas* by the mere Act of the mind abstracting from Circumstances, but from *Reason* and Consideration of the true Nature of things."[28] That individuals belong to the same kind depends, not on a nominal essence we arbitrarily put together, but on a real, common essence that is unchangeable. Another way to put this is that Aristotelian primary substances are also, individually, instances of Aristotelian secondary substances; Dobbin is an individual but also an individual horse. For Stillingfleet, the same rational necessity that bundles qualities as qualities of Dobbin bundles such things as Dobbin into kinds.

Locke responded to this in two ways. Stillingfleet's notion of a real essence he found inconceivable. "This, my lord, as I

---

26 As John Yolton points out, "Stillingfleet's description of our idea of substance is, of course, identical with Locke's except that he prefers to call it a necessary or rational idea 'because it is a Repugnance to our first conception of Things that Modes or Accidents should subsist by themselves.'" Yolton, *John Locke and the Way of Ideas* (Oxford: Oxford University Press, 1956), pp. 133–4. But even this nominal difference seems removed by Locke at one point: "I grant it to be a good consequence that to those who find this repugnance the idea of a support is very necessary; or, if you please to call it so, very rational." *Mr. Locke's Reply to . . . Worcester's Answer to His Second Letter* (1699), in *The Works of John Locke* (London: Thomas Tegg, 1823), 3:452–3.

27 *A Letter to the . . . Bishop of Worcester* (1697), in *The Works of John Locke*, vol. 3, esp. p. 11.

28 *A Discourse in Vindication of the Trinity* (1696), in *The Works of . . . Stillingfleet* (London: Mortlock, 1710–13), 3:511.

understand it, is to prove that the abstract general essence of any sort of things, or things of the same denomination, v.g. of man or marigold, hath a real being out of the understanding; which I confess, my Lord, I am not able to conceive."[29] This may mean a number of things: (i) Things have a real essence, but it is unknown to us. This *is* Locke's view, but I think his claim here is stronger, for it suggests the reason why Stillingfleet should be mistaken in his view that we apprehend real essences, namely, he has a mistaken view of what real essences are. (ii) That things should have a real essence of the sort described by Stillingfleet is inconceivable. This too is Locke's view. Things are not classifiable into kinds, as Stillingfleet thinks, because of real essences in which they are grounded. (iii) That an abstract general essence should exist outside the mind is inconceivable, because abstract general essences are ideas, which exist only within the mind. Thus Stillingfleet, with respect to substance, confuses the world with our way of looking at it; he confuses things with ideas.[30]

The denouement of the story concerning the confusion of things and ideas lies, of course, with Berkeley, who charged Locke with the same confusion. Having argued that the being of sensible things lies in their being perceived, that sensible things in this sense exist only in the mind, Berkeley felt constrained to defend himself against the charge of having advanced skepticism, of having changed "all things into ideas." For Berkeley's thesis is that the entire sensible world has the same status previously accorded only to dreams and illusions.[31] Berkeley's reply was that he was "not for changing things into ideas but rather ideas into things, since those immediate objects of perception, which, according to [the objection], are only appearances of things, I take to be the real things themselves."[32] What is real is all that, and only that, which immediately appears to the senses, and these are ideas.

29 *Letter*, in *The Works of John Locke*, 8:83.
30 Thus Locke, at the end of the controversy: "Your lordship here, I know not upon what ground, nor with what intention, confounds the idea of substance and substance itself." *Second Reply*, in *Works*, 3:451.
31 If I disagree with Popkin's interpretation of Gassendi's relation to skepticism, I largely agree with his interpretation of Berkeley's relation to skepticism. Popkin, "Berkeley and Pyrronhism," *Review of Metaphysics*, 5:2 (December 1951), 223–46, esp. p. 243. I am grateful to C. McCurdy for drawing my attention to this early article of Popkin's.
32 Berkeley, *Three Dialogues* III, in *Works*, 2:244.

Berkeley's atomism is the phenomenalist atomism of ideas, which, like physical atoms, are (1) inert; (2) ontologically isolated – that is, without internal relations; (3) separable from everything else but not further distinguishable; and (4) epistemologically transparent – that is, without hidden properties. These are the *minima perceptibilia*: *minima visibilia, minima tangibilia,* and so forth. Although he perhaps disagrees with Gassendi and Locke in claiming that atoms are known immediately and not by means of ideas, Berkeley agrees with them that the ultimate constituents of things are not of a metaphysically different kind from things as they appear. (Indeed, since for Berkeley *percipi* is not only necessary but also sufficient for *esse*, all appearances are real. Such distinctions as we make among reals – for example, between the color of this page and its afterimage – are purely contextual.) In this sense the Epicurean new way of ideas is, by contrast to the Cartesians, the new way of things.

14

The Stoic legacy in the early Scottish Enlightenment

M. A. STEWART

In Scottish thought in the eighteenth century we find something of a reenactment of the ancient debates between Stoics and Skeptics. The Stoic tradition is strongest among the teachers and preachers who find their way into the universities from about 1720 and who close ranks against the skeptical influence of David Hume from the 1740s onwards. Neither tradition survives in a pure classical form, and it would be misleading to suggest that these labels capture all that was at issue; but the adoption of classical models was not inadvertent, and both sides make use, where it suits them, of the language and sources of antiquity. When James Balfour of Pilrig in 1753 sided with the prevailing moralism against Hume's philosophy, Hume offered unsuccessfully to re-create the conditions of social intercourse "when Atticus and Cassius the Epicureans, Cicero the Academic, and Brutus the Stoic, could, all of them, live in unreserved friendship together, and were insensible to all those distinctions, except so far as they furnished agreeable matter to discourse and conversation."[1] But he was not always so conciliatory. In the first part of this essay I discuss Hume's explicit and generally hostile references to Stoic thought and practice. In the second part I describe the more

I am grateful to the British Academy and to the Calgary Institute for the Humanities for support which made possible the preparation and presentation of this essay; to Knud Haakonssen, Ian Kidd, and Richard Sher for advice; and to David Norton for posing me some awkward questions.
1 [James Balfour], *A Delineation of the Nature and Obligation of Morality* (Edinburgh: Hamilton & Balfour, 1753); *The Letters of David Hume*, ed. J. Y. T. Greig, 2 vols. (Oxford: Clarendon Press, 1932), 1:173. The latter work is hereafter cited as *HL*.

sympathetic reception accorded to Stoic ideas by some of his Scots contemporaries, who represent a Christian Stoicism which has lost much of its Roman rigor. Hume's own picture of Stoicism is neither Christian nor Roman, but this may be in part a deliberate literary tactic to present a less than flattering portrait of a tradition which, from experience, he did not consider wholly benign.

I

In 1741–2, roughly two years after the unsuccessful publication of *A Treatise of Human Nature*, Hume published two volumes of *Essays Moral and Political*.[2] Some of these essays dated from the summer of 1739, while he waited for better news of the first books of his *Treatise*.[3] At this stage in his career, he had ambitions to bridge the communication gap between "the Learned" and "the conversible World" by introducing into literary essays such philosophical content as might reasonably engage the minds of intelligent persons in company (*Essays*, pp. 533–4). The enunciation of this aim led him into some patronizing stereotyping, especially of women, in the tradition of the *Tatler* and *Spectator* which had no doubt served as his models; but several serious points remain – a mistrust both of "college" philosophy on the one hand and of "warm" feelings generated by emotive topics on the other, and a committed intellectualism. While the demanding studies of the learned "require Leisure and Solitude, and cannot be brought to Perfection, without long Preparation and severe Labour," they gain nothing from the "moaping recluse Method of Study" characteristic of "Colleges and Cells."[4] "The conversible World," for their part, "join to a sociable Disposition, and a Taste of Pleasure, an inclination to the easier and more gentle Exercises of the Understanding, to obvious Reflections on human Affairs, and the Duties of common Life, and to the Observation of the Blemishes or Perfections of the particular

---

2 My references to these works of Hume are to the following modern editions: *A Treatise of Human Nature*, ed. L. A. Selby-Bigge, rev. P. H. Nidditch (Oxford: Clarendon Press, 1978), cited as *T.*, and *Essays Moral, Political and Literary*, 2nd ed., ed. Eugene F. Miller (Indianapolis: Liberty Classics, 1987), cited as *Essays*.
3 *New Letters of David Hume*, ed. Raymond Klibansky and Ernest C. Mossner (Oxford: Clarendon Press, 1954), pp. 5–7: hereafter cited as *NHL*.
4 An echo of Shaftesbury (*Moralists*, I.i).

Objects, that surround them." For this they need some awareness
of "History, Poetry, Politics, and the more obvious Principles,
at least, of Philosophy" (p. 534). In this communication Hume
always identified himself with the Learned side of the divide, but
thought it important that the Learned should send out "Ambassa-
dors" and should not be "Men who never consulted Experience in
any of their Reasonings, or who never search'd for that Experience, where alone it is to be found, in common Life and Conversation" (p. 535).

This disenchantment with the higher reaches of abstractness
first came to Hume in the early 1730s. As an Edinburgh student
in the 1720s he would have devoted his last two years of study
to philosophy, following the set curriculum of the professors of
"logic and metaphysics" and "natural philosophy and ethics." He
would have had the option of additional lectures from the professor of moral philosophy, though there is no hard evidence that he
took it; but even if he did not, we know that the general climate
was one of presbyterian piety, theological and political conformity, and moral character training, and that whatever new learning
had established itself in metaphysics and the sciences had to be
compatible with those constraints.[5] Having "passed through the
ordinary Course of Education with Success," Hume found, when
he tried to prepare for a professional career, that he had "an
unsurmountable Aversion to every thing but the pursuits of
Philosophy and general Learning" (HL, 1:1); and by this point,
if not before, he had become intensely involved in ethical philosophy, which he conceived in the same practical terms that
his teachers would have intended. He embarked on an extended
program of reading in which he was later to recall the particular
impact of Cicero, Seneca, and Plutarch. Even if only one of these
is to be seen as a committed exponent of Stoicism, the others gave
considerable publicity to Stoic attitudes and ideals:

& being smit with their beautiful Representations of Virtue & Philosophy, I undertook the Improvement of my Temper & Will, along with
my Reason & Understanding. I was continually fortifying myself with

---

5 On the Edinburgh curriculum in the 1720s, see Richard B. Sher, "Professors of Virtue"
(pp. 87–126), and Michael Barfoot, "Hume and the Culture of Science in the Early
Eighteenth Century" (pp. 151–90), both in M. A. Stewart, ed., *Studies in the Philosophy
of the Scottish Enlightenment* (Oxford: Clarendon Press, 1990).

Reflections against Death, & Poverty, & Shame, & Pain, & all the other Calamities of Life. These no doubt are exceeding useful, when join'd with an active Life; because the Occasion being presented along with the Reflection, works it into the Soul, & makes it take a deep Impression, but in Solitude they serve to little other Purpose, than to waste the Spirits, the Force of the Mind meeting with no Resistance, but wasting itself in the Air, like our Arm when it misses its Aim. (HL, 1:14)

The solitary contemplation induced a mental breakdown. Hume recovered, but his philosophy was permanently marked by the experience; though in subsequent writing he does sometimes concede that Stoicism was a less sedentary philosophy than he had once tried to practice. As a student he would have been taught to disparage ancient science but to admire its associated ethics. He now rejected the utility of this distinction:

I found that the moral Philosophy transmitted to us by Antiquity, labor'd under the same Inconvenience that has been found in their natural Philosophy, of being entirely Hypothetical, & depending more upon Invention than Experience. Every one consulted his Fancy in erecting Schemes of Virtue & Happiness, without regarding human Nature, upon which every moral Conclusion must depend. This therefore I resolv'd to make my principal Study, & the Source from which I wou'd derive every Truth in Criticism as well as Morality. (Ibid., 1:16)

Hume continued to feel and to respond to this tension between the call of the study and the need to be active and to keep a sense of reality. In the shorter term, the problem was resolved by channeling the activity into writing, but steering clear of any more "beautiful Representations of Virtue & Philosophy" aimed at "the Improvement of my Temper & Will." The compromise attempt to write moral essays at the level of popular journalism was soon abandoned in favor of writings directed principally to his social and intellectual peers, but some of the earlier pieces which survived as part of the permanent canon throw interesting sidelights on his philosophical position, or the position he was promoting for discussion, around the time of the publication of the *Treatise*.

There is a cluster of four interacting essays which first appeared in the second collection of 1742 with the titles "The Epicurean," "The Stoic," "The Platonist," and "The Sceptic." The first three are subtitled, respectively "The man of elegance and pleasure," "The man of action and virtue," and "The man of contemplation, and *philosophical* devotion." On their first appearance, Hume wrote that in these essays "a certain Character is personated; and therefore, no Offence ought to be taken at any Sentiments contain'd in them" – allowing him to keep his distance, not only from the themes enunciated, but from the criticisms voiced by one character or another. In later editions he disclaimed any historical authenticity. His aim was rather "to deliver the sentiments of sects, that naturally form themselves in the world, and entertain different ideas of human life and happiness. I have given each of them the name of the philosophical sect, to which it bears the greatest affinity" (*Essays*, p. 138n).

A number of things strike one about these essays.[6] The first three are written in a high-flown style untypical of Hume's other writing, with extraordinary sections of purple prose[7] and an extensive but not consistent use of archaic forms, while the more sober essay on the Sceptic is longer than the others together. I read this as implying that the first three represent varying degrees of "enthusiasm," in the pejorative sense. The first three express attitudes whereas the fourth develops an argument, and it is an argument directed against an assumption which lies behind those attitudes. The details given for the different sects are not always as exclusive as Hume's strategy might suggest, but where they are the skeptic has the advantage of coming last in the sequence, so he can answer the others more readily than they can answer him. The impression conveyed, without pretending to any great historical sophistication, is that Epicureanism, Stoicism, and Platonism each represents one identifiable tendency, and collectively a gradation of tendencies, of the human mind, any one of which can be carried to excess; so what we have is a kind of Rake's Progress of philosophical enthusiasms, the third being for Hume the most

6 For recent commentary, see John Immerwahr, "Hume's Essays on Happiness," *Hume Studies*, 15 (1989), 307–24, and references cited there. The present essay was completed before the appearance of Immerwahr's thoughtful article. My main disagreements with it stem from a different view of Hume's attitudes to the essay genre and to the "easy" philosophy. See M. A. Stewart, "The Historical Significance of the First *Enquiry*," forthcoming in a collection on Hume's *Enquiry concerning Human Understanding*, ed. P. J. R. Millican.

7 Not to mention mixed metaphors: "O! for ever let me spread my limbs on this bed of roses, and thus, thus feel the delicious moments, with soft and downy steps, glide along" (*Essays*, p. 141).

extravagant. He himself, however, is partly to blame for the excesses of the exposition, because he detaches the moral doctrines from their historical roots in physics and metaphysics.[8] Skepticism is depicted as the reflective person's proper reaction to the tenets of the other schools, and in spite of the literary form the author's persona comes through as transparently here as it does in the character of Philo in *Dialogues Concerning Natural Religion.* "The Sceptic" is still an underused resource in Hume scholarship, showing more precisely and succinctly than the *Treatise* the nature and limits of moral inquiry within the framework of philosophical skepticism.[9]

The place of nature in the different philosophical schools is an important theme in Hume's characterization of their several systems, where he makes considerable play with the classical antithesis between nature and art. His own application of this antithesis in Book III of the *Treatise,* when he distinguishes between natural virtues and those which depend upon the existence of ordered conventions, is not found in the other traditions. It caused his contemporaries, and some of his later commentators, unnecessary difficulties.

1. The theme of "The Epicurean" is that art can only work upon nature as it finds it: There is nothing more ridiculous than to employ rules of art which run counter to human nature. The Stoic sage's attempt, therefore, by rules of reason and reflection to attain an "artificial" happiness of mind arising from a "consciousness of well doing," when our "sentiments and passions" owe their existence to our physical nature and well-being, is a kind of arrogant delusion (*Essays*, pp. 138–40). One who recognizes this will first cultivate the opposite extreme, a life of mind-less pleasure – graphically but somewhat vaguely portrayed as the satisfac-

---

8 Neither here nor elsewhere does Hume interest himself in Stoic epistemology, even as a topic for criticism. This is consistent with the impression that his skepticism was not modeled on Sextus.

9 Thus Hume's narrowing of the role of reason, which puts morals beyond the scope of knowledge or demonstration, makes him as much a sceptical moralist as a sceptical metaphysician. This has been recognized by Robert Fogelin, *Hume's Skepticism in the Treatise of Human Nature* (London: Routledge, 1985), Chap. 9. David Fate Norton has noted that Hume is not a "moral skeptic" in the sense of someone who challenges the existence of interpersonal moral standards: *David Hume, Common-Sense Moralist, Sceptical Metaphysician* (Princeton: Princeton University Press, 1982), Chap. 1. That is true, but not at issue.

tion of every sense and faculty, and literally roses, roses all the way – but the pace of such a life, and the saturation it induces, must destroy the pleasure of it. The true Epicurean recognizes a place for "virtue" in the good life, whereby he tempers sensual pleasures with those of friendship, personal relationships, and social harmony, and extends the pleasures of sense to include the accoutrements of culture (pp. 141–3). It is all somewhat like the Reverend Sydney Smith's idea of heaven: eating pâté de foie gras to the sound of trumpets. Within these constraints, the Epicurean self-consciously eschews "glory" and takes satisfaction in what he can, conscious of the ephemeral nature of all existence (pp. 143–5). The Epicurean life is tolerably civilized, unproblematic, and entirely superficial: His enjoyment of life does not appear to depend on his making any significant contribution to it.

2. "The Stoic" again begins with the distinction between art and nature, this time handled differently. Nature has conferred many benefits on us, both raw materials to be adapted to human ends and the intelligence – so different from the instincts of inferior animals – by which they can be adapted through our own industry. The Stoic is therefore more involved in the physical world than the Epicurean gave him credit for, but it is as a testing ground for his mind. The same art that has to be applied to develop our physical existence out of a state of savagery has to be applied also to our moral development. There is a progression from "wildest savage" to "the polished citizen, who, under the protection of laws, enjoys every convenience which industry has invented," and from the latter to "the man of virtue, and the true philosopher, who governs his appetites, subdues his passions, and has learned, from reason, to set a just value on every pursuit and enjoyment" (pp. 146–8). This latter is the progression from Epicurean to Stoic, where art, in the shape of reason, imposes upon our nature as creatures of passion. The "philosopher" learns the rules of practice from the observation of action and passion, studying the causes and remedies of his mistakes, but the person who systematically implements these rules is a "sage." Toil and fatigue are a "chief ingredient of the felicity" they bring, the business of life being that of integrating the parts of a complex machine into "just harmony and proportion" (pp. 148–9). "And cannot the same industry render the cultivating of our mind, the moderating of our passions, the enlightening of our reason, an agreeable

occupation; while we are every day sensible of our progress, and behold our inward features and countenance brighten incessantly with new charms?" (p. 149). Hume here repeats the vision he had personally repudiated, of the Stoic "charmed ' with the inward vision of virtue. While his Stoic shares the Epicurean rejection of mere sensuality, he is critical of the transitoriness of Epicurean pleasures: "Happiness cannot possibly exist, where there is no security; and security can have no place, where fortune has any dominion" (p. 150).

The sage will sometimes distance himself from the rest of humanity, but only to reinforce his view of the futility of others' lives without Stoic principles. In practice, he is regularly moved by sympathy and compassion. This leads to laudable action – for example, the disinterested action of parents for their children. It extends beyond the family to all our acquaintance, where we comfort those who are physically distressed, but we derive even greater joy from effecting moral improvement: "[W]hat supreme joy in the victories over vice as well as misery, when, by virtuous example or wise exhortation, our fellow-creatures are taught to govern their passions, reform their vices, and subdue their worst enemies, which inhabit within their own bosoms?" But even that is too cramping for the Stoic, who spreads his attention to the whole society, both now and for the future. Whatever the hardships along the way, the "glory" that accrues from meeting them offsets the danger and suffering, and the attainment of virtue is itself an adequate recompense for the life involved. This is seen to be the dispensation of "a being who presides over the universe; and who, with infinite wisdom and power, has reduced the jarring elements into just order and proportion." The sage achieves in his own person, like a master workman, the same harmony, proportion, and order as the deity achieves on a cosmic scale (pp. 152–4):

In the true sage and patriot are united whatever can distinguish human nature, or elevate mortal man to a resemblance with the divinity. The softest benevolence, the most undaunted resolution, the tenderest sentiments, the most sublime love of virtue, all these animate successively his transported bosom. What satisfaction, when he looks within, to find the most turbulent passions tuned to just harmony and concord, and every jarring sound banished from this enchanting music! If the contemplation, even of inanimate beauty, is so delightful; if it ravishes the senses, even when the fair form is foreign to us: What must be the effects of moral beauty? And what influence must it have, when it embellishes our own mind, and is the result of our own reflection and industry? (p. 153)

Hume's Stoic philanthropist is no doubt in some respects an improvement on the empty sociability of his Epicurean, but one is left with the impression that his happiness is not much different from smugness. Note, for future reference, the allusion to Stoic patriotism and the Stoic's susceptibility to the perception of moral "beauty."[10]

3. The Platonist, in Hume's brief depiction, carries further the incipient theism of the Stoic, and is more concerned than either of his predecessors to interpret human life against a backdrop of eternity. The rational soul is "made for the contemplation of the Supreme Being, and of his works," and its search for happiness will never be satisfied by the pursuit either of pleasure or of glory: The one degenerates in practice into remorse, the other into arrogance. The Platonist is particularly disparaging of what he sees as Stoic disingenuousness:

Thou seekest the ignorant applauses of men, not the solid reflections of thy own conscience, or the more solid approbation of that being, who, with one regard of his all-seeing eye, penetrates the universe. Thou surely art conscious of the hollowness of thy pretended probity, whilst calling thyself a citizen, a son, a friend, thou forgettest thy higher sovereign, thy true father, thy greatest benefactor. Where is the adoration due to infinite perfection, whence every thing good and valuable is derived? Where is the gratitude, owing to thy creator, who called thee forth from nothing, who placed thee in all these relations to thy fellow-creatures, and requiring thee to fulfil the duty of each relation, forbids thee to neglect what thou owest to himself, the most perfect being, to whom thou art connected by the closest tye? (p. 157)

To the Platonist, the relationship between art and nature is one neither of mere conformity nor reform: Rather, art copies nature. And since it does, it is appropriate to see in nature itself the hand of the unseen artist who deserves all our worship and adoration (p. 158). If the weakness of our faculties, or the brevity of life,

---

10 The Stoic connection between beauty of mind and moral virtue is made at Cicero, *De finibus*, III.75. In his previous, unsympathetic, references to "glory," Hume may be picking up on later Stoic attitudes condemned at *De fin*. III.57. *Gloria*, or *bona fama*, was more usually placed among things "indifferent," though it might still bring incidental advantage.

prevents our seeing all the beauty of the world and the perfect virtue of its just and benevolent maker, we shall make up for this in eternity: This gives the Platonist a long-term prospect of satisfaction which the Epicurean and Stoic cannot lay claim to – though by now the connection between these designations and any particular historical figures has disappeared from sight. Hume's "Platonism" is, at best, an imitation of the seventeenth-century pietism of the early latitudinarian divines, and distinguishable from the natural religion of classical Stoicism only by its confident expectation of an afterlife.

4. Turning finally to the Sceptic, as critic of these other tendencies, we find an initial complaint which might be thought to owe something to Hume's own oversimplified presentation in the preceding essays. Philosophers too often, he says, overlook the variety of human inclinations and carry some "favourite principle" to absurd lengths. "[I]f ever this infirmity of philosophers is to be suspected on any occasion, it is in their reasonings concerning human life, and the methods of attaining happiness. In that case, they are led astray, not only by the narrowness of their understandings, but by that also of their passions" (p. 160). Given any particular end in view, "common prudence" will dictate the means to gain it, but why should any inquirer turn to a philosopher to discover his (the inquirer's) own ends? As any skeptic can show, they vary between individuals, and on different occasions in a single life. Hume had already in correspondence criticized those who make "unphilosophical" assumptions about human ends,[11] and in the essay he defended this stance with a thesis that he must have expected at least that correspondent to share: "that there is nothing, in itself, valuable or despicable, desirable or hateful, beautiful or deformed; but that these attributes arise from the particular constitution and fabric of human sentiment and affection." This is as much true of sensory delights, of the pleasures of human relationships, and of aesthetic and moral sentiments (pp. 161–3). "You will never convince a man, who is not accustomed to ITALIAN music, and has not an ear to follow its intricacies, that a SCOTCH tune is not preferable" – a remark whose crassness is only mildly mitigated by the realization that it was initially written for a Scottish reading public.

11 Hume to Francis Hutcheson, 17 September 1739, in HL, 1:33.

Were I not afraid of appearing too philosophical, I should remind my reader of that famous doctrine, supposed to be fully proved in modern times, "That tastes and colours, and all other sensible qualities, lie not in the bodies, but merely in the senses." The case is the same with beauty and deformity, virtue and vice. This doctrine, however, takes off no more from the reality of the latter qualities, than from that of the former; nor need it give any umbrage either to critics or moralists. Though colours were allowed to lie only in the eye, would dyers or painters ever be less regarded or esteemed? There is a sufficient uniformity in the senses and feelings of mankind, to make all these qualities the objects of art and reasoning, and to have the greatest influence on life and manners. And as it is certain, that the discovery above-mentioned in natural philosophy, makes no alteration on action and conduct; why should a like discovery in moral philosophy make any alteration? (p. 166n.)

It thus becomes an empirical, not a speculative, question, which passions or sentiments are agreeable, and what it is that induces them; and while we can generalize, we cannot universalize. The "philosophical devotion" of natural religion (Platonism) is in general as transient as pleasure (Epicureanism), and many external goods are unattainable, others randomly distributed between the virtuous and vicious. In general, "the happiest disposition of mind is the *virtuous*; or, in other words, that which leads to action and employment, renders us sensible to the social passions, steels the heart against the assaults of fortune, reduces the affections to a just moderation, makes our own thoughts an entertainment to us, and inclines us rather to the pleasures of society and conversation, than to those of the senses" (p. 168). Although this appears to concede something to the earlier systems, those that are inclinable to such a life have little need of moral maxims to get them there: They are already subject to their "constitution and temper," and this imposes limits on how far they can, even by experience and practice, modify their habits; for someone born with no spark of the social virtues and no initially relevant responses cannot be trained by either reason or philosophy (p. 169). The skeptic therefore makes no great claims for philosophy, except as a study which familiarizes us with the consequences of actions and whose practice may serve as a constructive distraction from antisocial passions and help us develop greater sensibility. Its influence on life, whether profound or not, is essentially indirect.

A man may as well pretend to cure himself of love, by viewing his mistress through the *artificial* medium of a microscope or prospect, and beholding there the coarseness of her skin, and monstrous disproportion of her features, as hope to excite or moderate any passion by the *artificial* arguments of a SENECA or an EPICTETUS. The remembrance of the natural aspect and situation of the object, will, in both cases, still recur upon him. The reflections of philosophy are too subtle and distant to take place in common life, or eradicate any affection. (p. 172)

This remains Hume's point even at the end of a long and well-known footnote which at first seems to concede a little more to the moralist (pp. 177–9n.). At most, philosophy can help us to see ills and benefits with a due sense of proportion, so that we do not overreact to either.

Because of the difficulty in being sure whether the Stoicism Hume impugns in this essay is more than a literary fiction, it is worth noting a few other references to this philosophy in his writing.

(a) In the course of answering criticisms from Francis Hutcheson in 1739, Hume resorted to an argument from Cicero in Book IV of *De finibus*,

where you find him prove against the *Stoics*, that if there be no other Goods but Virtue, tis impossible there can be any Virtue; because the Mind woud then want all Motives to begin its Actions upon: And tis on the Goodness or Badness of the Motives that the Virtue of the Action depends. This proves, that to every virtuous Action there must be a motive or impelling Passion distinct from the Virtue, & that Virtue can never be the sole Motive to any Action. (HL, 1:35)

This is much more than just a disagreement with Hutcheson. It is a central plank in the development of Book III of the *Treatise*. It pervades Hume's argument about the motivation for justice in Part II, section I, and is thus intimately involved in establishing the distinction between natural and artificial virtues. If justice is a virtue, it must derive from a virtuous motive. That motive cannot, then, itself be that justice is a virtue; while the philanthropic feelings which the Stoic tradition offered as an external influence on action are too weak for the job (T., p. 483).

(b) Both in the *Treatise* and elsewhere in the *Essays* Hume sometimes identifies famous republican Stoics as paragons of virtue – Brutus (T., p. 582; *Essays*, pp. 30, 539) and Cato (T., p.

607; *Essays*, pp. 30, 272). There is a certain amount of conventionality in this; but equally, there is no reason why someone may not in practice come to the right conduct – or conduct which has the right social consequences – by the wrong theoretical route. In context, Hume seems always to be referring to their conduct in the public arena. Since, however, in Hume's view virtue depends importantly upon motive and Stoic theory with regard to motive is untenable, this leaves little room for conscientiously Stoic action to be virtuous at all: He would have to be claiming either that the great Stoics were just mistaken about their true motivation or that they fell into the category of those who conformed to duty without any genuine moral feeling (T., p. 479). I do not think that this was his true view, but rather that he regarded their patriotism and public service as reflecting a commendable social conscience: "*Brutus* riveted the Chains of *Rome* faster by his Opposition; but the natural Tendency of his noble Dispositions, his public Spirit & Magnanimity, was to establish her Liberty" (HL, I:35). For all that, the Stoics' was an essentially intellectual stance, one step from reality: "The virtue and good intentions of CATO and BRUTUS are highly laudable; but, to what purpose did their zeal serve? Only to hasten the fatal period of the ROMAN government, and render its convulsions and dying agonies more violent and painful" (*Essays*, p. 30).

(c) Hume's rather slight essay "Of Moral Prejudices" appeared in the same collection as the original printing of the four essays on the philosophical sects but was subsequently withdrawn. It was written by June 1739 (NHL, p. 7), so its anti-Stoic line cannot be attributed to the ill success of the *Treatise*. Hume begins by denouncing the moral nihilist – the person who rejects civilized values as "chimerical and romantic" – as distinct from the philosophical skeptic who, in investigating morality, merely rejects an illusory kind of philosophical activity in favor of a legitimate one. The nihilist will bring about the destruction of society. The Stoic, *malgré lui*, is another of the same breed, less pernicious to society but destructive of the individual.

I mean that grave philosophic Endeavour after Perfection, which, under Pretext of reforming Prejudices and Errors, strikes at all the most endearing Sentiments of the Heart, and all the most useful Byasses and Instincts, which can govern a human Creature. The *Stoics* were remarkable for this Folly among the Antients; and I wish some of more venerable Characters in latter Times had not copy'd them too faithfully in this Particular. The virtuous and tender Sentiments, or Prejudices, if you will, have suffer'd mightily by these Reflections; while a certain sullen Pride or Contempt of Mankind has prevail'd in their Stead, and has been esteem'd the greatest Wisdom; tho', in Reality, it be the most egregious Folly of all others. (*Essays*, p. 539)

The complaint is supported by anecdotes about individual Stoic heartlessness, contrasted with more normal (non-Stoic) and apparently commendable manifestations of human nature in which sentiment is superior to principle.

(d) In another essay published at this time, "Of the Rise and Progress of the Arts and Sciences," Hume with some special pleading identified the non-Skeptic sects, this time including Pythagoreanism, as schools which historically maintained their hold by the "tyrannical" attitude of their teachers and the "servile" attitude of their students, in a culture where "they sought for truth not in nature" but in their own cloisters. Skepticism is exempt from these strictures. In the modern era, "Upon the revival of learning, those sects of STOICS and EPICUREANS, PLATONISTS and PYTHAGORICIANS, could never regain any credit or authority; and, at the same time, by the example of their fall, kept men from submitting, with such blind deference, to those new sects, which have attempted to gain an ascendant over them" (p. 123). This historical claim is to be distinguished from the sociological one, previously quoted, that there is still a tendency to sectarianism along lines that reflect specific motifs which may be individually prominent in particular historic schools. But even the historical claim, I suspect, was targeted more at the pretensions of Scottish educational practice after "the revival of learning" than at classical schooling before it.

(e) In the first *Enquiry* (Sec. V, pt. I) Hume draws an unfavorable contrast between Stoicism and Skepticism. The former promotes, and the latter opposes, "the supine indolence of the mind, its rash arrogance, its lofty pretensions, and its superstitious credulity." Skepticism is the one philosophy which does not pander to the philosopher's "predominant inclination" (such inclinations had been previously noticed by the Sceptic at *Essays*, p. 160), while the philosophy of "Epictetus, and other *Stoics*" is "only a more refined system of selfishness":

While we study with attention the vanity of human life, and turn all our thoughts towards the empty and transitory nature of riches and honours, we are, perhaps, all the while flattering our natural indolence, which, hating the bustle of the world, and drudgery of business, seeks a pretence of reason to give itself a full and uncontrolled indulgence.[12]

A second reference to the Stoics (Sec. VIII, pt. II) comments on their view of suffering, which would convert human ills into cosmic goods. Hume sees this as too obviously the cloistered reaction of the pure theoretician and as ineffectual for the individual racked by pain.

(f) Some of his strongest comments on the Stoics of antiquity, as distinct from Stoic themes at large, are tucked away in *The Natural History of Religion*, where the charge of "enthusiasm" is explicit.

The STOICS bestowed many magnificent and even impious epithets on their sage; that he alone was rich, free, a king, and equal to the immortal gods. They forgot to add, that he was not inferior in prudence and understanding to an old woman. For surely nothing can be more pitiful than the sentiments, which that sect entertain with regard to religious matters; while they seriously agree with the common augurs, that, when a raven croaks from the left, it is a good omen; but a bad one, when a rook makes a noise from the same quarter. PANAETIUS was the only STOIC, among the GREEKS, who so much as doubted with regard to auguries and divinations. MARCUS ANTONINUS tells us, that he himself had received many admonitions from the gods in his sleep. It is true, EPICTETUS forbids us to regard the language of rooks and ravens; but it is not, that they do not speak truth: It is only, because they can foretel nothing but the breaking of our neck or the forfeiture of our estate; which are circumstances, says he, that nowise concern us. Thus the STOICS join a a philosophical enthusiasm to a religious superstition. The force of their mind, being all turned to the side of morals, unbent itself in that of religion.[13]

There are of course implicit Stoic undercurrents in Hume's other major writing on religion, the *Dialogues*. "Cleanthes" as the name of the protagonist for the Design Argument has Stoic associations,

---

12 Hume, *Enquiries concerning Human Understanding and concerning the Principles of Morals*, ed. L. A. Selby-Bigge, rev. P. H. Nidditch (Oxford: Clarendon Press, 1975), p. 40.

13 Hume, *The Natural History of Religion*, ed. H. E. Root (London: Black, 1956), pp. 62–3, citing Cicero, *De divinatione*, I.iii.vii; M. Aurelius, *Meditations*, i.17; Epictetus, *Enchiridion*, 18.

and this character is the counterpart to the Stoic Balbus in Cicero's dialogue which Hume took as his model. Significantly, when Cleanthes is forced out of his eighteenth-century Newtonianism, he regresses to the Stoic theory of the world as an organism. There is no reason to suppose that Hume was not equally aware of the Stoic roots of natural religion when he wrote his tendentious essays on the Stoic and the Platonist.

II

When Lord Bacon, at the start of the seventeenth century, excoriated the ancients and set out his program to rebuild learning upon a sound historical method, he inaugurated a rhetoric which soon came to obstruct the reforms he advocated. With the partial exception of the Cambridge Platonists, the principal English thinkers of the seventeenth and early eighteenth centuries aligned themselves with the Baconian "moderns" in a tug of war between ancients and moderns. In practice, the main casualty of Bacon's attack upon the ancients was Aristotelianism, and anyone who took sides against medieval and Renaissance scholasticism came to be seen as a modern. But the traditional language of learning in both England and Scotland continued to be Latin until well into the eighteenth century, and the traditional literature of learning in that language remained in circulation. If any one author came to fill the gap created by the downgrading of Aristotle and his Latin followers, it must be Cicero. In spite of Bacon's censure of his style, Cicero's eclectic philosophy, dominated by both Stoic and Skeptical elements, could be and was adapted to different and incompatible interests, while Epicurean theory was reestablished in expurgated form as a successor to Aristotelian physics. The seventeenth century, then, far from rejecting the ancients as Bacon had seemed to recommend, restored to the public curriculum the three schools of thought which had competed against Aristotelianism in the ancient world.

By the later part of the century Cicero's standing had become more ambiguous, not on doctrinal but on linguistic grounds. The movement towards plainer speech in science had the endorsement of the Royal Society and its spokesmen, and it was taken up by a new breed of theologians who found themselves having to fight increasingly persuasively to preserve the protestant cause in the decades after the restoration of the monarchy. Seneca replaced

Cicero as the approved model, so that Stoic thought could not help but come to the notice of aspiring preachers and writers, independently of its merits.[14] There was no major Stoic scholarship in Britain, so far as I have been able to judge, before Samuel Clarke, and whatever was adopted from Stoicism was assimilated into the Christian tradition. Those who wanted to read it for its own sake had Seneca, Epictetus (through Arrian), and Marcus Aurelius available as primary sources, and Cicero, Plutarch, Diogenes, and Simplicius as their principal secondary sources, and these were still widely read up to Hume's day. Besides this, indigenous traditions in political thought were developing both in protestant Europe and in Britain, both in Latin and in the national vernaculars. The result was a culture in which the principles of public and private virtue and the possibility of realizing them were much discussed, and there were pressures for radical educational changes to train people for citizenship. The impetus came from political writers like the Earl of Shaftesbury and Lord Molesworth, but it was in Scotland in the early eighteenth century that the movement developed furthest in Britain.

Hume was not an orthodox member of this culture. After receiving the standard education, he separated himself from it to pursue his own studies, which took him eventually abroad. The attempt to read Cicero and others for moral improvement was in his case, as we have seen, counterproductive: At least, it led him instead into the study of skeptical systems, fostered further by his sojourn in France. When he returned to Scotland in the early 1740s he made no headway with the new philosophical establishment, and only Adam Smith among his professional contemporaries was ever completely at ease – that does not mean he agreed with it all – in reading Hume's philosophy. Smith himself was more subject to Stoic influences, particularly in the doctrine of self-command; and two recent studies have identified a pronounced strain of Christian Stoicism in the teaching and preaching of Adam Ferguson and Hugh Blair, younger contemporaries who achieved the academic eminence at Edinburgh that eluded Hume.[15] Both moved freely

14 George Williamson, *The Senecan Amble* (London: Faber, 1951).
15 D. D. Raphael and A. L. Macfie, introduction to Adam Smith, *The Theory of Moral Sentiments* (Oxford: Clarendon Press, 1976), pp. 5–10; Richard B. Sher, *Church and University in the Scottish Enlightenment* (Edinburgh: Edinburgh University Press, 1985), Chap. 5; John Dwyer, *Virtuous Discourse* (Edinburgh: Donald, 1987), Chap. 2.

in the same social circle as Hume, yet we know that Hume was averse to Ferguson's philosophy and that he reached a point where he was unable to discuss his differences comfortably with Blair. At the beginning of his professional life, a similar breakdown in communication between Hume and Francis Hutcheson had far-reaching repercussions for Hume's career, and Hutcheson, who was professor of moral philosophy at Glasgow, shows many of the same signs of Christian Stoicism. Both the Christianity and the Stoicism posed difficulties for Hume, because both made more claims on human reason than he thought legitimate.

This is, of course, a historically imprecise categorization – the juxtaposition of the terms "Christian" and "Stoic" can hardly be anything but imprecise. I do not mean that Hutcheson accepted Stoic cosmology or was superstitious about omens, or even that he was on the more puritanical wing of the kirk. Not even his particular psychology of the passions is Stoic, though some Hutchesonian elements may have entered into Hume's account of Stoic philanthropy. The passions, in Hutcheson, are as integral to happiness and right action as to unhappiness and wrong action; and his cardinal concept of benevolence, whatever limited role one might find for it in Marcus Aurelius, hardly looms large in the classical literature.

What I have in mind is something more general. The epigraphs of Hutcheson's moral *Compend* are largely from Stoic sources; in the preface he acknowledges his debt to Cicero, and Cicero's to the Stoics. He has both a theoretical and applied interest in the nature of virtue – conceived as beauty of soul – and the means of attaining it, and believes that such an inquiry adds to our under-standing of the divinely ordained order of nature. Hutcheson also believes that the completed study will enhance our powers of self-control, an essentially rational function, and thereby promote better order in the individual and in society. As he advised in the preface to his students soon after his exchange with Hume, so possibly in implied criticism of him:

Let not philosophy rest in speculation, let it be a medicine for the disorders of the soul, freeing the heart from anxious solicitudes and turbulent desires; and dispelling its fears: let your manners, your tem-pers, and conduct be such as right reason requires. Look not upon this part of philosophy as matter for ostentation, or shew of knowledge, but as the most sacred law of life and conduct, which none can despise with impunity, or without impiety toward God.[16]

And according to biographical testimony from several sources, Hutcheson was highly effective in selling the message, promoting his moral aesthetic by a combination of analytic and affective skills. To Hume, this was a waste of good talents. There was much in Hutcheson's account of moral psychology that Hume could at least understand and debate. But that it could be used to *teach* morality was something he rejected, and his own handling of moral-sense theory suggests that Hume would have expected Hutcheson to reject it too. Hume's familiar dictum (T., p. 415) that reason, as a motivating force, "is and ought only to be the slave of the passions" must be a consciously anti-Stoic sentiment, just as Stoicism is the target in the notes to Bayle's dictionary article on Ovid from which the dictum is derived.[17]

Hutcheson and George Turnbull, who were the most articulate Scottish exponents of the moral-sense philosophy, cite Epictetus and Marcus Aurelius in their mature philosophy rather more than the Stoicism of the Roman republic. Clearly those authors offered a morality sufficiently congenial to the moderate Calvinist mind that they could be used to satisfy the pedagogical requirement that secular writings must not jeopardize the faith. This is most apparent in the translation of Aurelius which Hutcheson published with assistance from his former pupil, the Hellenist James Moor.[18] Whatever can be construed as an intimation of monotheism, and

---

16 Francis Hutcheson, *Philosophiae moralis institutio compendiaria*, 2nd ed. (Glasgow: Foulis, 1745), p. v; from the anonymous and posthumous English translation, *A Short Introduction to Moral Philosophy* (Glasgow: Foulis, 1747), p. iv. Already, in "The Sceptic," Hume had voiced his doubts that philosophy could be a "medicine of the mind," in the sense of a strong corrective to natural propensities (*Essays*, p. 169).

17 Desmaizeaux edition, note H. Christopher Bernard, "The Role of the Imagination in Hume's Science of Man," Ph.D. thesis, University of St Andrews, 1990, Chap. 3, notes that "slavery" to the passions was identified by Malebranche with the fall of Adam, in a work well known to Hume (*De la recherche de la vérité*, V.i). Bayle himself traced the image back to the abbé Esprit's *La Fausseté des vertus humaines* (1678), but it originated, I assume, in the hostile context of Plato's *Republic* (IV.442ab).

18 *The Meditations of the Emperor Marcus Aurelius Antoninus. Newly Translated from the Greek: With Notes, and an Account of his Life* (Glasgow: Foulis, 1742). It is Hutcheson's concurrent work on this edition, rather than any shift in philosophical views, that explains the increased references to Stoicism in his later publications, contrary to the thesis of W. R. Scott, *Francis Hutcheson* (Cambridge: Cambridge University Press, 1900), Chap. 12. On the pedagogical constraints, see Sher, "Professors of Virtue."

particularly of Christian doctrine or precepts – principally the Sermon on the Mount and the Epistles of Paul – is carefully annotated, while the students are cautioned against the Stoic line on mortality. Even more interesting than the theological annotation is the strong political angle to the prefatory Life. Hutcheson does more than argue the general compatibility of the Christian and Stoic systems. He turns the potentially embarrassing evidence of religious persecution in the reign of Aurelius against the fanatical sectarianism – the conflicts between presbyterianism and episcopacy, between Calvinism and Arminianism, and between credal subscription and nonsubscription – in his own age. To construe independence in religion as political rebellion is characteristic of all persecutors, "As we see in all the defences of the R. catholic persecutions in France, and the protestant persecutions in England and Scotland; when the clergy have once persuaded the legislator, impiously to invade the prerogatives of God, over the consciences of men, by penal laws about such religious opinions, and forms of worship as are no way hurtful to society" (pp. 38–9).

Let none make this objection to Antoninus, but those, who, from their hearts, abhor all Christian persecutions, who cannot hate their neighbours, or deem them excluded from the divine favour, either for neglecting certain ceremonies, and pieces of outward pageantry, or for exceeding in them; for different opinions, or forms of words, about some metaphysical attributes or modes of existence, which all sides own to be quite incomprehensible by us; for the different opinions about human liberty; about which the best men who ever lived have had opposite sentiments: for different opinions about the manner in which the Deity may think fit to exercise his mercy to a guilty world, either in pardoning of their sins, or renewing them in piety and virtue. (pp. 40–1)

This passionate toleration is strongly reminiscent of the Molesworth circle, to which Hutcheson had belonged in the first years of his professional career in Dublin, in the 1720s, along with some other Glasgow-trained Ulstermen. We know from surviving letters that a like-minded colony of young intellectuals in Edinburgh enjoyed the same rhetoric. These were some of the members of the Rankenian Club.[19] Turnbull was newly appointed

---

19 On these two groups see M. A. Stewart, "John Smith and the Molesworth Circle," *Eighteenth-Century Ireland*, 2 (1987), 89–102, and other documents cited there.

to Marischal College, Aberdeen; William Wishart was later principal at Edinburgh, where he subsidized the publication of small devotional works, including a re-edition of the Stoic meditation *De animi tranquillitate* by the sixteenth-century Scot Volusinus (Florence Wilson). They wrote to Molesworth in topical commonplaces advertising their devotion to virtue, extolling the inculcation of patriotism in ancient Rome, and looking forward to such time as these standards would replace clerical and political domination of the educational system in Scotland and bring true liberty with them. Ancient Rome can obviously mean more than Stoicism, but for these young firebrands Cato and Brutus are the names which most often recur at this time. What the latter stood for did not need to be made explicit: It is clear from the literature of the day that Cato and Brutus were habitually seen as bastions against self-interest, corruption, and absolutism in government, while the death of Caesar is frequently used as an allegory for the rightful overthrow of dictatorial government. A fictional sequence of "Cato's Letters" was serialized in *The London Journal* and *The British Journal* in the 1720s,[20] and dramatic portrayals of Cato, which had become popular on the London stage, caused unrest when they were imported into the strife-ridden halls of the College of Glasgow.[21]

A younger member of the Rankenian fraternity with Stoic leanings was John Pringle, a physician who held the moral philosophy chair at Edinburgh in the late 1730s and early 1740s, whom Hume hoped vainly to succeed in 1744. When Pringle was deputed to go off as medical officer with the British army in Flanders, the only reading he took with him was the *Meditations* of Marcus Aurelius, and he was chagrined to find that this did not equip him for the boorishness of mess-table conversation during a military campaign.[22] Some student dictates from his lectures on

20 C. B. Realey, "The London Journal and Its Authors, 1720–1723," *Bulletin of the University of Kansas*, Humanistic Studies, 5:3 (1935). For the impact of republican Stoicism on the work of Thomas Blackwell of Aberdeen around this period, see M. A. Stewart, "The Origins of the Scottish Greek Chairs," in E. M. Craik, ed., *"Owls to Athens": Essays on Classical Subjects Presented to Sir Kenneth Dover* (Oxford: Clarendon Press, 1990), pp. 391–400, and Howard D. Weinbrot, *Augustus Caesar in "Augustan" England* (Princeton: Princeton University Press, 1978).

21 [John Smith], *A Short Account of the Late Treatment of the Students of the University of G——w* (Dublin, 1722), p. 16. See further detail in Stewart, "John Smith and the Molesworth Circle."

22 British Library, Add. MS. 6861, fol. 176.

Cicero's *De officiis* survive. Much of it is uninspired paraphrase, but at two points he breaks away from the text he is expounding. In place of Cicero's defense of philosophy at the beginning of Book II, Pringle inserted some statutory exercises in natural religion, and even this turns out to be the fashionable Christian–Stoic synthesis:

You See what great use the light of Nature may be to illustrate the Scriptures the Naturall theology was greatly cultivate among the Ancients especialy among the Stoicks whose Sentiments are far from being a disgrace to Christianity Epictetus & Antoninus agree in what they Say with Relation to God & his atributes you may See them yourselves their is no man can be a good Member in Society without believing the being of a God & he can never be content with a Condition that may befall him except he be of an opinion that this god is good & that all that happens him happens him for the best.[23]

More interestingly, a chance remark by Cicero on suicide, in Book I, Chapter 31, sparks off two weeks' lectures to his teenage audience on the arguments for and against. Pringle finds the arguments against suicide "tedious" and comes down firmly on the Stoic side in defense of the practice, illustrating his advocacy with several improving stories from antiquity and strongly denying that there is any religious basis for condemning it.[24] This was at the time a touchy subject, and Hume was later to suppress an essay of his own that had reached as far as the press.[25] Hume's

---

23 Edinburgh University Library, MS Gen. 74D, p. 59.
24 Ibid., pp. 15–44. An older member of the Rankenian Club with fewer Stoic sympathies, the Rev. William Wallace, was defending suicide in youthful papers c.1720: Edinburgh University Library, MS La. II. 620/19.
25 *Essays*, pp. 577–89. It appeared posthumously. As a member of General St Clair's expedition to Brittany in 1746, Hume had come face to face with a sick officer who had decided to make a Stoic exit, but, confronted with the practical reality, Hume thought it his duty to intervene. "I immediatly sent for a Surgeon, got a Bandage ty'd to his Arm, & recoverd him entirely to his Senses & Understanding. He liv'd above four & twenty hours after, & I had several Conversations with him. Never a man exprest a more steady Contempt of Life nor more determind philosophical Principles, suitable to his Exit. He beg'd me to unloosen his Bandage & hasten his Death, as the last Act of Friendship I coud show him: But alas! we live not in Greek or Roman times. He told me, that he knew, he coud not live a few Days: But if he did, as soon as he became his own Master, he wou'd take a more expeditious Method, which none of his Friends cou'd prevent. I dye, says he, from a Jealousy of Honour, perhaps too delicate; and do you think, if it were possible for me to live, I woud now consent to it, to be a Gazing-Stock to the foolish World. I am too far advanc'd to return. And if Life was odious to me before, it must be doubly so at present." (HL, 1:97–8)

line, though sympathetic to the suicide, is not the Stoic line as Pringle read it, for Hume appears to allow the taking of one's life as a form of self-indulgence to those who are at the end of their useful existence. Pringle claimed to find such an attitude in Seneca but used that to argue that Seneca was not a serious Stoic, just a political opportunist who "traduced" the philosophy to which he claimed allegiance; indeed Pringle doubted the morality not only of Seneca's death but of those of Cato and Brutus as well. Pringle's view was that suicide is permissible when one's "dying would be of more advantage to the Publick than his living," a thesis he found in the writing of Epictetus and Aurelius, and exemplified in the death of Cleomenes. Even the wretched Lucretia fails to be absolved: "What was the Losing her Virtue" – for which she committed suicide – "in respect of her reputation," which she voluntarily allowed to be tarnished rather than face an honorable death for refusing to succumb? "Perhaps she was no philosopher & Some allowances are to be made the fair Sex."

One can add other samples to show that the general tradition of philosophy teaching in Scotland at this period was dominated by such concerns, and by the search for a theoretical grounding which will supposedly teach people to be Christian, virtuous, and pillars of society. Hume thought the project not only misguided but intellectually harmful, since it showed a profound misunderstanding of the place of reason in action and instruction, and of the nature of moral judgment. There was therefore little chance that he would obtain employment to teach a curriculum whose credentials he had effectively destroyed. Not that he did not try; and the first *Enquiry*, which contains the defense of the Skeptical philosophy against its latter-day Stoic rivals, was written on the rebound from his defeat at Edinburgh in 1745.[26] In the opening section of the *Enquiry* he distinguishes a fashionable "easy" philosophy from an "abstract" metaphysics founded in a science of human nature, to which he himself gives preference. And what is this easy philosophy to which the skeptical exercises of the abstract philosophy are to be preferred? It is a new synthesis of those aspects of the human temperament which appeal to the popular-essay reader and which have become distorted in the Epicurean and Stoic philosophies: the love of cultured discourse and the desire for maxims to

26 See Stewart, "Historical Significance."

guide an active person in the affairs of life. You can have these, Hume now thinks, if you are content with "polite letters." The effect upon your sentiments is achieved by scene painting, rather than by reasoning; and an exemplary exponent of the technique was Cicero.

INDEX

Cleanthes, 14, 235
clinamen (swerve of atoms), 156, 165, 165n,
    210n, 223
    Gassendi's objections to, 166
clock metaphor, 257
cognition, 251–2
cohesion of bodies, 213n–14n, 223–4
    Newton on, 225, 232
Colish, Marcia, 141
Collège Royal, 192
conservation of matter, 222
contemplative life, 178
contingency, 158, 161
conversos, 71, 79
corpuscularianism, 203, 226
    Boyle on, 199, 266
    Leibniz on, 203n
    in Locke, 266
cosmology
    Aristotelian, 147
    Stoic, 138–9, 145–7
Cratylus, 20
creation, 158, 161
    biblical, 222
    Epicurean denial of, 156, 222
    Newton on, 222
    Plato on, 166
    in Stoics, 166
Cudworth, Ralph, 202n, 204
Cyrenaics, 44

Dante, 50
Dassaminiato, Fra Giovanni, 63
Daudin, Jean, 63
Democritus, 163–4, 166, 202n, 203n, 261
    Boyle on, 201–2, 201n, 207, 211n
demons, 170–2
Descartes, René, 152–3, 176, 202n, 203n,
    260
    on animal spirits, 248, 248n
    Boyle on, 197n, 198–9, 201–2, 201n,
        205, 205n
    and cosmology, 154, 227
    dualism of, 265
    on ideas, 259–60
    on logic, 25–6
    on origin of the world, 205n
    on personal identity, 257
    on sensation, 245, 249
    on skepticism, 259
    on the soul, 242, 245–6, 253
    Willis on, 244
design, argument from, 3, 139, 141, 208–9,
    287
determinism, 6, 16, 162–3, 166
    attacked by Cicero, 41
Diaz, Pero de Toledo, 68–9

Digby, Kenelm, 201–2, 201n, 203n, 244
Dijksterhuis, E. J., 260
Diodati, Elié, 188, 188n
Diogenes Laertius, 2, 2n, 6–7, 28, 116, 122,
    126, 207
Dionysius of Syracuse, 75
divination, 59, 163, 170, 171, 171n
    Calvin on, 87
    Gassendi on, 155–74
    in Stoicism, 13, 19, 170, 172–3, 172n
divine foreknowledge and human freedom,
    168–70
Dominici, Giovanni, 90n
dreams, 59, 172–3, 243, 270
Dryden, John, 214n
Duarte, King of Portugal, 71
Dupuy brothers (Jacques and Pierre), 187

elements, 147, 154
    Aristotelian, 138
    in Stoicism, 139, 236
emotions, disturbances of, 40–66
Epictetus, 6, 16, 29, 65, 286, 289, 291, 295
    against Epicureanism, 177
Epicureanism, 1–9, 26, 89–134, 155–271
    Boyle on, 197–219
    and ethics, 2–3, 102, 125, 161, 175–95
    Gassendi on, 155–95, 260
    Hume on, 276–9, 286–8
    in Newton, 221–38
    and physics, 3, 125, 161, 177
    on public life, 125, 175
    reputation of, 122–3, 177n
    on the soul, 242, 246, 253
    unorthodoxy of, 3, 125, 156, 156n, 222
    Valla on, 89–114
Epicurus, 1, 3, 44, 91n, 107–8, 110, 123–4,
    164, 181, 203n, 261, 264n
    Boyle on, 197n, 201, 205–6, 211n, 212,
        213n
    Valla on, 107, 110, 112–14
Erasmus, Desiderius, 85–7, 121
essences, real and nominal, in Locke,
    266–70
eternity of the world in Epicureanism, 156,
    222
ethics
    Aristotelian, 69, 71, 76, 101
    Epicurean, 102, 125, 155, 175–95
    Stoic, 3, 5, 11–24, 45–115
Euclid, *Optics and Catoptrics*, 143
Eusebi, Mario, 68
exorcism, 172

fate, 163–9
    Gassendi on, 155–74
    Saint Jerome on, 55